SLOW SAND FILTRATION:
Recent Developments in
Water Treatment Technology

45.00

G

SLOW SAND FILTRATION:
Recent Developments in Water Treatment Technology

Editor:

N. J. D. GRAHAM
Lecturer in Public Health Engineering
Department of Civil Engineering
Imperial College of Science and Technology, University of London

ELLIS HORWOOD LIMITED
Publishers · Chichester

Halsted Press: a division of
JOHN WILEY & SONS
New York · Chichester · Brisbane · Toronto

First published in 1988 by
ELLIS HORWOOD LIMITED
Market Cross House, Cooper Street,
Chichester, West Sussex, PO19 1EB, England
The publisher's colophon is reproduced from James Gillison's drawing of the ancient Market Cross, Chichester.

Distributors:
Australia and New Zealand:
JACARANDA WILEY LIMITED
GPO Box 859, Brisbane, Queensland 4001, Australia
Canada:
JOHN WILEY & SONS CANADA LIMITED
22 Worcester Road, Rexdale, Ontario, Canada
Europe and Africa:
JOHN WILEY & SONS LIMITED
Baffins Lane, Chichester, West Sussex, England
North and South America and the rest of the world:
Halsted Press: a division of
JOHN WILEY & SONS
605 Third Avenue, New York, NY 10158, USA
South-East Asia
JOHN WILEY & SONS (SEA) PTE LIMITED
37 Jalan Pemimpin # 05–04
Block B, Union Industrial Building, Singapore 2057
Indian Subcontinent
WILEY EASTERN LIMITED
4835/24 Ansari Road
Daryaganj, New Delhi 110002, India

© **1988 N. J. D. Graham/Ellis Horwood Limited**

British Library Cataloguing in Publication Data
Slow sand filtration
1. Water supply. Slow sand filtration.
I. Graham, N.J.D. (Nigel Jonathon Douglas) *1953–*
628.1'64

Library of Congress Card No. 88–29292

ISBN 0–7458–0585–X (Ellis Horwood Limited)
ISBN 0–470–21374–4 (Halsted Press)

Printed in Great Britain by Hartnolls, Bodmin

Table of Contents

Preface   ix

Chapter 1   Filter Design, Operation and Management

1.1   Water treatment by slow sand filtration –   1
        considerations for design, operation and
        maintenance.
        J.T.Visscher.
1.2   Slow sand filtration: an approach to practical   11
        issues.
        I.P.Toms and R.G.Bayley.
1.3   Slow sand filtration in the United States.   29
        G.S.Logsdon and K.R.Fox
1.4   Improvement of slow sand filtration – application   47
        to the Ivry rehabilitation project.
        A.Montiel,B.Welte and J.M.Barbier.
1.5   Management of slow sand filters in respect to   91
        ground water quality.
        H.Sommer.

Chapter 2   Pretreatment Methods

2.1   Roughing gravel filters for suspended solids   103
        removal.
        M.Wegelin.
2.2   Upflow coarse-grained prefilter for slow   123
        sand filtration.
        L.Di Bernardo.
2.3   Pretreatment with pebble matrix filtration.   141
        K.J.Ives and J.P.Rajapakse.
2.4   Ozonation and slow sand filtration for the   153
        treatment of coloured upland waters – pilot-plant
        investigations.
        G.F.Greaves,P.G.Grundy and G.S.Taylor

Chapter 3   Biological Aspects

3.1   The ecology of slow sand filters.   163
         A.Duncan.
3.2   Some aspects of the filtration of water   181
         containing centric diatoms.
         K.B.Clarke.
3.3   Development of a slow sand filter model as a   191
         bioassay.
         C.Schmidt.
3.4   The removal of viruses by filtration through sand.   207
         D.Wheeler,J.Bartram and B.Lloyd.

Chapter 4   Process Performance

4.1   The effects of high-carbon and high coliform   231
         feed waters on the performance of slow sand
         filters under tropical conditions.
         J.M.Barrett and J.Silverstein.
4.2   Benefits of covered slow sand filtration.   253
         J.A.Schellart.
4.3   Comparisons between activated carbon and slow sand   265
         filtration in the treatment of surface waters.
         J.Mallevialle and J.P.Duguet.
4.4   Modifications to the slow rate filtration process   281
         for improved trihalomethane precursor removal.
         M.R.Collins and T.T.Eighmy.

Chapter 5   Process Developments

5.1   Pilot plant evaluation of fabric-protected slow   305
         sand filters.
         T.S.A.Mbwette and N.J.D.Graham.
5.2   Advanced techniques for upgrading large scale slow   331
         sand filters.
         A.J.Rachwal,M.J.Bauer and J.T.West
5.3   Developments in modelling slow sand filtration.   349
         C.A.Woodward and C.T.Ta.
5.4   The application of polyurethane to improve slow   367
         sand filters.
         P.Vochten,J.Liessens and W.Verstraete.

Chapter 6   Developing Country Case Studies

6.1   Dual media filtration for the rehabilitation of   379
        an existing slow sand filter in Zimbabwe.
        T.F.Ryan.
6.2   The performance of slow sand filters in Peru.   393
        B.Lloyd,M.Pardon and D.Wheeler.

Subject index   413

Preface

This book presents recent research and state-of-the-art information on the scientific basis, modes of use, and engineering developments of slow sand filtration. The information is loosely grouped into the the following themes: filter design, operation and management; pretreatment methods; biological aspects; process performance; process developments; and developing country case studies.

Slow sand filtration is well known for being an effective water treatment unit process provided appropriate pretreatment is incorporated . However, its low-technology image and perceived disadvantages (e.g. low throughput, high operation and maintenance costs) have resulted in it being considered generally inappropriate as a treatment alternative for industrialized countries. Such perceptions are now rapidly changing in the light of the rising need to meet higher drinking water quality standards. In particular, the use of slow sand filtration as a means of reducing dissolved organic carbon (e.g. trihalomethane precursor material, assimilable organic carbon) and pathogenic micro-organisms from polluted surface water sources is currently receiving considerable attention in Europe. In developing countries slow sand filters are well established as an appropriate treatment technology. However, many installations fail as a result of inadequate pretreatment facilities, lack of a proper understanding of the process, and poor operation and maintenance practise.

Although slow sand filtration has been applied widely as a unit water treatment process throughout this century, our knowledge of the scientific basis of the process remains extremely limited. This is in part the result of the lack of interest in the process in the industrialized countries, as well as due to the very complex nature of the treatment process.

The purpose of this book is to bring together the experience and knowledge from a broad spectrum of related disciplines, such as plant operators and designers, universities and research organisations to focus on and summarize the latest understanding and developments in this important process of water treatment. Each chapter of the book has been compiled from presentations made at the International Seminar "Advances in Slow Sand Filtration", held at Imperial College, London, between the 23rd and 25th November 1988. The contributions of the authors, session chairmen and conference organizers at Imperial College made this International Seminar and book possible, the editor wishes to express his thanks to them for their assistance.

Imperial College, London.   N.J.D. Graham
November, 1988.

x

1

Filter design, operation and management

1.1 WATER TREATMENT BY SLOW SAND FILTRATION CONSIDERATIONS FOR DESIGN, OPERATION AND MAINTENANCE

Jan Teun Visscher — IRC, International Water and Sanitation Centre, P.O. Box 93190, 2509 AD The Hague, The Netherlands.

ABSTRACT

Slow sand filters because of their advantage of simplicity, efficiency and economy are appropriate means of water treatment particularly for community water supply in developing countries. The basic elements of a slow sand filter have been briefly described. Planning, design and construction aspects are presented with special emphasis on maintenance. Some cost figures of systems in India and Colombia are included. Operational procedures for maintenance and re-sanding are set out.

INTRODUCTION

Slow sand filtration is a very appropriate method for making surface water safe to drink. When surface water is readily available, whilst groundwater is not, slow sand filtration will frequently prove to be the simplest, most economical and reliable method to prepare safe drinking water.

In a slow sand filter the water percolates slowly through a porous sand bed. During this passage the physical and biological quality of the raw water improves considerably through a complex of biological, bio-chemical and physical processes. In a mature bed a thin layer forms on the surface of the bed. This filter skin consists of retained organic and inorganic material and a great variety of biologically active micro-organisms which break down organic matter. When after several weeks or months the filter skin gets clogged, the filtration capacity can be restored by cleaning the filter, i.e. by scraping off the top few centimeters of the filter bed including the filter skin.

Slow sand filters are an essential element of water treatment works in various European cities, e.g. London and Amsterdam, as well as in many developing countries. The majority of these systems are giving satisfactory results, but problems are also being reported. It has been shown over and again that good design and adequate provision for maintenance are essential prerequisites for proper functioning of the system. Particularly these issues are being addressed here. The information is largely based on a recent IRC publication "Slow Sand Filtration for Community Water Supply; Planning, design, construction, operation and maintenance" (Technical Paper No.24), and it is very much appreciated that IRC granted permission to use several tables from their publication. Information is also included from IRC's Development and Demonstration Programme on Slow Sand Filtration (SSF project).

This project, which was financially supported by the Department of Research and Development of the Netherlands Ministry of Foreign Affairs, embraced applied research, demonstration programmes and the transfer of information. It was implemented in collaboration with institutes in developing countries including India and Colombia.

Laboratory research was followed by the installation of a number of village demonstration plants in selected villages. The communities in the villages were involved in the planning, construction and operation of the schemes, to increase the sustainability of the water supply and to ensure that it met the needs of the population. In some areas the community played an important role in the selection of the site and the caretakers, and established a water committee to manage the system. Some communities have also provided free labour which has cut down the construction cost of some of the systems by more than 30 percent (ref.1).

The SSF project proved the efficiency of slow sand filtration, helped developing country based organizations to gain experience with the technology and showed the importance of community involvement. This research also underscored the major drawback of slow sand filtration - its vulnerability to high turbidity, which will cause rapid clogging of the filters. Good progress has been made in identifying suitable and simple pre-treatment systems. Comparative testing of several of these systems is now underway in a project in Cali, Colombia, which is implemented by the University of Valle in collaboration with IRC. Support for this project is obtained from several organizations including the Netherlands Government, the International Reference Centre for Waste Disposal in Switzerland, Cali Water Authority and many individual advisors.

THE SLOW SAND FILTER SYSTEM

Basically, a slow sand filter consists of a box constructed of reinforced concrete, ferrocement and stone or brickwork masonry, containing:
- a supernatant layer of raw water;
- a bed of fine sand;
- a system of underdrains;
- an inlet and outlet structure;
- a set of filter regulation and control devices.

The water flow in a slow sand filter may be controlled at the outlet (Figure 1), or at the inlet of the filter (Figure 2), and the method chosen may slightly affect the structure, the control devices and the functioning.

Inlet structure

The inlet structure is intended to allow water to flow into the filter without damaging the filter skin on top of the sand bed. Usually, the inlet structure is a box which can also be used to drain the supernatant water quickly, which greatly facilitates the cleaning process.

Supernatant water layer

The supernatant water layer provides a head of water which is sufficient to drive the raw water through the bed of filter medium, while creating a detention period of several hours for the raw water. An outlet has to be provided to serve as an overflow for the supernatant water and to enable removal of scum which may be formed on the water surface.

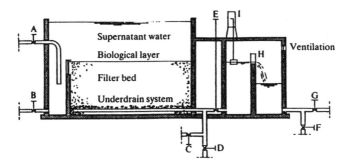

Figure 1. Basic components of an outlet-controlled slow sand filter
(source: IRC, 1987)

A: raw water inlet valve
B: valve for drainage of supernatant water layer
C: valve for back-filling the filter bed with clean water
D: valve for drainage of filter bed and outlet chamber
E: valve for regulation of the filtration rate
F: valve for delivery of treated water
G: valve for delivery of treated water to the clear-water reservoir
H: outlet weir
I: calibrated flow indicator

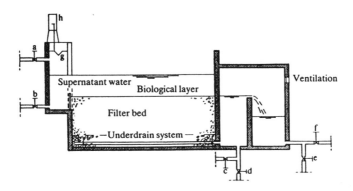

Figure 2. Basic components of an inlet-controlled slow sand filter
(source: IRC, 1987)

a: valve for raw water inlet and regulation of filtration rate
b: valve for drainage of supernatant water layer
c: valve for back-filling the filter bed with clean water
d: valve for drainage of filter bed and outlet chamber
e: valve for delivery of treated water to waste
f: valve for delivery of treated water to the clear-water reservoir
g: inlet weir
h: calibrated flow indicator

Filter bed

Although any inert, granular material can be used as the filter medium, sand is usually selected because it is cheap, inert, durable and widely available. When placed in the filter, the sand should be free from clay, soil and organic matter.

Sand used in slow sand filters should be relatively fine, have an effective size in the range of 0.15–0.30 mm, and a uniformity coefficient preferably below 3. It is important that the effective grain size of the sand should not be finer than necessary, because sand which is too fine will add to the initial head loss (ref.2) although the quality of the effluent will be improved.

Underdrain system

The underdrain system provides unobstructed passage of treated water and supports the bed of filter medium. Usually, it consists of a main and lateral drain constructed from perforated pipes, or a false floor made of concrete blocks or bricks, and is covered with layers of graded gravel. These layers prevent the filter sand entering or blocking the underdrains and ensure uniform abstraction of the filtered water.

An innovative underdrain system is being applied in Colombia, where corrugated PVC pipes of 6 cm diameter are placed one meter apart, and covered with 0.1 m layer of fine gravel. This development is interesting because it reduces the need for graded gravel, and also lessens the total height of the underdrain system compared to traditional systems, so that the total height of the filter box is lower.

Outlet chamber

The outlet chamber usually consists of two sections separated by a wall, on top of which a weir is placed with its overflow slightly above the top of the sand bed. This weir prevents the development of below-atmospheric pressure in the filter bed, and ensures that the filter operates independently of fluctuations in the level of the clear-water reservoir. By allowing the free fall of water over the weir, the oxygen concentration in the filtered water is increased and the weir chamber should therefore be suitably ventilated to facilitate aeration. Provision is often made in the outlet chamber to enable back-filling through the underdrains.

Flow control

In an outlet-controlled filter, the rate of filtration is set with the outlet valve. Daily or every two days this valve has to be opened a bit further to compensate for the increase in resistance in the filter skin. The outlet valve has to be manipulated on a regular basis, causing a slight variation in the rate of filtration. Thus, the operator is forced to visit the plant at least every day, otherwise the output will fall. The water is retained for 5 to 10 times as long as an inlet-controlled filter at the beginning of the filter run, which may make purification more efficient.

In an inlet-controlled filter, the rate of filtration is set by the inlet valve. Once the desired rate is reached, no further manipulation of the valve is required. At first the water level over the filter will be low but gradually it will rise to compensate for the increasing resistance of the filter skin. Once the level has reached the scum outlet, the filter has to be taken out for cleaning.

Inlet control reduces the amount of work which has to be done on the filter to just cleaning it. The rate of filtration will always be constantly with this method and the build-up of resistance in the filter skin is directly visible. On the other hand, the water is not retained for very long at the beginning of the filter run, which may somewhat reduce the efficiency of treatment.

Flow indicator

A flow indicator is required to continuously measure the flow and facilitate the job of the plant operator.

DESIGN CONSIDERATIONS

Based on the results of the SSF project and on information from the literature (ref. 2 and 3) general design criteria have been formulated for slow sand filter plants (Figure 3). Following these criteria a suitable plant can be designed provided the designer always keeps in mind that operation and maintenance is the crucial factor in producing safe water and should be facilitated as much as possible.

Design Criteria	Recommended level
Design period	10-15 years
Period of operation	24 h/d
Filtration rate in the filters	0.1-0.2 m/h
Filter bed area	5-200 m^2 per filter, minimum of 2 units
Height of filter bed:	
initial	0.8-0.9 m
minimum	0.5-0.6 m
Specification of sand:	
effective size	0.15-0.30 mm
uniformity coefficient	<5, preferably below 3
Height of underdrains including gravel layer	0.3-0.5 m
Height of supernatant water	1 m

Figure 3. General design criteria for slow sand filters in
          rural water supply (Source IRC, 1987)

Design Period

As there is hardly any economy of scale in the cost of slow sand filter construction, short design periods of 10-15 years can be adopted. This will prevent over-design of facilities which is unnecessarily eroding available resources in many countries.

Filtration rate

Controlling the rate of filtration is the key to adequate functioning of a slow sand filter. For surface water, operation at a rate between 0.1 and 0.2 $m^3/m^2/h$ is usually satisfactory, because the filter tends to clog within a shorter period of time using higher rates of filtration. However, the rate may be increased to 0.3 $m^3/m^2/h$ for short periods of one or two days without undue harm, for example, while another filter is being cleaned.

Even higher rates can be applied if very good quality water is being treated as for example is the case in Amsterdam were the slow sand filters as one of the last of a series of treatment steps, operate at 0.6 $m^3/m^2/h$.

Mode of operation

It is most effective to operate a slow sand filter continuously because good quality effluent is ensured, and the smallest filter area is required. Continuous operation of the filter is feasible where raw water can be fed into the filters by gravity flow, but in many cases the raw water has to be pumped. If continuous pumping cannot be guaranteed, continuous operation at a constant rate may be ensured by constructing a raw water balancing tank. Water is pumped into this tank at certain intervals before being continuously fed into the filters by gravity flow. Alternatively, declining rate filtration can be applied in outlet-controlled filters. These, however, would require a larger filter area or a higher filter-box and therefore may be more costly.

Intermittent operation in which the filtration process is stopped at intervals should not be permitted, because it has been shown conclusively that an unacceptable breakthrough of bacteriological pollutants occurs four to five hours after filters recommence operation (ref.4).

Number of units

At least two slow sand filter units are required to ensure uninterrupted supply and facilitate maintenance. In larger plants the number of units may be increased to ensure greater flexibility, often at little additional cost. A first indication of a suitable number may be obtained with n = 0.5 3 A, where A is total area of filter beds in m^2. No additional beds need to be provided as standby. When one of the beds needs to be cleaned the rate of filtration in the other bed(s) may be gradually increased. A temporary increase of the rate up to 0.3 $m^3/m^2/h$ does not have an adverse effect on the effluent quality.

Filter area per unit

For rural areas, it is advisable to restrict the area per filter unit to 200 m^2 to facilitate filter cleaning.

Height of filter bed

The minimum depth of sand (depth before re-sanding) should be 0.5-0.6 m. Selecting a minimum depth of 0.6 m is advisable if slow sand filtration is the only treatment. The initial depth should be 0.8-0.9 m to allow for a sufficient number of scrapings before re-sanding is needed. Sand found locally can often be used as filter sand and will cut down the cost considerably. A sieve analysis is required to confirm its suitability.

COST ASPECTS

The construction cost of a filter excluding pipes and valves is made up of two components: the total cost for floor, underdrains, sand, and gravel; and the cost of walls of the filter box. The cost of labour and land is usually of lesser importance. In rural areas in India, the cost of land rarely exceeds 1% of the total construction cost. However, the somewhat larger areas required for slow sand filter plants can be a problem in densely populated areas.

The following quotation can be used to estimate the construction cost of a slow sand filter excluding cost of pipes and valves:

$$C_t = C_a \times A + C_1 \times L_w$$

where:

C_t = total construction cost, excluding pipes and valves
C_a = combined costs per square metre of filter bed area of floor, underdrains, gravel, filter sand and excavation
C_1 = cost of the walls per running metre of wall length
A = total surface area (m^2)
L_w = total wall length (m)

Whereas the required area A can not be reduced it is important to minimize the total wall length. For rectangular filter units having common walls, the total wall length becomes smallest when:

$$A = \sqrt{\frac{2A}{n+1}} \quad \text{and } b = \frac{(n+1)a}{2n}$$

where:
A = wall length of unit (m)
A = total surface area (m^2)
n = number of units
b = breadth of unit (m)

The construction cost of small and medium sized slow sand filters often will be less than that of other types of treatment. Figure 4 gives an example of a comparison of the construction cost of slow sand filters and rapid sand filters in India (1983 prices). Construction of a slow sand filter in India is less expensive than a rapid sand filter up to a capacity of 3,000 m^3/day. However, if recurrent costs for operation and maintenance are taken into account, the balance shifts to 8,000 m^3/day. When cheaper materials are used, for instance ferrocement and cheaper drainage systems, the cost of SSF will fall (ref.5).

Recent developments in Colombia, with 20 new slow sand filters in the offing, are even more interesting. Not only are new slow sand filtration plants competitive with conventional treatment, but even reconstruction of conventional plants into slow sand filtration plants has been found to be cost effective. For example, in a sub-urban settlement near Cali, Colombia, operation and maintenance of a conventional treatment plant was costing US$ 500. Because of the increasing cost of chemicals, it was decided to reconstruct the plant and adopt slow sand filtration as the main treatment. Most of the existing structure could be used, holding the cost of reconstruction as low as US $ 7000. Operation and maintenance costs are now amounting to US $ 50 per month.

Capacity in m^3/day	Cost in US $ 1,000	
	SSF	RF
1,000	50	60
1,500	60	80
2,000	80	90
3,000	120	120
4,000	150	140
5,000	190	160
7,000	250	190

Figure 4. Construction cost of slow sand filters
(Source, NEERI)

OPERATION AND MAINTENANCE

One of the most attractive aspects of slow sand filtration is its simplicity of operation. This makes its application particularly appropriate in rural areas where local people can do the job including normal maintenance. Provided that the plant has been well designed and constructed, the performance of the filter will depend on the conscientiousness of the operator carrying out the daily routine. An example of a maintenance schedule is set out in Figure 5. For each plant the schedule will be different, as it depends on site specific variables such as how much suspended solids there are in the raw water, size of the plant, and type of supply. It is very important to draw up the maintenance schedule in co-operation with the caretaker(s) (ref.7).

Proper training and supervision are essential, as the caretaker will have the responsibility of ensuring that the water supplied to the community is safe and attractive in appearance. The caretaker should not only understand the technical aspects but also the concept of community participation and health education. Caretakers can be selected from the village or be employees from the water agency. Formal education is an advantage, but not really necessary to operate most village water supply systems. Other factors such as their being likely to stay in the job for a decent length of time, having the respect of the community, receiving sufficient reward in cash, kind or in a rise in status, and being trained, are far more important. Also, it is necessary to give the caretaker back-up and supervision when needed.

Frequency	Activity

Daily	– Check the raw-water intake (some intakes may be visited less frequently)
	– Visit the slow sand filter
	  * check and adjust the rate of filtration
	  * check water level in the filter
	  * check water level in the clear well
	  * sample and check water quality
	– Check all pumps
	– Keep the logbook of the plant

Weekly	– Check and grease all pumps and moving parts – Check the stock of fuel and order, if needed – Check the distribution network, and taps, repair if necessary – Communicate with users – Clean the site of the plant
Monthly or less frequently	– Scrape the filter bed(s) – Wash the scrapings and store the retained sand
Yearly or less frequently	– Clean the clear-water well – Check the filter and the clear-water well for watertightness
Every two years or less frequently	– Re-sand filter unit(s)

Figure 5.   Schedule of activities for caretakers

(Source IRC, 1987)

Initial commissioning of the filter bed

With all outlet valves closed, filtered water is admitted from the bottom to flow upwards through the drainage system and the gravel and sand bed until it reaches 10-15 centimeters above the sand bed. This method of charging ensures that air accumulated in the system, especially in the pores of the sand bed, is driven out. When no filtered water is available for backfilling, a temporary connection can be made with the raw water.

Next, the inlet valve is gradually opened and water is allowed to flow on top of the bed. The rate of filling is initially low to prevent scouring of the sand around the inlet. With the increasing layer of supernatant water the rate of filling can be increased. When the normal working depth of the supernatant is reached the outlet valve is opened and the effluent is run to waste at the design rate of the filter.

The filter must now be run continuously for a few weeks to allow for so-called "ripening". The ripening period has ended when bacteriological analysis indicates that the effluent quality reaches local water quality standards. From then on water can be passed into the water supply system. Often it is not possible to wait so long; in that case the water can be supplied after a few days provided there is adequate chlorination of the effluent during the ripening period.

Daily Operation

The daily operation is limited to checking and possibly adjusting the rate of filtration and to monitoring of the plant performance and the effluent quality. Only occasionally are larger inputs required for cleaning the filters.

Cleaning

To clean the filter it is necessary to lower the water level in the bed to about 0.1-0.2 metres below the top of the sand. The inlet valve should be closed and both the supernatant water drain valve and the valve on the underdrains should be opened to quickly drain the filter to the required level of about 0.1-0.2 metres below the surface of the sand.

Subsequently cleaning of the filter bed is accomplished by scraping the top 1-2 centimeters of the bed. The cleaning operation must be completed as quickly as possible in order to minimize the interference with the life of the micro-organisms in the remaining part of the filter bed.

Floating matters such as leaves or algae should be discharged over the scum weir by first raising the level of water in the unit to carry the floating matter over the weir. If this floating matter is not removed it becomes a nuisance when cleaning takes place and may make it difficult to drain the bed. When one unit is shut down the others are run at a slightly higher rate to maintain the output of the plant.

After cleaning, the unit has to be refilled with water through the underdrains. In the absence of an overhead storage tank, this backfilling may be effected by using filtered water from an adjacent filter. If this method of operation leads to a temporarily reduced output of the plant, the population should be informed in advance.

When the filter is put back into service a period of at least 24 hours is required to allow for re-ripening of the bed. After that period, the bacteriological flora has been sufficiently re-established to be able to produce a safe effluent which can then be put back into the supply. In cooler areas this ripening period may have to be extended to a few days.

Re-sanding a filter

Re-sanding becomes necessary when successive scrapings have reduced the thickness of the sand bed to 0.5-0.6 m. Fortunately this rather lengthy operation only has to be done at two or three years intervals. Assuming, for example, an initial thickness of the filter medium of 80 cm and an average of 6 scrapings a year (or 6 x 2 = 12 cm total scraping a year) then re-sanding would only be necessary after 2 1/2 years of operation.

Before re-sanding, the filter bed must be cleaned and the water level in the sand must be lowered to the gravel layer. Then a layer of "new" sand (accumulated from earlier scrapings) has to be placed underneath the sand layer in the bed.

The method of handling the sand is as follows: The old sand which has to be replaced is moved to one side, the new sand is placed and the old sand replaced on the top of the new. This "throwing over" process is carried out in strips.

Excavation is carried out on each strip in turn. The removed material from the first strip is stacked to one side in a long ridge, the excavated trench is filled with washed or new sand and the adjacent strip is excavated, throwing over the removed material from the second trench to cover the new sand in the first. When the whole of the bed has been re-sanded, then the material from the first trench is used to cover the new sand in the last strip. By doing so, the layer of "old" sand (which is rich in microbiological life) is replaced at the top of the filter bed which will enable the re-sanded filter to become operational within a minimum re-ripening period.

REFERENCES

1.　Visscher, J.T., Paramasivam, R., Raman, A., Heijnen, H.A. (1987). Slow Sand Filtration for Community Water Supply; Planning, design, construction, operation and maintenance. Technical Paper No.24. IRC.

2.　Ellis, K.V., (1985). Slow Sand Filtration; CRC Critical Reviews in Environmental Control. Department of Civil Engineering, University of Technology, United Kingdom. Volume 15, Issue 4.

3.　Huisman, L. Prof. dr. ir. (1986). Slow Sand Filtration, Delft University of Technology, The Netherlands.

4.　Sundaresan, B.B. and Paramasivam, R. (1982). Slow Sand Filtration Research and Demonstration Project - India. NEERI, Nagpur, India.

5.　Paramasivam, R., Mhaisalkar, V.A., and Berthouex, P.M. (1981). Slow Sand Filter Design and Construction in Developing Countries. NEERI, article in AWWA, USA.

6.　Filtración Lenta en Arena y Pretratamiento (1987). Technologia para Potabilizacion. A joint publication by Universidad del Valle, Colombia. Ministerio del Salud, Colombia. IRC, The Netherlands.

7.　Visscher, J.T., Veenstra, S. (1985). Slow Sand Filtration; Manual for Caretakers. Training Series No.1, IRC, The Netherlands.

1.2 SLOW SAND FILTRATION: AN APPROACH TO PRACTICAL ISSUES

I. P. Toms and R. G. Bayley — Regional laboratory Services, Thames Water Authority, U.K.

ABSTRACT

All London's surface derived water is slow sand filtered. The paper summarises the water quality criteria for modern filters operated at various rates. The relationships between head-loss, flow, run time and water quality are quantified. Modern approaches to resanding and weed growth are discussed.

INTRODUCTION

Since the mid-nineteenth century, all surface derived potable water, supplied to Greater London, has been treated by slow sand filtration. Although the original benefits were seen as good aesthetic qualities and improved public health, scientific proof was not provided until Koch was able to identify water borne bacteria in 1885. That discovery initiated the concepts of quality control and quality assurance in water supply.

Eighty five per cent of London's potable water is river derived, and all that water continues to receive slow sand filtration. The present process has developed and evolved so that its technology and application is now very different. There have been three major changes in raw water quality which have increased the capacity of the process. Firstly, a pre-filtration stage was introduced which decreased the particulate and ammonia load on the slow sand filters (1). Secondly, there was an improvement in river water quality which decreased the loading of ammonia and organic contamination. Thirdly, there were major developments in the design and management of reservoirs which have increased their capacity for self-purification and decreased the concentrations of phytoplankton (2). As these developments occurred, scientists responded by redefining the design and operational criteria for the slow sand filtration process.

In advising their engineering colleagues, scientists have had to address the day-to-day management of existing processes and planning for future systems and unit processes. They developed a suite of analytical techniques which not only advise on the consumer acceptability of the product, but also advise on the economic and technical reliability of the process.

This paper addresses some of the technical issues raised by operational engineers in the existing and future management of slow sand filters.

TREATMENT OBJECTIVES

Within the treatment systems operated for the London area, the treatment objectives can be split into five groups:-

1.   To make the water safe to drink, i.e.
        a)   free from harmful bacteria and viruses, and
        b)   no significant   concentrations of hazardous   or toxic
              substances.

2.   To make the water palatable and aesthetically pleasing
        a)   visually acceptable with no obvious colour, turbidity or
              particles, and
        b)   free of objectionable tastes and odours.

3.   To minimise subsequent treatment problems
        a)   minimum of organic carbon which could encourage the growth
              of slimes etc., and
        b)   minimum amount of residual ammonia and other   substances
              likely to react with chlorine.

4.   To minimise subsequent problems in the distribution system
        a)   minimum dissolved   and particulate organic   carbon to
              minimise aftergrowth of bacteria, fungi or animals,
        b)   chemical stability to prevent precipitation (e.g. lime) in
              pipes and plumbing,
        c)   a balance of ions and pH to prevent corrosion of plumbing
              and dissolution of metals such as copper and lead.

5.   To meet the statutory obligations of water suppliers.

                PRACTICAL ISSUES IN SLOW SAND FILTRATION

Slow sand filters are extremely complex in their physical, biological,
chemical and hydraulic behaviour. The search for a "unified theory", which
would provide answers to all questions, is illusive. However the questions
asked of scientists are usually simple:-

1.   What quality of water should be expected from the process?

2.   How fast should filters be operated?

3.   How often should they be cleaned?

4.   How quickly should they be cleaned, and how soon after cleaning can they
      be operated at full rate?

5.   What conditions determine the need for trenching, resanding or complete
      sand changes ('deep skimming')?

6.   How far into the future can one predict the behaviour of a filter run?

7.   Can weed growth be controlled?

8.   What causes sudden increases in head-loss ('jacking-up'), and can this
      problem be predicted?

9.   How clean is clean sand?

10.   What is the ideal specification for sand?

11.   What are the limitations of slow sand filtration?

12.   Is slow sand filtration really a treatment process for the future?

13.   How should new, or uprated, filters be designed and managed?

These are the basic questions to which Thames Water scientists (and their predecessors) have addressed themselves over the past 20 years. It will be possible to describe some of the work which has been done in providing some answers.

FILTERED WATER QUALITY

The effective size of the medium used in most of London's slow sand filters is about 0.32mm. This grade of sand should not physically permit particles much larger than 40µm to pass through the filter (3). In practice, the biological and physical processes that take place within the filter mean that the maximum size of particles in the filtrate should be less than 40µm. However, the presence of particles much larger than 40µm is indicative of faults in the filter shell or the sand matrix. Figures 1a and 1b show typical size distributions for particles entering and leaving a filter operating in Summer.

The operational sampling programme includes the biological examination of large volumes (up to 10m³) of filtrate to determine the presence, taxonomy and size of larger organisms. The presence of animals characteristic of still stored water (e.g. Daphnia spp.) or the schmutzdecke (e.g. chironomid larvae) is usually indicative of bypassing because of structural failures or holes in the filter medium. Similar problems would be indicated by filamentous algae such as Melosira spp. or Cladophora sp.. Other animals, such as Asellus aquaticus or Nipharqus sp. can live in underdrains, and are expected to be endemic at low numbers. Iron bacteria indicate that the filter is being operated too slowly for too long, or that water flow is not evenly distributed over the surface of the filter.

Certain unicellular algae are able to penetrate the sand and appear in the filtrate. Typically problems are experienced with small pennate diatoms, unicellular green algae, and single cells from blue-green algae. The problems are aggravated at low filtration rates (less than 15cm.h⁻¹) in summer when conditions in the supernatant water encourage the growth and formation of planktonic unicellular algae. Simple turbidity measurements are not sufficiently adequate, or technically appropriate, to detect the onset of the problem. Thames Water scientists use particulate organic carbon (POC), chlorophyll a and particle size analysis to measure the breakthrough of general organic and algal derived material.

Details of the seasonal performance of slow sand filters, on different works, are shown in Table 1. Works A is a works of modern construction filtering at rates which are often in excess of 25 cm.h⁻¹, whilst works B has good structural engineering but rates usually less than 15 cm.h⁻¹. Works C is an old works, scheduled for closure, where rates are usually less than 10 cm.h⁻¹.

Table 1: Seasonal performance of slow sand filters (mean values)

Season	Works A		Works B		Works C	
	POC $\mu g.l^{-1}$	Chl.a $\mu g.l^{-1}$	POC $\mu g.l^{-1}$	Chl.a $\mu g.l^{-1}$	POC $\mu g.l^{-1}$	Chl.a $\mu g.l^{-1}$
Jan-Mar	37	0.27	30	0.25	30	0.19
Apr-Jun	35	0.27	75	0.90	40	0.36
Jul-Sep	39	0.22	75	0.49	67	0.35
Oct-Dec	24	0.04	23	0.02	26	0.04
Year	34	0.20	50	0.42	41	0.23

From the analysis of many years of data from a variety of works, it has been possible to define the following standards as shown in table 2:-

Table 2: Operational water quality standards for particulate material.

	excellent	good	poor	action needed
POC	$<50\mu g.l^{-1}$	$50-100\mu g.l^{-1}$	$100-150\mu g.l^{-1}$	$>150\mu g.l^{-1}$
Chlor	$0.1\mu g.l^{-1}$	$0.1-0.5\mu g.l^{-1}$	$0.5-1.0\mu g.l^{-1}$	$>1.0\mu g.l^{-1}$
TPN	$<200.cm^{-3}$	$200-500.cm^{-3}$	$500-900.cm^{-3}$	$>900.cm^{-3}$
TPV	$<0.10ppm$	$0.10-0.25ppm$	$0.25-0.40ppm$	$>0.40ppm$

where TPN is Total particle numbers, TPV is Total particle volume and ppm is parts of particle volume per million of sample volume.

A mature filter bed, operating at temperatures above 10°C, will normally produce filtered water which is free of coliform bacteria in a 100cm³ sample. However, immediately after cleaning or resanding, there may be significant numbers of faecal indicator organisms in the filtrate. Under normal conditions of surveillance, it is Thames Water's policy not to return a filter to supply, if the Escherichia Coli concentration in the general well (the well where all the filtered water is mixed before disinfection) would exceed 10CFU.100cm⁻³ (CFU are colony forming units). In practice, this means that from March to December, filters can be returned to supply immediately after cleaning. However, after resanding, filters must be run to waste until the bacteriological criteria for the general well can be met. It is an unfortunate paradox that the waste filtration rates are often too slow to produce optimum conditions for bacterial removal (4).

In recent full scale prolonged trials at filtration rates in excess of 30cm.h^{-1}, it has become clear that there may be a low temperature limitation on the capacity of a filter to remove faecal indicator organisms. When water temperatures were below 4°C, it was not possible to sustain average concentrations of less than 50 E. coli.100cm^{-3} in the general wells in a works filtering at over 30cm.h^{-1}, whereas the older works which filtering at less than 20cm.h^{-1} were usually within the criterion of 10 E. coli.100cm^{-3}. These low temperature effects are specific to bacterial removal, and overall particulate quality in terms of TPV (total particulate volume), TPN (total particle numbers), POC and chlorophyll a appear to be independent of flow rate in winter. Under these conditions, increased monitoring in necessary with greater attention to the reliability of subsequent disinfection. This paper does not purport to be a general review of all aspects of particle removal, and other parameters, such as virus removal, are well reviewed elsewhere (5).

HYDRAULIC BEHAVIOUR AND SAND MANAGEMENT

1. Background

The present operation of London's slow sand filters is constrained by many factors which include rate of flow, structural integrity, maximum pressure differential (head-loss) across individual filters and the works, weed formation, run duration and performance of cleaning plant. Most of these factors depend on the quality and quantity of particulate matter which is trapped in and on the sand.

During the run of a filter, most of the restriction to flow arises from material which originates from either a) the raw water, or b) growth within the filter. Operated over winter at 20cm.h^{-1}, the approximate ratio of allochthonous to autochthonous material is about 1:3, in spring it may be 1:2 whereas in summer it may be in the range 6-8:1. The situation is further complicated by the nature of the autochthonous material e.g. many of the reservoir filamentous diatoms will continue to grow on the sand surface whereas the blue-greens will not. Although the precise figure varies widely, a filter will usually require cleaning when its content of trapped particulate carbon is in the range 100-300gC.m^{-2}.

After cleaning, the filter bed is put back into supply and continues to be operated until either a) its head-loss exceeds a predetermined value, or a predetermined run time is exceeded, or there are operational reasons (such as the availability of manpower and plant) which override head-loss and run-time. At the end of the run, the surface layer (about 2.5cm) is removed by mechanical skimming. The bed is then back-charged, and returned to supply at a rate which depends on both the advice of scientists and local working practices. The "conditioning period" (the time to establish full rate) can be very short if beds are cleaned quickly.

The Operational Science group has analysed the hydraulic and quality information from several hundred filter runs on a range of treatment works. It has been possible to characterise filter runs, and build a model to assist at times of crisis.

The repeated skimming of a bed brings about a progressive decrease in the depth of the filtration medium until it becomes too thin to be certain that there will be adequate removal of particles including bacteria. For London's filters, this minimum depth is 30cm. When this minimum sand depth is reached, it is necessary to restore the bed to its full thickness. This is done in one of three ways:-

1)   Resanding. The bed is double-skimmed, and clean sand is then put on top until the full depth is restored

2)   Trenching. After final skimming, a longitudinal trench is dug down as far as the support gravel and the displaced sand is put in a long mound alongside the trench. The trench is then filled with clean sand. The next trench is dug alongside the previous one, and its contents are placed on top of that previous trench. The process is repeated until the bottom layer of the bed contains entirely clean sand while the top layers contain the older sand (6).

3)   Deep-skimming. When the sand depth falls to its minimum, all the old sand is completely removed down to the support gravel. This is achieved by repeated skimming.

If a bed is only resanded, the lower layers of the medium progressively become less permeable as they become clogged by silt, and concreted by precipitated calcium carbonate. This deterioration is slow, and with low overall rates of filtration, its effects may not be perceived over periods of up to 10 years.

Scientists have worked with engineers to compare the various resanding processes, and advise on their consequences.

2.   Characterising filter runs

Because, at any one time, head-loss (h) is directly proportional to flow, the head-loss can be corrected to a standard flow to produce a "normalised head-loss", h_v (in cm) where the subscript, v, is that standardised flow rate in cm.h^{-1}. When hv is plotted against total quantity of water treated per unit area of sand (Q in m), the following observations are made about the normalised head-loss:-

(a)   it usually falls in the first few days of the run.

(b)   thereafter, it usually increases exponentially at rates which depend on season.

(c)   its rate of increase (slope) may sometimes change very rapidly, and the probability of this occurring is dependent on season, being very low in winter, and very high in late spring.

(d)   its value extrapolated to the beginning of the run ("starting head-loss") is a function of sand quality and water temperature.

From observation (b), a general model of filtration is:-

Equation (1):-

$$(h_v)_Q = (h_v)_o . e^{BQ}$$

where h   is the head-loss (cm)

   h_v is the normalised head-loss (cm)

   $(h_v)_o$ is the starting normalised head-loss (cm)

   v   is the standardised flow rate (cm.h^{-1})

   Q   is the total quantity of water filtered per unit area of sand (m)

   B   is the rate of increase of head-loss

The normalised head-loss (NSH) is calculated using a standardised flow rate of 20cm.h^{-1} unless stated otherwise.

The principles of this model are used to define a graphical representation for filter runs. Typical filter behaviour is shown in Figures 2(a), 2(b) and 2(c). Three aspects of each run can be used to describe it:-

   (a)   the normalised starting head-loss

   (b)   the rate at which head-loss is increasing at a particular time

   (c)   the pattern of head-loss change over the entire run.

Under (c), it is possible to discern three main classes of head-loss development over the entire run. These classes are called, for convenience, "standard" (S-type), "jacked-up" (J-type), and "jacked-up-recovered" (R-type).

A typical S-type run is shown in Figure 2(a). The normalised head-loss increases exponentially with time (and water filtered provided the rate is reasonably constant). Any deviations from the trend are not significant. This pattern is most frequently observed in winter when biological activity and rates of change are low. Typically, the head-loss increases by 0.4 - 0.5 per cent for each metre of water filtered, i.e. the head-loss doubles for each 140 - 175m of water filtered.

A typical J-type run is shown in Figure 2(b). In the early stages of the run, it behaves as an S-type. However, there is a sudden change in which the rate of increase in head-loss itself increases by a factor between 5 and 10. The period from the start of the run to this discontinuity depends mainly on season varying from over 200 days in winter to less than 12 days in spring. Although the fundamental reasons for J-type runs are associated with raw water qualities and changes in the ecology of the filter skin, it is not yet possible to use fundamental science to predict the occurrence of such events. However, the probability of "jacking-up" depends on run time and season, and the statistical analysis of filter runs has enabled scientists to identify the critical times of year.

An R-type run is shown in Figure 2(c). The run starts as an S-type, then changes to a J-type, but then there is a further series of changes until a point is reached where head-loss begins to decrease with time and continues to fall until it reaches a point it would have reached had the run been an S-type run. The rate of change of head-loss then changes to the rate it had at the beginning of the run.

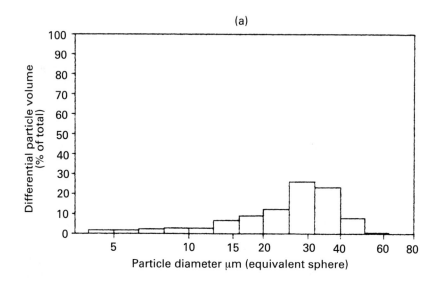

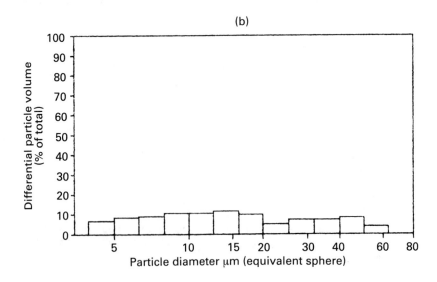

Figure 1:   Typical size distributions for (a) particles entering, and (b)
particles leaving a filter operating in summer.

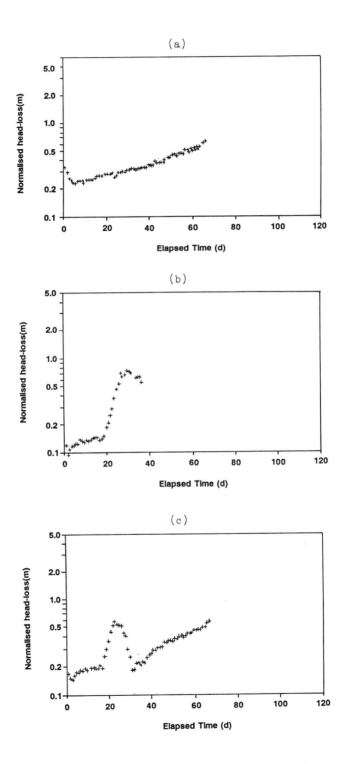

Figure 2:   Typical filter behaviour showing (a) S-type run, (b) J-type run, and (c) R-type run.

It has been possible to prepare a generalised diagram which incorporates the seasonal pattern of the rate of change of head-loss ('B' in equation (1)), with the run classification. This diagram is shown in Figure 3.

3. Resanding strategies

From equation (1), it can be seen that the maximum run length of a filter is determined by the permeability of the sand at the start of the run (starting head-loss), the final head-loss and the rate at which head-loss increases with time.

Following a major crisis, in which no filters were cleaned for 13 weeks, scientists analysed all filtration data in order that they could recommend a optimum cleaning strategy to the works management. This analysis revealed that many of the beds had starting head-losses which indicated that the lower layers of sand were compacted and silted up. From this exercise two strategies emerged. The first was an order of bed cleaning, and the second was a recommendation that the deep layers of sand, in key beds, should be replaced, and the results studied to compare the performance of those beds with resanded beds.

In most cases, the effect of deep-skimming was apparent from the nature of the filter run before and after the process. Figures 4(a) and 4(b) show this process reduced the normalised starting head-loss (NSH) from 0.5m before to less than 0.2m after the process: in effect the sand had 30cm less head-loss assuming the bed was operated at $0.2m.h^{-1}$.

A year's data was summarised into monthly units, and the results are summarised in Table 3. Over the year, the mean NSH prior to deep-skimming was 0.59m, whilst afterward, it was 0.28m. Differences in length of run are not significant, because cleaning was often based on run length itself. However, the more permeable sand allowed more water to be filtered, and deep-skimmed beds produced an average of 24 per cent more water than the others.

The data were further analysed to determine the rate at which the deep layers lost their permeability in the runs following deep-skimmimg. Although there was some variation, the average rate at which NSH increased was about $0.2m.y^{-1}$. At the filtration rates operated in the uprated London works, it would appear that the bottom layer should be removed when the sand depth is down to 30cm.

During conventional resanding, there is considerable disturbance of the medium. It was often observed that this process brought about a reduction in NSH in the immediate succeeding runs. However, the effect was short-lived, and within a few runs, the NSH could exceed 0.5m.

A similar analysis of data has been carried out on a works in which trenching is practised instead of deep-skimming. Trenching also produces a marked and robust reduction in NSH, indeed the NSH's were often lower than on the works using deep-skimming. This difference was not significant for comparative purposes, because it may have arisen from different gradings of sand, and different standards for clean sand.

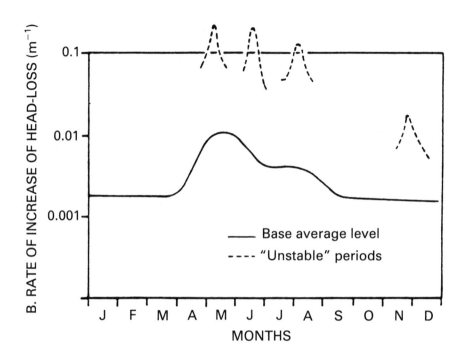

<u>Figure 3</u>:   Seasonal rate of change of head loss.

(a)

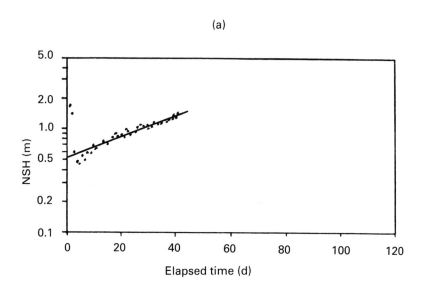

(b)

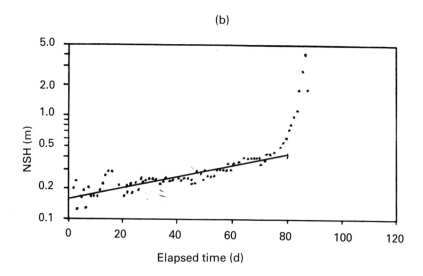

Figure 4: Reduction in NSH (a) prior to, and (b) post deep-skimming.

Table 3: Comparison of slow sand filter behaviour prior to and post deep-skimming.

Month	Prior				Post			
	NSH (m)	Flow (m.d^{-1})	Run (d)	OBS (n)	NSH (m)	Flow (m.d^{-1})	Run (d)	OBS (n)
Aug. 1984	0.39	2.68	58	4	0.21	3.13	72	4
Sep. 1984	0.48	2.72	72	8	0.10	3.37	91	3
Oct. 1984	0.46	2.35	90	3	0.28	3.05	120	6
Nov. 1984	0.88	2.47	94	3	0.25	3.32	100	2
Dec. 1984	0.36	2.81	85	4	-	-	-	0
Jan. 1985	-	-	-	0	-	-	-	0
Feb. 1985	1.22	2.76	49	2	0.46	3.16	46	4
Mar. 1985	0.68	2.75	44	12	0.43	2.94	51	8
Apr. 1985	1.03	2.81	31	3	0.29	3.65	27	12
May 1985	0.40	2.64	56	8	0.29	3.33	46	14
Jun. 1985	0.66	2.66	49	4	0.22	3.60	50	10
Jul. 1985	0.46	3.05	76	3	0.22	3.47	78	6
Total	31.77	146.04	3291	54	19.66	230.85	4009	69
Mean (weighted)	0.59	2.70	61		0.28	3.35	58	

When filters are operated at rates in excess of 25cm.h^{-1}, Thames Water recommends that filter beds are deep-skimmed or trenched when they have a sand depth of 30cm. The choice between the two processes is at the discretion of local management. On other works, where rates are lower, we recommend that the state of each bed should be reviewed before deciding how to resand.

Thames Water has found that there is recent evidence that filters operated at high filtration rates accumulate organic material relatively rapidly within the deeper layers of the sand column. Figure 5a shows the POC concentration shortly after trenching whilst figure 5b shows the results after some nine runs. Particle size analysis has shown that filters that contain accumulations of trapped material at depth may output some of this material in the filtrate, some of which may form a silt layer in the contact tank.

4. Sand washing and cleaning standards

Sand, which is removed from beds, is reclaimed and washed. It has been necessary to set standards for new sand washing plant, and to ensure quality control on existing systems. This work has shown how plant performance can deteriorate in the short and long term. Thames Water's current standards are given in Table 4.

(a)

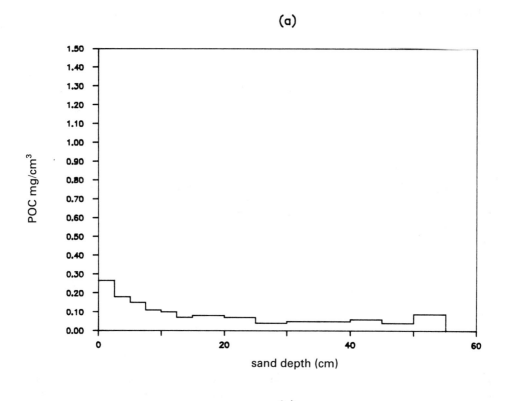

sand depth (cm)

(b)

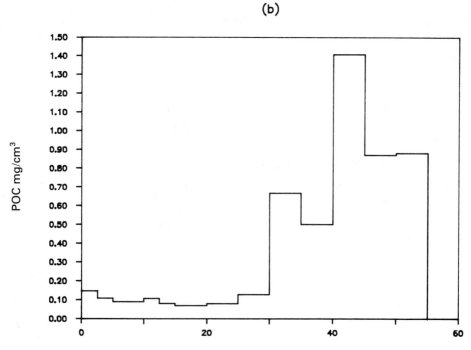

Figure 5:   POC concentration in sand column (a) shortly after trenching, and
            (b) after nine filter runs.

Table 4: Standard of sand cleanliness

POC   0.1kg.m^{-3}

Silt   1.0kg.m^{-3}

These standards may be changed with different sand washing techniques.

The standards in Table 4 are arbitrary, and reflect the technological capabilities of well operated systems. Thames Water scientists have regularly measured the silt and carbon content of columns of sand in filters, but it is not yet possible to quantify the relationship between silt content and sand permeability. However, we know that clean sand, meeting the criteria in Table 4, will produce a sand column which has an NSH of less than 0.2m, and there is no serious build-up of sand derived silt in the contact tanks which follow the filters.

FILAMENTOUS GREEN ALGAE ON FILTER BEDS

The algal flora of slow sand filters has been extensively reviewed by Bellinger (7) and Brook (8). In winter and spring, the autochthonous algal flora is usually dominated by filamentous diatoms. These algae are easily removed in conventional skimming, and are amenable to washing.

In summer and autumn, the algal flora is dominated by filamentous green algae, usually Cladophora sp.. This alga can cause serious operational and water quality problems. It forms a blanket which is too tough to be skimmed off, and the blanket must be manually raked into piles to allow the bed to be skimmed. These piles of weed are one of the few sites where E. coli are known to multiply naturally outside the mammalian gut. If the piles are left as long as 24h, a culture of E. coli can form and this seeps into, and contaminates, the underlying sand (9).

The dynamics of the growth of Cladophora sp. on filter beds have been investigated using operational data. The results of this work have been used to prepare the control diagram (Figure 6), and this indicates that filters cleaned before 1st June with run times of less than 50 days will not have significant crops. Thereafter, starting data and run time determine the crop. If the run time is longer than 20 days, a crop is established which doubles every 7-10 days reaching levels up to 100m^{3}.ha^{-1} (2). In terms of filter bed management, there are clear advantages in having runs of less than 30 days in summer, and these advantages would have to be offset against the disadvantages of not running a filter to its maximum head-loss. During uprating of filter beds, runs have often been shorter than 30 days in summer, and as a consequence, there has been a significant reduction in the problems associated with blanket weed.

SUMMARY AND CONCLUSIONS

Scientists, in the Thames Water Authority and it's predecessor bodies, have over fifty years experience in advising on the use and design of slow sand filters. As filtration rates have increased, the traditional water quality

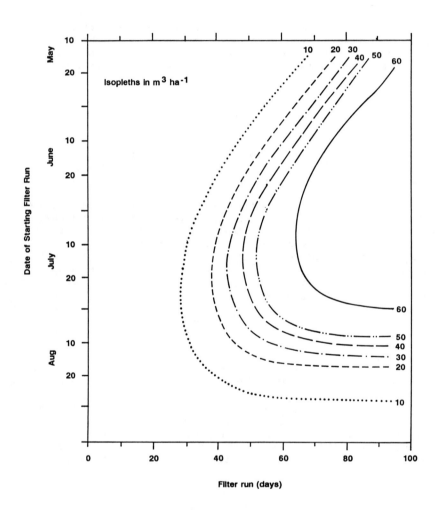

Figure 6:   The dynamics of the growth of Cladophora sp. on filter beds.

determinants of bacteria and turbidity have had to be supplemented by other determinants which are more sensitive to process behaviour. The paper has shown how POC, chlorophyll $\underline{a}$, TPN, TPV and microscopical examination can be used.

Statistical analysis of hydraulic data has been used to build up models of filtration. Such analysis has indicated the importance of minimising the normalised starting head-loss particularly when filters are uprated. This, in turn, requires that the sand medium should have minimum standards of cleanliness. The hydraulic and analytical data should be used in advising on the management of resanding and sand washing.

Problems, from filamentous green algae, can be minimised by optimising the run-length at critical times of the year. Similar solutions can be applied to managing other natural problems.

For most water quality determinants, there have been few problems in uprating from approximately 20cm.h^{-1} to 40cm.h^{-1}, and this is in agreement with previous studies (9). The most serious problem has been in the break-through of $\underline{E.coli}$ in very cold water conditions. With appropriate attention to the reliability of the disinfection process, under these conditions, the problem is manageable.

Uprating the slow sand filtration process has been a very complex task, but in its uprated form it will continue to be a key unit process in the development of London's water treatment systems.

ACKNOWLEDGEMENTS

The authors wish to thank Mr.J.A.P.Steel for initiating much of the work in this paper, and for the provision of figure 6. They wish to thank their colleagues at the Wraysbury laboratory who have analysed much of the basic data.

The paper is published by kind permission of Mr.G.A.Thomas (General Manager, Technology and Development, Thames Water), but the views expressed herein are entirely those of the authors.

REFERENCES

1.   Ridley, J.E., (1967): Experiences in the use of slow sand filtration, double sand filtration and microstraining. - Proc.Soc.Wat.Treat.Exam. 16: 170-191.
2.   Steel, J.A., (1975): The management of Thames Valley reservoirs. - Proc. WRC Symp. "The Effects of Storage on Water Quality", publ. by WRC, Medmenham, U.K.: 371-419.
3.   Toms, I.P., (1987): Developments in London's water supply system. - Arch.Hydrobiol.Beih.: Stuttgart.: 28: 149-167.
4.   Burman, N.P., (1962): Bacteriological control of slow sand filtration. - Effluent and water treatment journal. 2:12: 674-677.
5.   Poynter, S.F.B. & Slade, J.S., (1977): The removal of viruses by slow sand filtration. - Prog.Water.Technol.:9: 75pp

6.   Van Dijk, J.C. & Oomen, J.H.C.M., (1978): Slow sand filtration for community water supply in developing countries: a design and construction manual. Technical paper No.11, WHO, Geneva, 177pp.

7.   Bellinger, E.G., (1979): Some biological aspects of slow sand filtration – J.Inst.Wat.Eng.Sci.:33: 19-29.

8.   Brook, A.J., (1955): The attached algal flora of slow sand filter beds of waterworks. – Hydrobiol:7: 103-117.

9.   Windle Taylor, E., (1974): Forty-forth rep. results bact.chem.exam. London's waters for the years 1971-1973. – Metropolitan Water Board, London.

Authors address:

I.P.Toms/R.G.Bayley, Thames Water Authority, Wraysbury Reservoir Laboratory, Coppermill Road, Wraysbury, Nr.Staines, Middlesex, England. TW19 5NW.

1.3   SLOW SAND FILTRATION IN THE UNITED STATES

G. Logsdon and K. Fox — Drinking Water Research Division, Water Engineering Research Laboratory, U.S. Environmental Protection Agency, Cincinnati, Ohio, USA 45268

ABSTRACT

Interest in slow sand filtration has increased dramatically in the United States in the past ten years. Research conducted to evaluate removal of Giardia cysts and bacteria, showed that slow sand filtration is very effective in removal of these contaminants. Slow sand filters are much simpler and easier to operate than plants that employ coagulation and rapid filtration. They are very well suited for treatment of low-turbidity, unfiltered surface waters and would be ideal for small utilities serving from 25 to 3000 persons. The U.S. EPA estimates that about 1000 slow sand filters may be built as a result of proposed EPA regulations on surface water treatment.

INTRODUCTION

The history of slow sand filtration in the United States has been one of slow and reluctant acceptance, as contrasted to the European experience. Installation of both slow rate and rapid rate filtration plants took place in the 1890's and 1900's, but early in this century, rapid filters gained popularity. By 1940, the United States had about 100 slow sand filtration plants, whereas about 2,275 rapid rate plants had been constructed (1). Rapid sand filters, rather than slow sand filters, became the overwhelming choice of design engineers and water utilities because of some important advantages. Treatment consisting of coagulation, flocculation, sedimentation and filtration at 4.9 m hr^{-1} (m^{3}m^{-2}hr^{-1}) was shown by Fuller (1) at Louisville, Kentucky to be capable of treating a wide range of raw water qualities. In particular, river water that is muddied by runoff from farm land or from dry land susceptible to erosion because of a lack of vegetation can be successfully treated to reduce turbidity from hundreds of Nephelometric Turbidity Units (NTU's) to less than 1 NTU by means of a properly designed and operated rapid rate filtration plant. Such waters would quickly clog slow sand filters. Additional advantages for medium sized and large water utilities were the reduced land requirements in populated areas and the lower labor requirements in operation/maintenance for rapid filters vs. slow filters.

Because of perceptions about disadvantages of slow sand filtration, some water utilities that had originally used this process later abandoned the filters and replaced them with rapid sand filters. Pittsburgh and Philadelphia, Pennsylvania; Albany, New York; Indianapolis, Indiana; and Washington, D.C. are in this category. At Denver, Colorado the Kassler slow sand filtration plant was removed from service after a new rapid rate plant was brought on line in recent years.

Some water utilities have continued to use existing slow sand filters but built rapid rate plants when additional treatment capacity was needed. Included in this category are Springfield, Massachusetts and Auburn, New York. Sites such as these, with both processes treating a common source water, offer the possiblity for full scale comparative evaluations.

RENEWED INTEREST

In the late 1970's and early 1980's, the potential for application of slow sand filtration in the United States was reconsidered, in part because of successful applications in other parts of the world. Limitations for use by large water utilities were recognized, but the process was considered to be a strong candidate for use by small systems, where requirements for land and labor would not be a serious drawback. In addition, slow sand filtration's efficient removal of bacteria and viruses was the basis for expectations that it also might be effective for removal of <u>Giardia</u> cysts.

Results of a survey of slow sand filtration practice were published by Slezak and Sims (2). They received completed survey questionnaires from 27 slow sand filter plants, most of which were located in the eastern United States. The largest number of plants was in the State of New York. Slezak and Sims reported that the plants included in their survey served small to medium sized communities. Eleven percent served populations exceeding 100,000. Communities of up to 1000 persons were served by 31 percent of those responding, and another 31 percent served those in the 1,000 to 10,000 range. Slightly over half of the plants had been built between 50 and 100 years ago, but one third (2) had been built within the past 25 years. About three fourths of the plants used reservoirs or lakes, and 22 percent used streams for the raw water source. Only one of the 27 used a ground water source. The mean filter cycle length at these plants ranged from 42 days in spring to 60 days in winter. About half of the plants surveyed had raw water turbidity that averaged 2 NTU. Filtered water turbidity was 1 NTU or lower at 85 percent of the plants. Total coliforms exceeded 20 organisms per 100 mL in the raw source water at half of the plants. After slow sand filtration, total coliform densities were reduced to less than 1 organism per 100 mL in approximately 90 percent of the facilities. Overall, most of the plants responding to the survey appeared to be doing a good job of water treatment.

RECENT RESEARCH

An apparent increase in outbreaks of waterborne giardiasis in the USA throughout the 1970's played an important role in the renewed interest in

slow sand filtration. Most giardiasis outbreaks had occurred at places where
the raw water was of high quality, i.e. low turbidity and not much evidence
of sewage contamination. The quality of raw waters in these areas appeared
to be suited for treatment by slow sand filtration. This process had a
demonstrated record of excellence with regard to removal of bacterial contaminants, and research in the United Kingdom (3,4) had developed data on
the removal of viral contaminants. Slow sand filtration was expected to be
quite effective for removal of Giardia cysts, but no data were available to
verify this. In the early 1980's, the U.S. EPA sponsored four projects to
develop more information on the capabilities of slow sand filtration.

The Drinking Water Research Division started a research project at the
U.S. EPA's Environmental Research Laboratory in Cincinnati in November, 1980
when a pilot-scale filter (Filter A) was constructed and placed into operation (5). Table 1 contains information on this and subsequent filters.
The sand source for Filter A was "fine builders's sand" from a local sand
and gravel quarry that was mining glacial deposits. The raw water used in
this pilot plant study was a low turbidity surface water (gravel pit water)
with algae and other biological life typical of a clean surface water. To
increase the concentration of bacteria in the source water, raw municipal
sewage (approximately 1 part in 2,500) was added to the raw water stored in
the holding tank.

The research results were encouraging. The initial ripening period for
the new sand bed lasted about 40 days. During this time, a schmutzdecke
developed, effluent turbidity decreased and total coliform counts in filtered
water declined steadily and approached 1 per 100 mL, even though the concentration in the spiked raw water ranged from 1 to 10,000 per 100 mL. During
the first 40 days of run 1, all 12 samples of filtered water tested for coliforms exceeded 1 per 100 mL. The remainder of run 1 and the next 9 runs,
lasted for a total of 751 days. In this time period, 163 filtered water coliform samples were analyzed, and only 9 were positive. Of these, 8 occurred
in run 2, in which raw water coliforms ranged from 1 to 25,400 per 100 mL.
The other positive coliform sample was noted in run 6, in which the geometric
mean raw water coliform count was 4,060 per mL. This filter was capable of
removing about 90 percent of the influent particles in the 7 to 12 um size
range (similar to size of Giardia cysts). The initial erratic effluent plate
count results were found to be related to bacterial growth on clear plastic
flexible tubing through which filtered water was discharged. When the tubing
was replaced with stainless steel, the heterotrophic plate count results in
the process water improved dramatically. Then bacteriological examination
for heterotrophic bacterial population revealed removal of about 99 percent
of the plate count bacteria.

These results were encouraging, so another pilot filter (Filter B) was
constructed. Details are also given in Table 1. Some important differences
in behavior were noted. After seven weeks of treating a reservoir water,
Filter B was switched to Ohio River water having turbidity that ranged from
0.4 to 23 NTU. This, coupled with the larger effective sizes of sand Filter
B media, resulted in poorer turbidity control than observed in Filter A.
Throughout the study, effluent turbidity levels gradually increased as Filter

Table 1. Pilot-Scale Slow Rate Filter Description*

	Sand Filter "A"	Sand Filter "B"	Activated Carbon Filter "B"
Inside Dimension m^2(sq ft)	0.21	0.1	0.1
Media	Sand	Sand	12 x 40 GAC
Effective Size, mm	0.17	0.29	0.55 - 0.65
Uniformity Coefficient	2.1	1.8	1.6 - 2.1
Depth, m	0.76	0.82	0.82
Mean Reservoir Height (above media) m	1.25	1.09	1.09
Flow Rate m^3/day	0.58 0.87*	0.29	0.29
Loading Rate m/hr	0.12 0.18*	0.12	0.12
Mean Reservoir Detention Time hr	10.6 7.07†	9.21	9.21
Empty Filter Bed Contact Time hr	8.69 5.79†	6.87	6.87

*From: Fox et al. (5)
†Flow rate increased 50 percent at start of fifth cycle.

B became clogged. After 5 filter scraping cycles, the filter was ineffective for meeting the 1 NTU MCL. Turbidity profiles developed during cycle 6, prior to taking the filter out of service, demonstrated 1.0 NTU, or greater, at each depth sampled. The effect of excessive turbidity loading of a slow sand filter is also seen in Table 2. Cycle length decreased and startup head loss increased while mean effluent turbidity increased with successive cycles. Clearly, the filter was clogged. Even a 5 cm scraping that returned the filter to low head loss at the start of cycle 6 yielded a disappointingly short run (Table 2). This was reminiscent of problems encountered with slow sand filtration in the Ohio River Valley several decades ago.

During its first two cycles, Filter A was examined for total organic carbon (TOC) and 7-day trihalomethane formation potential (THMFP) control. Removals averaged 19 and 18 percent, respectively. Because gravel pit water TOC was relatively high ($7.1 - 11.1$ mg L^{-1}) a capillary column, gas chromatographic profile was conducted and the water found to be very heavily laden with low-to-mid-molecular weight compounds relative to typical surface water profiles. This was believed attributable to a nearby oil dump in the gravel pit. For this reason gravel pit water was not treated in Filter B. The reservoir water and, later, Ohio River water influent to Filter B had a lower TOC ($1.5 - 3.2$ mg L^{-1}). Treating these waters, sand Filter B averaged 15 percent removal of both TOC and 7-day THMFP. On a percent removal basis, the two sand filters behaved similarly, but on a weight removal basis, sand Filter A was more effective.

Activated carbon Filter B's control of organics was very effective, but time dependent, as expected. Mean TOC and 7-day THMFP removals were 88 and 97 percent, respectively. Periodic profile sampling demonstrated a time-dependent wave front movement of organics and TOC through the adsorbent.

While research was underway at EPA, a filtration research project was undertaken at Iowa State University (6). A slow sand filter 2.74 m high, with a diameter of 0.76 m, was constructed. Media depth was 0.94 m. The effective size of the sand was 0.32 mm, and the uniformity coefficient was 1.44. This filter was operated from October, 1981 until December, 1982. During this time, 11 filter runs, ranging in length from 9 to 123 days, were conducted, with the filter operating at 0.12 m hr^{-1}. The raw water source was a gravel pit north of Ames, Iowa. Turbidity of this water often ranged between 2 and 6 NTU, but it was less than 1 NTU when the water surface was frozen, and exceeded 30 NTU at times during the summer of 1982.

Cleasby et al. (7) reported that the pilot slow sand filter produced water well below 1 NTU in all filter runs, even during the first two days of each run. This was when filtrate quality was somewhat inferior as compared to the quality of water produced during the remainder of each run. The turbidity of the last seven runs, conducted in May to December, (excluding the first two days of each run) averaged near 0.1 NTU. Filter performance gradually improved throughout the series of runs, with respect to turbidity, particle counts, total coliform bacteria, and chlorophyll-a. When results from the initial two days (re-ripening period) of each filter run are omitted

Table 2.   Slow Sand Filter B Operation*
e.s. = 0.29 mm
Rate = 0.12 m/hr

Cycle Number	Cycle Start Headloss cm	Cycle End Headloss cm	Cycle Length Days	Mean Turbidity, NTU		Media Scraped cm
				Influent	Effluent	
1	1	44	98	7.6	0.59	1.0
2	3	43	62	2.4	0.63	1.0
3	6	38	40	2.7	0.73	1.0
4	10	39	28	4.5	0.77	1.0
5	24	41	6	16	2.4	5.0
6	2.5	38	16†	16	3.1	†

*From: Fox et al. (5)
†Filter taken out of service

from calculations, reductions of turbidity, particle count and total coliform bacteria were always 90 percent or greater, and were often close to 99 percent. During the final seven runs (omitting data from the initial two days of operation after scraping) average turbidity removal was 97.8 percent or more, 7 to 12 um particle removal was 96.8% or higher, coliform bacteria removal was 99.4 percent or greater, and average chlorophyll-a removal was 95 percent or more. Filter run length was 41 days or less in 9 of 10 runs in which the filter was operated to a terminal head loss of 1.35 m of water. A winter run, part of which occurred when the source water was under ice cover, lasted 123 days. Serious algal blooms occurred in summer, and one run then was only 9 days long.

Cleasby et al. (7) concluded that using turbidity measurements by themselves would not be adequate for predicting filter run length. In addition to turbidity, the concentration of algal cells had a strong influence on run length. Chlorophyll-a concentrations of less than 5 mg m^{-3} were measured when runs were longer than 30 days. Under these conditions, raw water turbidity averaged 4 to 5 NTU, but short term excursions as high as 16 NTU were noted. They suggested that chlorophyll-a of less than 5 mg m^{-3} and turbidity of 5 NTU or less were appropriate upper limits for application of slow sand filtration. They also suggested that collecting raw water data on turbidity, suspended solids, and chlorophyll-a for at least one year, with inclusion of seasonal extreme conditions, would be essential to a sound decision-making exercise in which filtration alternatives were considered. Results of this project showed that slow sand filtration could be an effective treatment technique for small water supplies if the raw water quality was appropriate.

Another slow sand filter project was funded at Colorado State University for evaluation of removal of turbidity, total coliform bacteria, and Giardia cysts (8). During the first phase of this work (9), three 0.30 m diameter formity coefficient = 1.46) over 0.46 m of coarse sand and gravel support media. These filters were operated from August, 1981 until December, 1982 at rates of 0.04, 0.12 and 0.40 m hr^{-1}, using a common raw water source. Raw water from Horsetooth reservoir was stored in a 1,400 L holding tank at temperatures of 5° or 15°C. Giardia cysts were spiked into the raw water at concentrations ranging from 50 cysts L^{-1} to 5,075 cysts L^{-1}. Primary settled sewage was added to the raw water in the holding tank to raise the concentration of coliform bacteria.

Results of the first phase of this research focused on the effects of the filtration rate and the condition of the filter bed (new, ripened, just scraped). The relationship of these factors to total coliform removal is presented in Table 3. Comparisons are made in Table 3 by normalizing data to the equivalent of 1.00 x 10^6 coliforms per 100 mL applied in the raw water, with filtration results shown in terms of numbers of coliforms in the filter effluent. The data show trends for decreased removal (more organisms passing) as filtration rate increases and as the condition of the sand bed is disturbed. In contrast to these results, Giardia cyst removal did not appear to be influenced by the presence or absence of a schmutzdecke. Bellamy et al. (9) reported cyst removals ranging from 98.7 to 100 percent for a mature sand bed with no schmutzdecke, and 99.77 to 100 percent for a

Table 3.   Influence of Slow Sand Filtration Rate and
Filter Bed Condition on Coliform Passage*

Filter Bed Condition	Coliforms in Effluent if Influent Contains 10^6 per 100 mL		
	Filtration Rate		
	0.04 m hr^{-1}	0.12 m hr^{-1}	0.40 m hr^{-1}
Mature sand, Established schmutzdecke	40	1,000	4,000
Mature sand cleaned, No schmutzdecke	300	10,000	47,000
Mature sand cleaned and disturbed, No schmutzdecke	10,000	28,000	62,000

*From: Bellamy et al. (9)

mature sand bed with a schmutzdecke. Cysts were detected in filter effluent
in 4 of 15 of the runs with no schmutzdecke, and in 12 of 24 runs with a
schmutzdecke developed. Filtration rate did seem to have some influence on
Giardia cyst removal. When all data were used to calculate average removal
for each filtration rate, the percentage exceeded 99.99 for both the 0.04
and 0.12 m hr^{-1} rates, but it was just over 99.98 for the 0.40 m hr^{-1} rate.

One aspect of the results was completely unexpected. Turbidity of the
raw water ranged from 2.7 to 11 NTU, but removals were poor -- from 27 to 39
percent. Horsetooth Reservoir contained fine particulate matter, as indicated
by passage of 30 percent of the turbidity through a 0.45 um pore size membrane filter. Further investigation of the nature of the turbidity problem
was carried out in the second phase (10) of this project.

To obtain information about the effects of temperature, bed depth, media
size, and the extent of the influence of biological activity in the filter
bed, six 0.3-m diameter filter columns were set up and operated at a rate
of 0.12 m hr.$^{-1}$. The control filter had a bed depth of 0.97 m, 0.29 mm
effective size sand (uniformity coefficient = 1.5) and was operated at 17°C.
Other filters had different bed depth (0.48 m), sand sizes (0.13 mm and
0.62 mm) and temperature (5° and 2°) during the 11 month operating period.
Finally, to develop data on how nutrient loading and the biological community
within the filter influenced filtered water quality, two filters like the
control filter were operated in different ways. One was chlorinated to
maintain a residual of 5 mg L^{-1} between test runs. This reduced the extent
of biological growth in the media in the filter. The other filter treated
Horsetooth Reservoir water supplemented with sterile synthetic sewage (a
modified version of Piper's synthetic sewage) sufficient to cause a dissolved
oxygen drop of about 4 mg L^{-1} through the filter. The nutrient material,
described in the project report (8), was added to promote biological growth
within the bed.

Results for removal of total coliform bacteria were what would be expected with regard to bed depth, media size, and temperature. Removal declined as bed depth decreased, and as temperature decreased. Results of
nutrient augmentation and biological growth inhibition indicated the importance of the biological community in the treatment process (Table 4). The
extent of development of the biological community strongly influenced removal of total coliform bacteria. Reduction of turbidity also improved cón-
siderably. An X-ray diffraction analysis (11) of the residue on a 0.22 um
pore size membrane filter indicated that it consisted largely of kaolinite,
illite, and montmorillonite clays, and Pavoni (12) has shown that natural
exocellular polymers can flocculate clays. Bellamy et al. (10) suggested
that higher levels of production of exocellular polymer by the enhanced
biological community in the filter receiving supplemental nutrients may have
contributed to that filter's more efficient reduction of the particulate
matter present in the raw water.

While the project at Colorado State University was being operated under
controlled conditions to define the influence of various factors on slow sand
filter performance for control of turbidity, bacteria, and Giardia cysts, a

Table 4.   Effect of Condition of Filter on Removal
of Turbidity and Bacteria*

Parameter	Filter Chlorinated Between Runs	Control Filter	Filter with Nutrient Addition
	Percent Reduced	Percent Reduced	Percent Reduced
Total Coliform	60.1	97.5	99.9
Standard Plate Count	− 89	− 41	58
Turbidity	5	15	52

*From: Bellamy et al. (10)

project was conducted at McIndoe Falls, Vermont (13) to develop data on the
efficacy of a full-scale slow sand filter for controlling those same contaminants. This facility had two filter beds, each with an area of 37 m^2.
The plant operated at a rate of 0.08 m hr^{-1}. Because Giardia cysts and
sewage were spiked into the raw water in order to define the limits of treatability, all of the water except that needed for sampling purposes was wasted.
At this plant, cyst removal efficacy was influenced by water temperature.
When raw water temperature ranged from 7.5°C to 21°C, cyst removal was 99.98
to 99.91 percent. During one experiment at 0.5°C, both Giardia cysts and
sewage were spiked in the raw water at the same time. The biological
capacity of the filter seemed to have been stretched to the limit, as cyst
removal dropped to 93.7 percent. Treatment of both sewage and Giardia cysts
would not be expected to be a problem for plants treating pristine, unpolluted waters, but locating slow sand filters downstream from sewage treatment
plants, sewer mains, or septic tanks could lead to drinking water quality
problems if a treatment failure or main break took place. This supports the
concept that a sanitary survey ought to be undertaken in conjunction with
plant design studies.

Turbidity reduction under ambient water quality conditions was quite
good. Influent turbidity ranged from 0.2 NTU to 59 NTU and averaged 1.4 NTU.
Filtered water turbidity averaged 0.22 NTU, and 90% of the samples were 0.5
NTU or lower. The highest value recorded during the study was 8 NTU.

For ambient raw water, the average total coliform reduction was 87 percent, whereas the average was 89 percent for heterotrophic plate count bacteria. When sewage was spiked into the raw water, coliform removal averaged
98 percent.

One aspect of the slow sand filter performance that was not encouraging
was its inability to reduce trihalomethane precursors. The water source for
this filter was a small brook that flowed from a marsh (about 20 hectares)
and Coburn Pond (about 4 hectares). In a previous water quality study (14),
color ranged from 0 to 50 color units, and averaged 24 color units. Raw
water sampled by Pyper in February, 1982 had a color of 5 units. He reported
that after 10 days of exposure to free chlorine, the raw water and the slow
sand filter effluent contained similar amounts of trihalomethanes. Obviously,
some types of the materials that can form disinfection byproducts upon chlorination are not removed efficiently by slow sand filtration.

Another evaluation of an operating slow sand filter verified the Giardia
cyst removal capability of this process. Seelaus et al. (15) reported that a
slow sand filter was built and operated at Empire, Colorado in the mid-1980's,
following an outbreak (110 cases) of waterborne giardiasis caused by unfiltered
water in this small community in the fall of 1981. The raw water source, Mad
Creek, is about 2,700 m above sea level, does not exceed 8°C in summer and is
near 0°C in winter, and generally has a turbidity less than 1 NTU. The plant
was designed to operate at a rate of 0.26 m hr^{-1}, with a sand bed 1.2 m deep.
A local sand having an effective size of 0.21 mm and a uniformity coefficient
of 2.7 was used. A Giardia cyst sampling program was conducted and Giardia
cysts were collected between March 16 and April 27, 1985. The cyst members

ranged from 14 to 48 cysts, in raw sample test portions ranging from 1892 to 3860 liters. Filtered water samples collected on the same days contained no Giardia cysts. In four of five cases, filtered water sample volume was similar to or greater than raw water sample volume. These winter time results showed that even when the water was so cold that a thin layer of ice formed on the water surface within the filter plant, cysts were effectively removed from this clear mountain water.

Bryck et al. (16) reported on pilot testing, design, construction and evaluation of a slow sand filter at 100 Mile House, British Columbia, Canada. The facility has three filters, each 43 m long and 6 m wide. This plant operated at an average rate of 0.08 m hr^{-1} and a maximum of 0.19 m hr^{-1} during its first year of operation (November 1985 to November 1986). In a water quality evaluation program, Giardia cysts were detected in 11 of 21 raw water samples. No Giardia cysts were detected in any of the 22 filtered water samples. This confirmed the earlier work indicating slow sand filtration was very effective for Giardia cyst removal.

In addition to conducting a survey of slow sand filtration practice in the USA, Slezak and Sims (2) also conducted research with a slow sand filter having a capacity of 75 m^{3}d^{-1}. This plant, which had a filter bed 0.81 m deep, with 0.18 mm effective size sand, reduced turbidity from 5 NTU (raw) to 0.28 NTU (filtered), while operating at a rate of 0.20 m hr^{-1}. Their test filter, located at Logan, Utah, used locally available unsieved sand with a uniformity coefficient of 4.4.

The authors generally observed about a one $\log_{10}$ decrease in turbidity over a six month period of operation. Seven pairs of raw and filtered water samples were analyzed for total coliforms. Influent concentrations ranged from 4 to 50 colonies per 100 mL, whereas one coliform or none were detected in 100 mL concentrations of the effluent. Samples were collected for analysis of particle count. The pilot filter consistently reduced the concentration of 7 to 12 um particles by about three $\log_{10}$.

Cullen and Letterman (17) conducted a field evaluation of slow sand filter maintenance and the influence of maintenance practices on water quality. Seven filter plants located in the State of New York were evaluated. During the study period, the researchers collected data on a total of ten schmutzdecke scraping or resanding operations. In four of the ten maintenance operations, some evidence of a filter ripening period, or initial improvement period, was noted, as indicated by improvement in filtered water turbidity or particle counts, or both. The ripening periods ranged in length from 6 hours to 2 weeks. The quantity and nature of the particulate matter being removed by the slow sand filter appeared to influence the length of the ripening period. Recently scraped filters were less successful in reducing high concentrations of influent particles than filters that had been operating long enough to be in a ripened condition.

The authors (17) also investigated maintenance costs and reported that scraping a filter required about 5 hours of work per each 100 m^2. Resanding that involved replacement of 15 to 30 cm of sand required about 53 hours of

work per 100 m^2. The plants visited were small (up to 23,000 m^{3}d^{-1}) and
all employed shovels for scraping. Only one used shovels and wheelbarrows
and no mechanically powered or hydraulic equipment to remove sand from the
filter box.

Letterman and Cullen (18) reported that one water utility included in
the survey was operating both a rapid sand filter plant and a slow sand
filter plant. In July, 1984, both plants were cleaned. Both were returned
to service, treating the same water, because the rapid filter plant was not
employing coagulation at that time. The effluent turbidity from the slow
sand filter was 0.3 NTU, whereas from the rapid filter it generally ranged
from 1.3 to 2.0 NTU, with a raw water in the 2 to 3 NTU range. The particle
count (2 to 60 um range) in the slow sand filtered water was about one tenth
of that in the rapid filter effluent. Whereas slow sand filters can be very
effective without chemical pretreatment of the water to be filtered, rapid
rate filters are not.

Schuler and Ghosh (19) recently presented data on the capability of a
pilot scale slow sand filter to remove Giardia cysts and Cryptosporidium
oocysts. Their research at Pennsylvania State University showed that slow
sand filters are capable of removing >99.83 percent of Giardia cysts and
>99.98 percent of the Cryptosporidium oocysts. Initial use and backwashing
of the pilot filter in an attempt to remove fine clay particles may have
partially stratified the sand media and coated the top layer of the sand
with fines. The layer of fines caused the slow sand filter to build up
excessive headloss very rapidly. When the fines were removed, the filter
removed more than 99.99 percent of the Giardia cysts.

The removal of trihalomethane precursor material by slow sand filters
was evaluated at the University of New Hampshire (20). Six pilot filters
were set up and compared, each filter containing a different media. The
first filter contained 38 cm of sand, the second contained 38 cm of anthracite, the third contained 8 cm of clinoptilolite on top of 30 cm of sand,
the fourth filter had 8 cm of aluminum oxide over 30 cm of sand, 8 cm of
granular activated carbon (GAC) on 30 cm of sand were in Filter 5 and Filter
6 had 8 cm of an anionic exchange resin supported by 30 cm of sand.

The amended filters containing GAC and anionic resin consistently
achieved higher removals for organic precursor material than the other filters (>75 and >90 percent respectively). The other filters achieved 5 to 25
percent removal of the precursor material. One major drawback to the use of
the GAC and the anionic resin was excessive headloss buildup.

RECENT PILOT PLANT STUDIES AND NEW CONSTRUCTION

The attention focused upon slow sand filtration by the appearance of
three papers on this topic in the Journal American Water Works Association
in December, 1984, and the interest generated by numerous technical presentations on the subject since the early 1980's have resulted in a renewed
willingness to consider use of slow sand filtration. State drinking water
regulatory agencies and designers are now including slow sand filters among

the options to consider for small systems. Some water utilities have expressed strong interest in the process. A number of plants have been built,
or are in the planning stage. Table 5 presents information on some of these.
While not exhaustive, it does convey the impression of renewed interest in
slow sand filtration in the United States after several decades of little
application.

Of the 39 facilities listed in Table 5, 23 have a design capacity of
$1,000 \text{ m}^3 \text{ day}^{-1}$, or less. Fifteen of the 39 were built before 1980. The
number currently proposed and under study is 11. This indicates a higher
level of interest in slow sand filters at present.

Another indication of the change in attitude in the USA, and of the
likely application for the process, is seen in the the American Water Works
Association Research Foundation's (AWWARF's) funding of a project to develop
a design and construction manual for slow sand filtration. This project involves developing a set of plans that would allow modular construction of
slow sand filters for small communities. Once the basic engineering has
been done for a single module, a group of modules can be built side by side,
with a common wall separating each module. By preparing designs for two
or three different sizes of modules, a wide range of treatment capacities
could be provided. Although some engineering services would be needed to
adapt the modular design to each site where it was applied, the basic design
plans would provide savings in the customizing process. Availability of
carefully prepared construction plans may encourage engineers who serve
small communities to use slow sand filtration. The AWWARF has also funded
a slow sand filtration research project to explore techniques for enhancing
the removal of aquatic organic matter that may serve as trihalomethane precursor or as a food source for microorganisms in the distribution system.
Pretreatment of source water with ozone and addition of benzoate to the
source water will be studied over a 14 month period.

The Environmental Protection Agency expects slow sand filtration plants
to be built by numerous small communities in the future. On November 3, 1987,
the EPA proposed the Surface Water Treatment Rule in the Federal Register (21).
The proposed rule would require that surface waters be evaluated to determine
if filtration is needed. About 2,900 water systems are estimated to be providing surface waters to users without filtration at the present time. Of
these, the EPA (21) estimated that about 1000 of the systems needing to install filtration would select slow sand filtration. This process is likely
to be appropriate because many unfiltered systems have raw water that is of
uniform quality and apparently has been adequately treated by disinfection
to provide an effective barrier to pathogenic agents. The number of waterborne disease outbreaks, however, provided a strong incentive for improvement
of the present situation. The strong regulatory effort is likely to result
in a substantial increase in treatment plant construction in small communities.

SUMMARY

Slow sand filtration has been widely accepted in many parts of the world,
but in the USA, interest in this process was weak for many years. The search

Table 5.   Recent Slow Sand Filters in USA

State	Water Utility	Daily Capacity, m^3	Status
Idaho	Mission Creek	270	Built 1964
"	Skin Creek	330	" 1966
"	Paradise Valley #1	270	" 1967
"	Paradise Valley #2	270	" 1967
"	Fernwood	530	" 1967
"	Monrovia	110	" 1968
"	Twenty Mile	45	" 1968
"	Beeline	570	" 1968
"	East Hope	950	" 1975
"	City of Salmon	11,000	" 1976
"	Yellow Pine	160	" 1976
"	City of Challis	5,300	" 1981
"	Colburn	830	" 1985
"	Schweitzer Basin	120	" 1985
"	Rocky Mountain Academy	110	" 1986
"	Cavanaugh Bay	160	Built 1988
"	Rockford Bay	3,300	Proposed 1988
"	Harbor View Estates	1,100	" 1988
Idaho	Sunnyside	160	Proposed 1988
Oregon	Salem	150,000-170,000	Built 1958
"	Salem	190,000-260,000	" 1970
"	Stayton	15,500	" 1975
"	Stayton	15,500	" 1987
"	Westfir	540	" 1986
"	Wickiup Water District	650	Built 1987
"	Cape Meares	380	Proposed 1988
"	Detroit	1,100	" 1988
"	Idanha	1,100	" 1988
"	Kernville	2,300	" 1988
"	Oaklodge Water District	11,400	" 1988
"	Panther Creek Wtr. Dist.	760	" 1988
Oregon	Astoria	3,800	Proposed 1988
Washington	Blue Spruce Wtr. Dist.	910	Built 1987
"	Eatonville	3,300	Built 1988
"	Doe Bay--Orcas Island	760	Built 1988
Washington	Cashmere	14,200	Proposed 1988
Colorado	Empire	950	Built 1984
Vermont	McIndoe Falls	82	Built about 1974
New York	Waverly	4,500	Built 1982

for processes appropriate for control of Giardia cysts in surface waters led
to a program of research in slow sand filtration. The success attained in
the research, which revealed the capability of slow sand filtration to remove
a very high percentage of Giardia cysts present in raw water, is expected to
cause design engineers and regulatory officials to reconsider their previous
position on this simple treatment process. The EPA's regulatory proposal for
surface water treatment is likely to stimulate the construction of several
hundred new slow sand filters in small communities that are presently providing only disinfection for their surface water supplies.

ACKNOWLEDGEMENT

    The authors with to thank Steve Tanner (Idaho), D. W. Liechty (Wash-
ington), and David Leland (Oregon) for the information on the status of slow
sand filters in those states, and Elise Creamer for typing the manuscript.

REFERENCES

1.   M.N. Baker. The Quest for Pure Water, Vol. 1, Second Ed., American
      Water Works Assoc., 147-148 (1981).

2.   L.A. Slezak and R.C. Sims, J. American Water Works Assoc., 76(12),
      38-43 (1984).

3.   S.F. Poynter and J.S. Slade, Prog. Water Tech., 9(1), 75-88 (1977).

4.   J.S. Slade, J. Inst. Water Engineers and Scientists, 32(6), 530-536
      (1978).

5.   K.R. Fox, R.J. Miltner, G.S. Logsdon, D.L. Dicks and L.F. Drolet.
      J. American Water Works Assoc., 76(12), 62-68 (1984).

6.   J.L. Cleasby, D.J. Hilmoe, and C.J. Dimitracopoulos, J. American
      Water Works Assoc., 76(12), 44-45 (1984).

7.   J.L. Cleasby, D.J. Hilmoe, C. Dimitracopoulos, and L.M. Diaz-Bossio,
      EPA-600/2-84-088, U.S. Environmental Protection Agency, Cincinnati,
      Ohio, (1984).

8.   W.D. Bellamy, G.P. Silverman, and D.W. Hendricks, EPA-600/2-85-026,
      U.S. Environmental Protection Agency, Cincinnati, Ohio, (1985).

9.   W.D. Bellamy, G.P. Silverman, D.W. Hendricks, and G.S. Logsdon,
      J. American Water Works Assoc., 77(2), 52-60 (1985).

10.   W.D. Bellamy, D.W. Hendricks, and G.S. Logsdon, J. American Water Works
      Assoc., 77(12), 62-66 (1988).

11.   E.R. Baumann, Private Communication to D.W. Hendricks, (1980).

12. J.L. Pavoni, M.W. Tenney, and W.F. Eichelberger, Jr., J. Water Pollution
     Control Federation, 44(3), 414–431 (1972).

13. G.R. Pyper, EPA/600/2-85/052, U.S. Environmental Protection Agency,
     Cincinnati, Ohio (1985).

14. R.J. Eberhard, Water Supply and Distribution System, McIndoe Falls Fire
     District No. 3 Evaluation Report, McIndoe Falls, Vermont (1976).

15. T.J. Seelaus, D.W. Hendricks, and B.A. Janonis, J. American Water Works
     Assoc. 78(12), 35–41 (1986).

16. J. Bryck, B. Walker and G. Mills, in Proceedings of AWWA Seminar on
     Coagulation and Filtration: Pilot to Full Scale, American Water Works
     Association, Denver, Colorado, pp. 49–58 (1987).

17. T.R. Cullen and R.S. Letterman. J. American Water Works Assoc. 77(12),
     48–55 (1985).

18. R.D. Letterman and T.R. Cullen. EPA/600/2-85/056, U.S. Environmental
     Protection Agency, Cincinnati, Ohio, (1985).

19. P.F. Schuler, M.M. Ghosh, and S.N. Botros. Comparing the Removal of
     Giardia and Cryptosporidium Using Slow Sand and Diatomaceous Earth
     Filtration. Presented at American Water Works Assoc. Ann. Conf.,
     Orlando, Florida, June 21, 1988.

20. M.R. Collins, T.T. Eighmy, S.K. Spanos, and J.M. Fenstermacher.
     Modifications to the Slow Rate Filtration Process for Improved Removal
     of Trihalomethane Precursors. American Water Works Assoc. Research
     Foundation Research Report, In press July, 1988.

21. U.S. Environmental Protection Agency, Federal Register, November 4,
     42178–42222 (1987).

1.4   IMPROVEMENT OF SLOW SAND FILTRATION
Application to the Ivry rehabilitation project — Part I —

A. Montiel – B. Welte – J. M. Barbier — Société Anonyme de Gestion des Eaux de Paris, 9 rue Schoelcher – 75014 PARIS

Summary of part I :

Slow sand filtration is one of the most ancient techniques. It was gradually abandoned between 1960 and 1975 and has sice come back into use.

We reviewed the advantages and disadvantages of this technique and examined the different means of improving this process before and after biological filtration.

1 - INTRODUCTION

Slow sand filtration is one of de most ancient water treatment techniques. Although it had already been described in ancient times, it was only in the 17th, 18th and 19th centuries that the principles of this treatment for making water fit for drinking were dealt with in scientific literature or in the form of patents. (13)(8)

It was then that these treatment techniques were implemented in major cities to produce drinking water : London (1830), Hamburg (1890), Paris (1898). (8) (5)

As early as 1929, this treatment method wad incuded on the list of techniques approved by the Frenc Ministry of Health. During the sixties, however, the tred was to replace slow sand filtration by physico-chemical processes known as "rapid filtration". These nex processes developed and phased out biological treatments due to the excellent results obtained from the standpoint of the quality considerable avantage in industrialized countries and especially in urban areas.

Three key dates explen the renewed interest in biological treatment and slow filtration in particular.

1974 - Rook showed the secondary effectys of chlorine with organic matter found in water ; this led to the gradual phasing out of the ammonium removal process using break-point chlorination. (11)(6)

1980 - The UN declares the International Drinking Water Supply and Sanisation Decade open. The only treatment considered reliable and to be recommended for developing countries is slow sand filtration. (9)

1985 - Pursuant to numerous epidemics in the U.S.A., especially giardia cysts, American laws imposed filtration and disinfection as a minimum treatment to be applied to surface water.

This led to renewed interest in slow biological filtration dealt with in a large number of studies conducted at international level.

An exhastive bibliographic survey of all studies concerning this treatment process reveals that it is not outdated and is at least as efficient if not better than physico-chemical processes known as rapid filtration (9). The most important sutdies in this area were conducted in England, France, Germany and the U.S.A.

A prospective survey on the future of this type of treatment has even revealed that, since 1975, new slow sand filtration units have been built or their production capacity has been increased.

In spite of this renewed interest, it is indispensable to understand the reasons why this process was gradually abandone after the second world war.

2 - <u>REASONS WHY THE PROCESS WAS GRADUALLY ABANDONED</u>

Three possibilities ar to be considered :

1°) as certain authors suggest, this process was no longer suited to
    water quality requirements ;

2°) the process necessitates a considerable labour force and is not
    easily automated ;

3°) the process uses up large surfaces.

We shall now examine these three hypotheses.

2.1 - <u>Is this treatment ill-adapted as regards water quality</u> ?

To examine this hypothesis, we will define the principles of slow
filtration, on the one hand, and on the other hand, in the light
of the most recent research on this treatment, we will explain
the mechanisms of this surface water treatment. (9)

Slow sand filtration consists in percolating water through a
filtering medium composed of a 0.6 to 1 meter thick and bed. Sand
usually has an effective grain size structure varying from 0.5 to
1.2 mm at speeds of 2 to 12 m/per day.

After a few days, a complex biocenosis composed of algae,
bacteria and zooplankton develops in the top layer of the
filtering media. Extremely complex phenomena take place,
involving a large number of organisms living in symbiosis and as
mutual predators.

This biofiltration process must clarify the water and remove
organic and mineral micropollutants as well as microorganisms
contained in the water to be treated.

2.1.1 - Clarification

Clarification consists in removing suspended matter and colloids. As soon as this process was introduced, it was considered extremely efficient for removing and decreasing turbidity. It even served as a reference treatment when physico-chemical processes were first implemented.

Based on the current state of the art, not only can we explain the effect of biological filters on water clarification but we also know the limitations of such filters.

Suspended matter in removed by filter screening. The larger the volume of syspended matter presents, the more quickly the filter is clogged.

Colloids are neutralized and coagulated by microorganisms fixed on the sand (algae, bacteria, etc) and which secrete polysaccharides secreted in insufficient. This leads to a high turbidity of water treated.

It may be considered that water arriving on these filters with a turbidity of more than 10 NTU will not be adequalety treated (turbidity of filtered water exceeding 1 NTU).

2.1.2 - Removal of micropollutants

2.1.2.1 - Removal of micropollutants :

Mineral micropollutants are removed by precipitation - coprecipitation. Adsorption and bioconcentration exist in physico-chemical as well as biological treatments but play as small part in removing mineral micropollutants. Trace elements will therefore be precipitated or coprecipitated either in the form of hydroxides or carbonates, or after an insoluble compound is formed with a reagent added to the water.

Cd, Pb, Hg, Zn, FeIII, MnIV, Al, CrIII, Ni, and Co are removed by the formation of a hydroxide or carbonate and coprecipitation of these compounds.

The most important parameters to be considered to ensure removal of such compounds are :

- pH : the more alkaline it is the better it is removed, except in the case of amphoteric elements such as aluminium.

- turbidity of water treated : low turbidity indicates extent of precipitation and coprecipitation.

The valency of CrVI, MnII and FeII must correspond to their precipitation - coprecipitation in the water to be treated. (10)(7)

Hexavalent chromium must be made trivalent by ferric sulfate. (14)

Bivalent iron an manganese may be biologically oxidized to a valency of 3 or 4, respectively, on the biological filters.

Oxyanions   (arsenates,   vanadates,   selenates   and phosphates) may react with ferric salts to produce insoluble substances which will then be coprecipitated. (10)(7)

It is shown that chemical reagents must be added, in the case of the last two groups of compounds, to ensure adequate retention of these groups of elements.

2.1.2.2 - Removal of organic micropollutants

Organic micropolluants may be removed using two different processes :

a) abiotic process : hydrolysis, photolysis, evaporation, precipitation, coprecipitation, adsorption, oxidation.

b) biotic process : biodegradation, hydrolysis, adsorption, bioconcentration.

In the abiotic process, all elements adsorbed on suspended matter will already be removed. A long retention time in the biological filters facilitates hydrolysis, photolysis and evaporation. The biological membrane has a considerable adsorption capacity in the case of hydrophobic molecules. (2)

In the biotic process, biodegradation is the chief means of degrading organic molecules. We may therefore consider that certain non-biodegradable molecules, wihich were not retained by abiotic means, may pass through this processing stage.

2.1.3 - Removal of microorganisms

The most recent as well as theearliest studies show that this process may be considered as a biological barrier. This is one of the reasons why is was selected for the International Drinking Water Supply and Sanisation Decade. (9)

Studies carried out in England, Germany and the U.S.A. show that this method was a good means of removing pathogenic bacteria, viruses, giardia cysts and cryptosporidia. (9)(1)(4)

2.2 - Is this treatment easily automated ?

In France, the fact that biological filters are washed manually is a major handicap for this type of treatment. Surveys, prospective studies (3) and missions abroad (England, Belgium and Switzerland) have shown that it is possible to automate - and this has actually been done - washing of slow biological filters.

2.3 - Does this treatment require much space ?

When compared with purely phosico-chemical processes, it is clear that this treatment requires fifteen to twenty times more space.

The choice of treatment method therefor depends of specific needs. In Paris, biological filtration is carried out in two out of three plants. In this case, biological filtration is no longer hindered by the amount of space required but it is at a disavantage ar regards improvement of the plant's reliability from the standpoint of quantity and quality of water produced.

We therefore examined the different aspects which call for improvement.

3 - <u>IMPROVEMENTS TO BE MADE TO SLOW BIOLOGICAL FILTRATION</u>

<u>Advantages of slow filtration</u> :

According to the most studies (8), biological filtration is a reliable method which is not subject to human error since all purification processes are physico-chemical and biological and do not necessitate input or dosing of chemical reagents.

This process is increasingly considered as a microbiological barrier for bacteria, viruses and parasites (giardia cysts, cryptosporidia, etc). (1)(4)

Low filtration speeds (0.1 to 0.5 m/h) result in a considerable water storage (3 to 4 h) making it possible, in case of accidental pollution of short duration, to cease supplying the plant with raw water without modifying delivery. Furthermore, this mass of water constitutes as signficant buffer.

We considered it very important to be able to continue using this processing stage with improvements upstream and downstream and to eliminate the aforementioned disadvantages while preserving all advantages.

The different problems to biofiltration are as follows :

1°) does not allow excessive turbidity ;

2°) it is difficult to remove certain mineral micropollutants requiring an emergency reagent ;

3°) it is difficult to eliminate certain organic micropollutants which
must be removed by an emergency reagent or which may escape
treatment or be generated by the treatment itself (metabolite) ;

4°) excessive proliferation of algae in the summer.

Improvements will therefore concern treatment either upstream or
downstream of biological filters.

3.1 - Improvements upstream

3.1.1 - Effect of clarification

This pretreatment consists in introducing a processing
stage between raw water and biological filters, eanbling
adequate stabilization of turbidity at the lowest
possible value.

This may be achieved through several means (9)(3) :

1°) storage of raw water with a retention time of more
than fifteen days ; (3)

2°) microstraining ;

3°) microstraining - direct filtration ;

4°) double filtration : roughing filtration - primary
filtration ;

5°) direct filtration - coagulation on filter ;

6°) contact coagulation - coagulation on filter ;

7°) physico-chemical coagulation - flocculation -
sedimentation.

3.1.2 - Micropollutants

3.1.2.1 - Effect of mineral micropollutants

Certain micropollutants require the use of emergency reagents (ferric sulfate, ferric chloride, aluminium sulfate, aluminium prepolymers).

These emergency reagents must be used in combination with a physicochemical treatment (coagulation - flocculation - sedimentation, direct filtration with emergency reagents).

3.1.2.2 - Effect on organic micropollutants

The removal of certain organic micropollutants requires the use of emergency reagents (powdered activated carbon, clays, etc). These emergency reagents must be used in combination with a physico-chemical treatment (coagulation - flocculation - sedimentation, direct filtration with chemical reagents).

3.1.2.3 - Effect on microorganisms

Part I of this study shall only deal with treatments enabling reduced algae proliferation on biological filters.

Algae proliferation on biological filters may be reduced upstream in two ways :

- by reducing phosphates through precipitation induced by adding ferric salts ; or

- by exterminating algae using copper salts.

These two treatments require retention of ferric phosphates and/or copper carbonate and a physico-chemical treatment must therefore be used upstream (coagulation - flocculation - sedimentation or direct filtration with chemical reagent).

3.2 - Improvements downstream

The aim is to improve the quality of water obtained from biological filters and this mainly concerns organic micropollutants and microorganisms.

3.2.1 - Organic micropollutants

It is now internationally acknowledged that water for human consumption is purified through the combined use of ozone and granular activated carbon.

Ozone is used to destroy or transform certain molecules which may be aromatic, etc. Certain molecules which resist biodegradation are made biodegradable, while other molecules which are more polar and soluble may be generated : aldehydes, Ketones, etc.

As a result, the subsequent use of activated carbon will make it possible on the one hand, to remove molecules which were not treated previously, and, on the other hand, to remove molecules produced by previous treatment stages (metabolites, secondary effects, etc).

Furthermore, carbon is itself a biological reagent and makes it possible to eliminate molecules by biodegradation. For years, ozone and activated carbon have been known for their beneficial action on the taste of water and it would therefore be superfluous to deal with this aspect.

3.2.2 - Microorganisms

Water ozonisation at 0.4 ppm for 10 minutes has long been known for tis bactericidal and virucidal action. This has been confirmed in a large nomber of publications and may be considered as an established fact.

Final disinfection in used not only to carry out disinfection but also for the bacteriostatic effect obtained by adding chlorine or chlorine dioxide.

As is the case of ozone, it has long been established that a 0.5 ppm concentration in water for 2 hours eliminates pathogenic germs and that a remaining 0.1 to 0.2 ppm of chlorine avoids bacterial post-proliferation.

Furthermore, elimination of bio-assimilable organic carbon using granular activated carbon also helps to reduce bacterial post-proliferation.

4 - CONCLUSION OF PART I

The purpose of this study was to show the strong and weak points of slow biological filtration. Although it is an extremely flexible and reliable process, it has certain restrictions concerning use : turbidity, organic micropollution.

We have shown that these treatments could be improved by upstream protection and downstream refining.

Downstream refining is well-known and as already been reviewed. It is currently used to refine water treated by means of rapid filtration.

The second part of this study will deal with upstream protection and we will try to show the advantages and disadvantages of all alternative processes.

BIBLIOGRAPHY

(1) Anonymous
ROUND TABLE : CRYPTOSPORIDIUM
JAWWA - 1988 - Vol. 80, n° 2, pages 14 to 27

(2) M. DREWS - K. HABERER
Wirksamkeit verschieden betriebener langsamfilter zur entfernung
organischer wasser inhaltstoffe
VOM WASSER - 1986, band 66, pages 255 to 275

(3) D. DROUET
Etude diagnostique du marché potentiel de rénovation des unités de
traitement des eaux par filtration lente en Grande Bretagne - Note de
synthèse
ETUDE SAGEP 1988 - SAGEP Document

(4) J.B. HORN - D.W. HENDRICKS - J.M. SCANLAN - L.T. ROZELLE - W.C. TRNKA
Removing giardia cysts and other particules from low turbidity waters
using dual stage filtration
JAWWA - 1988, vol 80, n° 2, pages 68 to 77

(5) LE MARCHAND
Commission Lemarchand.
La distribution des eaux à Paris
Historical account
Official Municipal Bulletin of the city of Paris - 1923

(6) A. MONTIEL
Livre : les halométhanes dans l'eau. Mécanismes de formation et
élimination
Book published by CIFEC - 1980

(7) A. MONTIEL
Importance du choix du coagulant pour l'élimination des micropolluants
minéraux
CHA - AGHTM days at Clermont Ferrand - 1983

(8) A. MONTIEL
    Historical background of water treatment
    Rennes workshop - French Ministry of Health
    "Evolution de la pensée sanitaire"
    2/3 june 1987

(9) A. MONTIEL - B. WELTE
    La filtration biologique à faible vitesse dans le monde
    Principales études récentes
    Etude SAGEP - 1988 - SAGEP Document

(10) A. MONTIEL - B. WELTE - M. COLMONT
     Influence du coagulant lors de l'élimination des micropolluants
     minéraux lors des traitements de clarification
     Journal Français d'Hydrologie - 1984 - 15, Fasc. 2, pages 119 to 129

(11) J.J. ROOK
     Formation of haloforms during chlorination of natural waters
     Water Treatment Exam. - 1974 - Vol. 23, Pages 234 to 243

(12) SCEVP - City of Paris
     Etude sur la dégradation de la qualité de l'eau dans le réseau de
     distribution
     Study financed by the "Agence de Bassin Seine Normandie" - 1987

(13) G.E. SYMONS
     Water treatment through the age
     JAWWA Centennial - 1981 - Spécial issue, pages 28 to 33

(14) B. WELTE - A. MONTIEL - J. OUVRARD
     Elimination du chrome hexavalent par les traitements de potabilisation
     des eaux
     Journal Français d'Hydrologie 1984 - Vol 15, fasc 2, pages 145 to 153

1.4   IMPROVEMENT OF SLOW SAND FILTRATION
Application to the Ivry rehabilitation project — Part II —

A. Montiel – B. Welte – J. M. Barbier — Société Anonyme de Gestion des
de Paris, 9 rue Schoelcher – 75014 PARIS

Summary of part II :

A previous bibliographic survey in part I showed the improvements which
should be brought to bear on slow sand filtration. In situ studies made it
possible to select, on the one hand, contact coagulation and coagulation on
filter prior to slow filtration and on the other hand, the combined use of
ozone and activated carbon downstream.

This study concerns the rehabilitation of the slow sand filtration plant at
Ivry which meets 15 percent of water needs for the city of Paris.

1 - INTRODUCTION

In part I, we defined the advantages and disadvantages of slow sand
filtration for producing drinking water and, on the other hand,
improvements to be made in order to ensure reliability as regards the
quantity and quality of water produced, beargin in mind the various
currently applicable or future standards : EEC guidelines (2), WHO
guidelines (11) and future French standards (7).

Part I of this article dealt with general considerations while par II will describe the situation at the Ivry plant which supplies approximately 15 percent of drinking water consumed in the city of Paris. (figure 1, table 1).

This is an extremely important plant for the city of Paris as it is the only plant which can supply all sections of Paris. This is not the case for the other plants. It is considered as an essential water supply point.

The treatment network must be removated and enhanced in order to improve the reliability of slow filtration from a quantitative and qualitative standpoint.

2 - DESCRIPTION OF THE CURRENT NETWORK (Figure 5)

The Ivry plant is located on the river Seine upstream Paris. Raw water is pumped from the Seine, screened and then passes through two filtration stages (one known as roughing filtration and the other as prior filtration) before reaching the biological filters. Ozonisation is then carried out virucidally and the water is disinfected by chlorine before being fed into the Parisian network.

The Ivry plant has a treatment capacity of 175 000 m³ for distributing good quality water. Annual output is 45 000 000 m³.

Rehabilitation of the Ivry plant must bear on several aspects :

- improving Paris water supply quality,

- improving the quality of water treated at the Ivry plant.

3 - IMPROVING PARIS WATER SUPPLY SECURITY

Spring water is the main source of water supply (an average of 60 percent of overall water supply) distributed to the city of Paris. The rest (40 percent) is obtained from surface water treated at the Orly and Ivry plants on the Seine and Saint-Maur on the Marne.

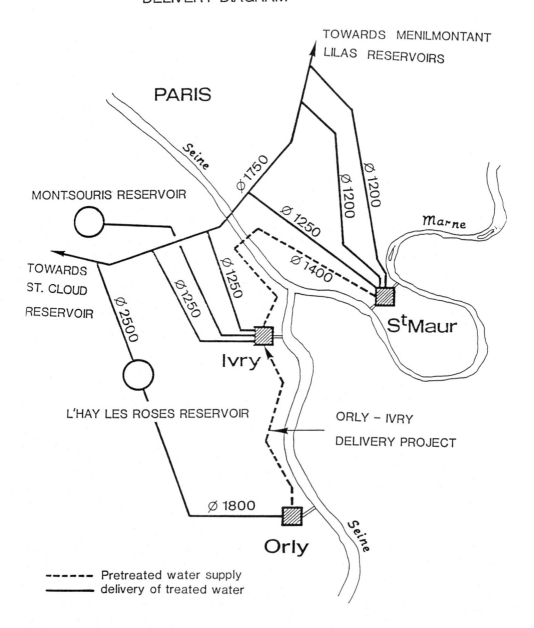

IVRY PLANT

WATER SUPPLY DIAGRAM

DELIVERY DIAGRAM

Figure 1

TABLE 1

	Annual production in millions de m3	Percentage
SPRING WATER		
Avre	60	19,5
Vanne + Loing ...	130	42
TREATED SURFACE WATER :		
Orly	30	9,5
Ivry	45	14,5
Saint-Maur	45	14,5
TOTAL	310 m3	100

PLANTS	Distribution to the different parts of Paris				
	North	North-East	East	South	West
ORLY	NO	NO	NO	YES	YES
IVRY	YES	YES	YES	YES	YES
St-MAUR	YES	YES	YES	YES	NO

As pointed out above, the Ivry plant is an essential feeding point in the Paris water supply system. This implies that there is a very high demand for good quality water in sufficiently large quantities with a need to ensure dependability as well as flexibility since the plant may be required to satisfy a need at any time.

3.1 - Quality of raw water at water intake

- Chronic pollution

Over the past ten years, the quality of raw water has deteriorated significantly with regard to background micropollution.

A comparison between current findings and results obtaindes some years ago reveals that, concerning mineral micropollutants, raw water has progessed from class A3 in 1980 to class A1 in 1986 (1975 EEC guidelines for raw water).(1)(10)

This improvement is due to the increased efficiency of the Agence de Bassin Seine-Normandie whose pollution control policy prompted a significant drop in direct river discharges (water treatment works at Valenton). It is also due to a decrease in the number of small industries along the Seine which used to discharge wastes directly without any prior treatment.

Organic micropollution is always present and we must devise a means of reducing it within the framework of the Ivry plant rehabilitation project.

- Accidental pollution

Accidental pollution is a problem which is common to all water treatment works. Average plant shutdown time due to accidental pollution over the past ten years is 2.5 days per year.

3.2 - Improving water supply security

Several studies have been conducted to improbe the security of Paris water supply. (figure 2 - table 2)

WARNING STATIONS BY THE SEINE

Figure 2

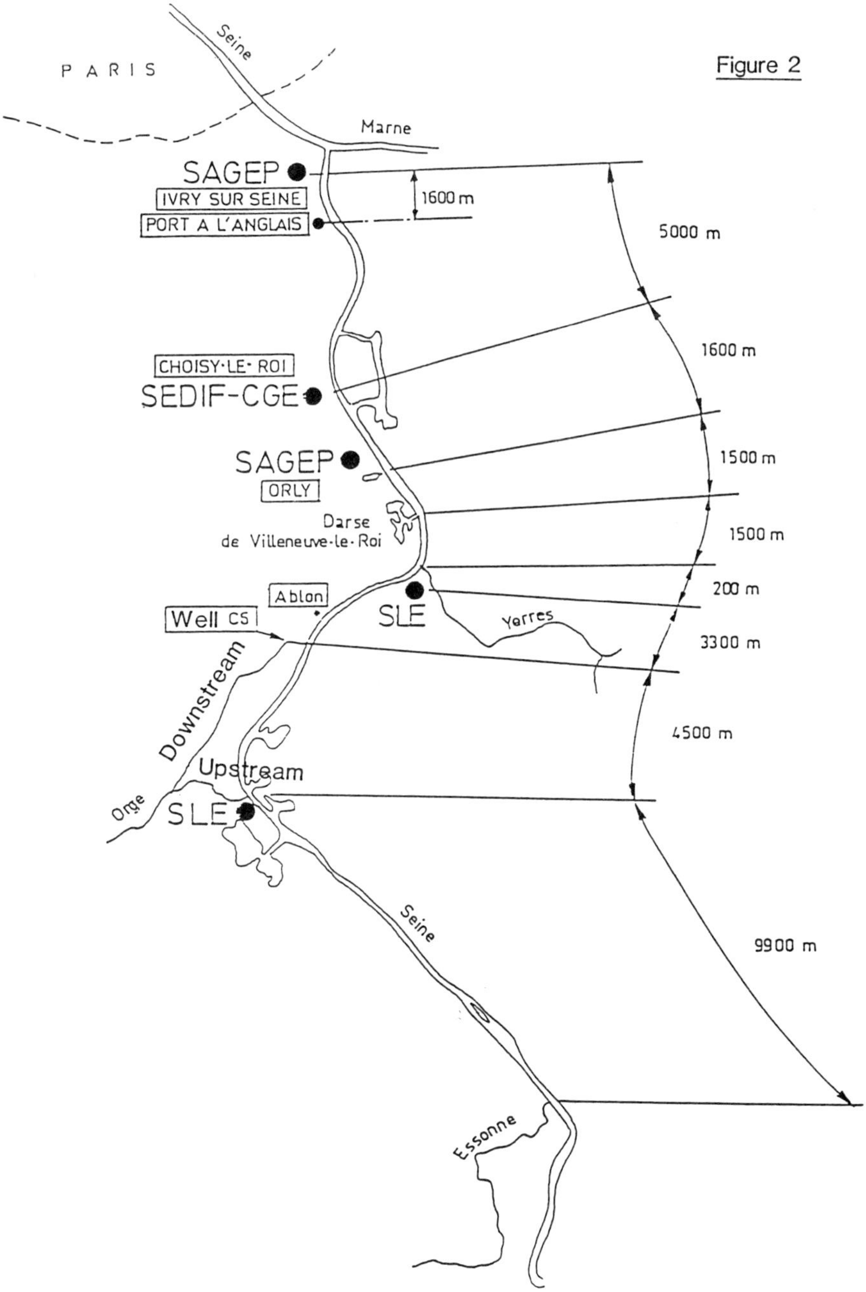

TABLE 2

WARNING STATIONS

SEINE PLANTS

	ABLON	ORGE	ORLY (PE)	PORT A L'ANGLAIS	IVRY (PE)
Temperature	X		X		X
pH	X		X	X	X
Conductivity	X		X	X	X
Redox Potential	X		X	X	X
Dissolved oxygen	X		X	X	X
Turbidity	X		X	X	X
COT	X	X	X	X	X
CH2 rate	X	X		X	
Ammonium	X		X	X	X
Nitrates	X		X		X
Fluorometry			X		X
UV absorption					
Metals :					
Cd	X			X	
Cu	X			X	
Zn	X			X	
Pb	X			X	
Cr VI	X			X	
Microtox	X			X	

3.2.1 – A network of warning stations has been set up by the
Seine to improve information transmission speed. These
stations constantly detect the most important parameters
concerning pollution and may trigger of alarms if an
abnormal concentration is detected.

The station are situated above the water intake and a
warning station for the Ivry plant will be set up ˎ at the
Port à l'Anglais dam, 1.6 km upstream. THis allows
intervention within 30 minutes in the event of pollution
and enable rapid input of emergency reagents, for
example, if necessary.

3.2.2 – If the Seine is considerably polluted at Ivry and this
requires shutdown of the plant for a long period, two
other water supply sources are available : the Saint-Maur
plant on the Marne and the Orly plant on the Seine. These
two sources may thus enable the Ivry plant to continue
producing good quality water in the event of significant
water pollution at Ivry.

3.3 – <u>Water quality problems in relation to the current network at the
Ivry plant</u>

In addition to shutdowns due to pollution, the Ivry plant must
cease or slow down water production in the event of excessive
turbidity in the Seine (attaining up to 100 NTU).

These are the findings of a statistical survey based on the past
ten years (6). The problem is due to the biological filters which
do not accept excessive turbidity as pointed out in part I of
this study. (figure 3)

Bearing in mind these problems, we considered it indispensable to
improbe two aspects, in particular, at the Ivry netword :

– upstream improvement of biological filters so as to avoid
turbidity peaks noted and which may therefore "protect" the
biological filters ;

– downstream improvement so as to enhance the quality of water,
especially concerning organic micropollution and in order to
reduce, as far as possible, any changes in quality during
distribution.

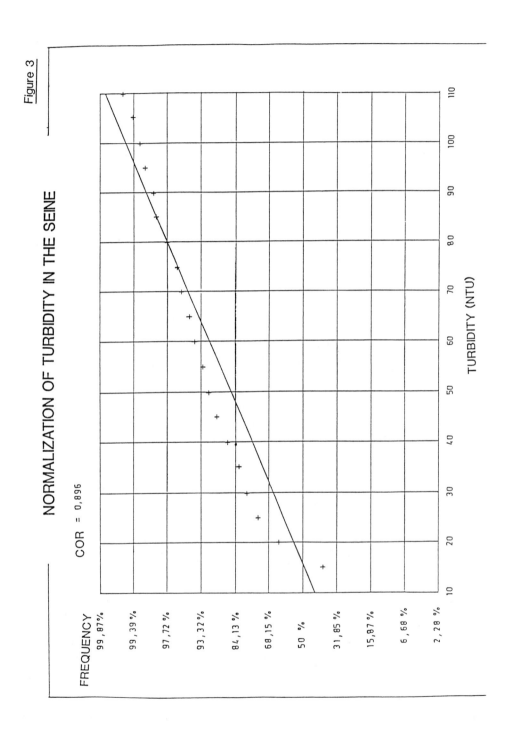

Figure 3

One may therefore consider (6) that a raw water turbidity of more than 30 to 40 NTU is the maximum which may be treated at these preliminary stages to ensure that the water reaches the biological filters with a turbidity of 10 NTU.

We imagined a treatment wherein reagents are added and which, in our opinion, has two advantages :

- improved turbidity of the water pretreated,

- possibility of using emergency reagents.

Our twenty-five years of experience at another plant supplying the city of Paris (the Saint-Maur plant) and previous studies on combined physico-chemical and biological treatment (12) at the Saint-Maur plant served as a basis for devising a new treatment.

At the Saint-Maur plant, we add 10 to 20 mg/l of ferric chloride at the source of a 671 meters long canal where mean water velocity varies between 0.1 to 0.2 m/s depending on plant operation. This gives rise to a sort of coagulation – flocculation – sedimentation and microstraining in then carried out before the water is prefiltered at a rate of 5 m/h and passes on the biological filters. The Saint-Maur pilot system has a full-scale physico-chemical treatment : coagualtion – flocculation – sedimentation – rapid filtration at the rate of 5 m/h before passing on the biological filters.

In view of preserving the structure of the Ivry plant, we imagined a pretreatment by contact coagulation followed by coagulation on filter. This type of treatment has been known since 1979 but, to date, it has only been used for low turbidites. (3)(4)(5)

A recent American study shows that this process is superior to rapid filtration treatments for removing giardia cysts. (5)

4.1 – Principles of contact coagulation – coagulation on filter

Contact coagulation – coagulation on filter is carried out on two different works which are filters, with each work playing a special role.

A small quantity of coagulation reagent (e.g. ferrice chloride) is added before the first filter. The usual dose is 1/5 or 1/10 of the optimal dose necessary in the case of physico-chemical treatment comprising coagulation - flocculation - sedimentation - rapid filtration.

The first filter uses a high granulometry (effective size > 2 mm) medium, thus forcing the water to pass between the grains and causing partially neutralized colloids to agglutinate and flocculate. This incoherent floc is filtered on the second filter which has a lower granulometry (effective size > 0.95), thereby initiating a coagulation process on this filter.

The first filtration stage diminishes raw water turbidity and has a certain buffering action. This allows good coagulation at the second filtration stage.

After this preliminaty treatment, we believe the water may have a turbidity of less than 10 NTU which is suitable for the proper operation of slow biological filters.

In order to verify this theory, we tried out this pretreatment using contact coagulation - coagulation on filter at the Ivry plant with its current structure.

First filtration stage :

- gravel 30 cm high
- effective size : 12 mm
- filtration rate is about 3 m/h at the plant's current normal production rate.

Second filtration stage :

- gravel 60 cm high
- effective size : 0.95 mm
- filtration rate is about 2 m/h at the plant's normal production rate.

4.2 - <u>Contact   coagulation - coagulation   on fiter tests   at the Ivry</u>
<u>plant</u>

4.2.1 - Description of facilities

Figure 4 give a detailled view of plant facilities.

Ferric chloride was added to the raw water channel of
approach at a point enabling an immediate and homogeneous
mixture.

4.2.2 - Optimization tests for ferric chloride dosing

A first series of tests was conducted on a half of plant
so as to determine the optimal dose of reagents to be
added and to verify the efficiency of this treatment and
compare it with the existing pretreatment. Findings are
presented in tables 3 to 4.

The following conclusions may be   drawn from these
findings :

- during the test period, raw water turbidity gradually
  increased from 7 to 30 NTU ;

- a beneficial effect was obtained for all parameters
  examined (turbidity, permanganate oxidation, NH4, NO2,
  PO4, COT)with a coagulant additive. The 5 ppm dosing
  seems to be quite sufficient for all parameters.

- two phenomena must be noted concerning turbidity :

  . increase in turbidity in raw water,
  . increase in coagulant dose.

  It is   therefore clear that differences   are not
  significative.

As a result, we considered 10 ppm of ferric chloride to
be satisfactory.

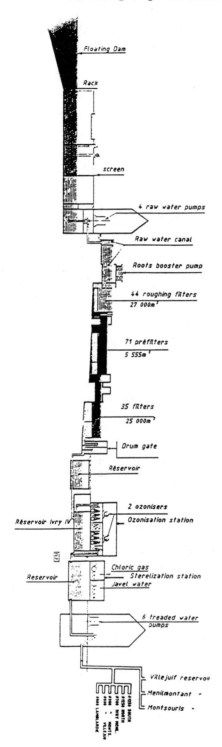

Floating Dam

Rack

screen

4 raw water pumps

Raw water canal

Roots booster pump

44 roughing filters
27 000m²

71 préfilters
5 555m²

35 filters
25 000m²

Drum gate

Réservoir

2 ozonisers
Ozonisation station

Réservoir Ivry IV

Chloric gas
Sterelization station
Javel water

Reservoir

6 treaded water
pumps

Villejuif reservoir
Menilmontant
Montsouris

CITY OF PARIS-IVRY FILTRATION PLANT

Figure 4

TABLE 3

RESULTS OF THE COAGULATION ON FILTER TESTS AT THE IVRY PLANT

Comparison between the current network and a netword using coagulation on filter

	5 g/m3			7,5 g/m3			10 g/m3		
	Raw	Préfiltered Without additive Fe	Préfiltred with additive Fe	Raw	Préfiltered Without additive Fe	Préfiltred with additive Fe	Raw	Préfiltered Without additive Fe	Préfiltred with additive Fe
Turbidity NTU	7,2	1,28	0,92	12,1	1,3	1,1	30,6	3,3	1,7
Oxydability KMnO4 mg/l O2	4,35	2,38	1,97	4,56	2,30	1,98	4,94	2,76	1,86
Ammonium mg/l NH4	0,69	0,11	0,04	0,68	0,072	0,035	0,45	0,072	0,05
Nitrites mg/l NO2	0,24	0,15	0,085	0,26	0,10	0,072	0,21	0,085	0,08
Phosphates mg/l PO4≡	1,28	0,96	0,60	0,94	0,94	0,55	0,84	0,77	0,30
TOC mg/l C	3,6	2,63	2,26	5,12	2,38	2,02	4,92	3,29	2,49

TABLE 4

POLLUTION REDUCTION RATE

	5 g/m3			7,5 g/m3			10 g/m3		
	Raw water content	Reduction %		Raw water content	Reduction %		Raw water content	Reduction %	
		Préfiltered without additive Fe	Préfiltered with additive Fe		Préfiltered without additive Fe	Préfiltered with additive Fe		Préfiltered without additive Fe	Préfiltered with additive Fe
Turbidity NTU	7,2	82,0	87,0	12,1	89,0	90,0	30,6	89,2	94,5
Oxydability KMnO4 mg/l O2	4,35	47,0	55,0	4,56	49,5	56,5	4,94	44,0	62,0
Ammonium mg/l NH4	0,69	84,0	94,0	0,68	89,5	95,0	0,45	84,0	89,0
Nitrites mg/l NO2	0,24	37,5	65,0	0,26	61,5	72,3	0,21	59,5	62,0
Phosphates mg/l PO4≡	1,28	25,0	53,0	0,94	0,00	41,5	0,84	8,3	64,0
TOC mg/l C	3,6	27,0	37,0	5,12	53,5	60,5	4,92	33,0	49,5

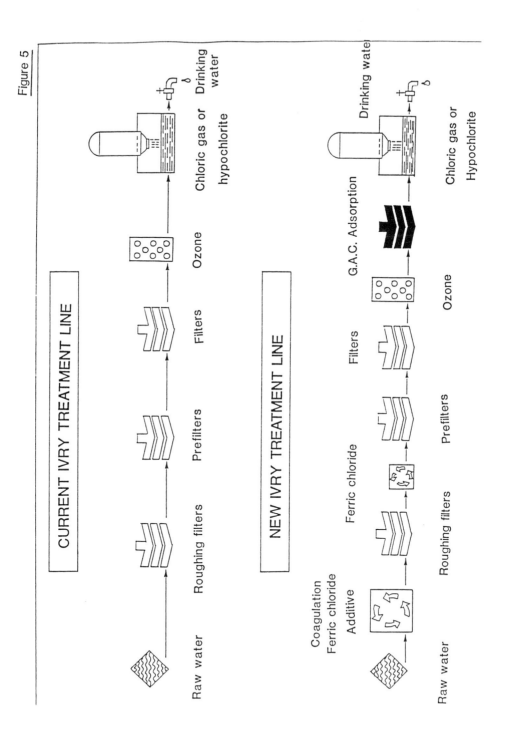

Figure 5

Based on these results, two additional series of tests
were planned at a pretreatment rate of 10 ppm considered
to be optimal for removing turbidity, phosphates and for
potassium permanganate oxidability.

One of these series of tests was conducted during the
warm season (august 1987) and the other during the cold
season (january/february 1988).

4.2.3 – Tests conducted during the warm season

Average results obtained during the test period are
presented in table 5. Reduction rates are grouped in
table 6.

The findings from the first tests are confirmed :

- reduction in turbidity after contact coagulation –
  coagulation on filter,

- reduction in organic matter measured by $KMnO_4$ and TOC
  oxidability,

- decrease in phosphate content of the water.

Furthermore, this series of tests confirmed findings for
the preliminary tests : washing of pretreatment stages
was facilitated in spite of the fact that a coagulant was
added.

4.2.4 – Tests conducted during the cold season

A third series of test was conducted during the cold
season, corresponding to cold temperatures and high
turbidity in raw water.

The results are presented in table 7. Reduction rates are
grouped in table 8.

TABLE 5

Average results for contact coagulation - coagulation on
filter tests during the warm season

	Raw water	Prefiltered water	Filtered water
Turbidity NTU	14,8	0,88	0,25
pH	8,1	7,80	7,85
Oxydability mg/l O2	3,37	1,94	1,29
Ammonium NH4 mg/l	0,30	0,036	< 0,02
Nitrite mg/l NO2	0,11	0,03	< 0,01
Phosphates mg/l PO4≡	0,68	0,33	0,33
Iron mg/l Fe	0,24 + 1,4*	0,025	< 0,02
TOC mg/l C	4,4	2,47	1,6

* 1,4 mg/l corresponds to the quantity of iron added by treatment.

TABLE 6

Reduction rates for contact coagulation - coagulation on filter
tests during the warm season

	Raw water content	Reduction %	
		Préfiltered water//raw water	Filtered water // raw water
Turbidity NTU	14,8	94	98,3
Oxydability mg/l O2	3,37	42,5	61,7
Ammonium NH4 mg/l	0,30	88	100
Nitrite mg/l NO2	0,11	63	100
Phosphates mg/l PO4≡	0,68	51,5	51,5
Iron mg/l Fe	0,24 + 1,4*	(89,6) 98,5	100
TOC mg/l C	4,4	43,9	63,6

* 1,4 mg/l corresponds to the quantity of iron added by the treatment.

TABLE 7

Average results for contact coagulation - coagulation on
filter tests during the cold season

	Raw water	Prefiltered water	Filtered water
Turbidity NTU	41	4,7	0,25
pH	8,23	7,96	7,91
Oxydability mg/l O2	8,45	4,88	3,62
Ammonium NH4 mg/l	0,13	0,018	< 0,01
Nitrite mg/l NO2	0,12	0,016	< 0,01
Phosphates mg/l PO4≡	0,84	0,16	0,126
Iron mg/l Fe	0,91 + 1,4 *	0,3	< 0,02
TOC mg/l C	6,72	3,9	3,9

* 1,4 mg/l corresponds to the quantity of iron added by the
treatment.

TABLE 8

Reduction rates for contact coagulation - coagulation on
filter tests during the cold season

	Raw water content	Reduction %	
		Prefiltered water // raw water	Filtered water // raw water
Turbidity NTU	41	88,5	99,5
Oxydability mg/l O2	8,45	42,5	67
Ammonium NH4 mg/l	0,13	86	100
Nitrite mg/l NO2	0,12	87	100
Phosphates mg/l PO4≡	0,84	81	85
Iron mg/l Fe	0,91 + 1,4*	(67,1) 87,1	100
TOC mg/l C	6,72	42	42

* 1,4 mg/l corresponds to the quantity of iron added by the treatment.

Interpretation of results during the cold season : the test confirmed a substantial rduction in turbidity, KMnO⁴ oxydability linked to total organic carbon as well as the removal of phosphates.

4.2.5 - Comparison between tests conducted during the warm and cold season

This comparison concerns pretreated water on the one hand and biologically filtered water on the other hand.

For all parameters, there is no difference in efficiency of the pretreatment stage between the warm and cold season.

If we now compare the results obtained after biological filtration, three parameters stand out : TOC, oxidability on the one hand and phosphates on the other hand.

Concerning oxidability, there is no difference between the warm and cold season as regards removal. A significant difference is noted, however, in the case of total organic carbon. Indeed, elimination of total organic carbon during the warm season is comparable to that of oxidability, whereas this is not the case during the cold season.

One explanation is that, during the warm season, biodegradation of molecules leads to their elimination. This explains a concomitant decrease in oxidability and organic carbon. In winter, biodegradation would lead to the formation of oxidized organic molecules causing a fall in the reducing power of waters without decrasing total organic carbon.

Removal of phosphates during the warm season is not a good as during the cold season. We have shown in previous studies that phosphates are removed from the water by reaction with ferric ions and formation of an insoluble ferric phosphate which is retained by the treatment.

This removal increases as the quantity of iron present in
the water increases. The initial quantity of iron in the
raw water (9) must therefore be taken into consideration.
This gives us 0.24 + 1.4 (added in the form of FeCl3) =
1.64 mg/l during the warm season and 0.91 + 1.4 (added in
the form of FeCl3) = 2.31 mg/l during the cold season.
This explains the difference noted.

4.2.6 – Comparison with cold season without adding ferric
        chloride

We took period around the test period and water
temperature was similar. Results obtained are shown in
tables 9 an 10. The most significative improvements
(table 11) concern the following parameters : turbidity,
oxidability, total organic carbon and phosphates.

4.2.7 – Test findings

The findings of these three test periods were as follow :

1) a 10 mg/l ferric chloride dosing enhances the plant's
realiability as regards high turbidity ;

2) elimination of turbidity was also accompanied by
improved removal of organic matters and phosphates.

5 – <u>DOWNSTREAM IMPROVEMENT OF BIOLOGICAL FILTERS</u>

The Ivry plant rehabilitation project takes into consideration the
improvement of the quality of water and the removal of organic
micropollution in particular.

Furthermore, as pointed out in part I, certain metabolites may be
generated by biological filtration and we believe that it is important
to be able to remove them.

It is now acknowledged that the combined use of ozone and granular
activated carbon is very efficient in removing organic matter (study on
carbon).(8)

TABLE 9

Average obtained using the current network at the Ivry plant
during the cold season

	Raw water	Prefiltered water	Filtered water
Turbidity NTU	10,3	1,3	0,48
Oxydability mg/l O2	5,11	3,58	2,66
Ammonium NH4 mg/l	0,12	0,01	< 0,01
Nitrite mg/l NO2	0,12	0,022	< 0,01
Phosphates mg/l PO4≡	0,57	0,47	0,48
Iron mg/l Fe	0,28	0,03	< 0,02
TOC mg/l C	4,75	3,49	2,75

TABLE 10

Reduction rates obtained using the current network at the
Ivry plant during the cold season

	Raw water content	Reduction %	
		Prefiltered water // raw water	Filtered water // raw water
Turbidity NTU	10,3	87,5	95
Oxydability mg/l O2	5,11	30	48
Ammonium NH4 mg/l	0,12	91,5	100
Nitrite mg/l NO2	0,12	81,5	100
Phosphates mg/l PO4≡	0,57	17	17
Iron mg/l Fe	0,28	89	100
TOC mg/l C	4,75	26	42

TABLE 11

Comparison between reduction rates in relation to raw water

	Prefiltered water		Filtered water	
	With addition of Fe during pretreatment	Without Fe	With addition of Fe during pretreatment	Without Fe
Turbidity NTU	88,5	87,5	99,5	95
Oxydability mg/l O2	42,5	30	67,0	48
Ammonium NH4 mg/l	86,0	91,5	100	100
Nitrite mg/l NO2	86,5	81,5	100	100
Phosphates mg/l PO4≡	81	17	85	17
Iron mg/l Fe	87	89	100	100
TOC mg/l C	42	26	42	42

Post-ozonization was eliminated in certain countries such as England and Belgium as it was not followed by filtration on activated carbon. These countries consider that activated carbon was redundant when biological filters were used.

We consider that these two stages (ozone - activated carbon filtration) allow for very good water purification ane may therefore avoid bacterial post-proliferation.

6 - <u>CONCLUSION</u>

The tests carried out at the Ivry plant allowed us to make a clear-cut classification of the different possibilities in the field of water pretreatment. Table 12 (summary) shows that, bearing in mind the problems to be solved (high turbidity, the need to be able to add emergency reagents in case of accidental pollution, need to reduce algae proliferation where necessary), the only suitable treatments are rapid filtration or dual-stage filtration (contact coagulation - coagulation on filter).

A previous study conducted at Saint-Maur by the city of Paris showed the advantages of the first pretreatment (12). This study shows the possibilities of this new pretreatment.

More highly specialized studies followed and made it possible to define the different parameters of this treatment : granulometry of filtering media, filtration rate, height of filtration beds. Studies conducted on carbon previously allowed us to enhance the reliability of downstream improvements of biological filtration.

All these enabled us to design a project for a new treatment line at Ivry composed of the following elements :

- pretreatment by contact coagulation followed by coagulation on filter at rates of 8 to 10 m/1, respectively, during the first stage, and at 5 to 8 m/h, respectively, during the second stage ;

- slow sand filtration at a rate of 5 to 10 m/day ;

- ozonization ;

- filtration using granular activated carbon with a water-carbon contact time of 6 to 15 minutes, depending on plant operation ;

- final chlorination.

TABLE 12

Summary of the advantages and disadvantages of the different pretreatment
stages prior to biological filtration

	Storage reservoir	Coagulation flocculation sedimentation filtration	Coagulation filtration	Contact coagulation coagulation on filtre
Advantages :				
Buffering action	X			
Pollution security	X			
Turbidité removal	X	X	X	X
Toxicity removal		X	X	X
Emergency reagents		X	X	X
Automation	X	X	X	X
Chemical reagents			X	X
Operating costs			X	X
Flexibility			X	X
Passivity			X	X
Disadvantages :				
Algae problems	X			
Space	X			
Chemical reagents		X		
Sludge treatment		X		
High turbidity			X	(X)

BIBLIOGRAPHY

(1) EEC - Guidelines concerning raw water - 1975

(2) EEC - Guidelines concerning drinking water - 1980

(3) CLEASBY J.L - SALEH F.M - Field evaluation of culligan filters for
    water treatment. Rept. ISU. ERI. AMES 79120 - Dep of civil engng.
    I.O.W.A STATE UNIV. AMES - March 1979

(4) HORN J.B - HENDRICKS D.W
    Removal of giardia cysts and particles from low turbidity water using
    the culligan multi Techn. Filtration System
    Envir. Engng. Techn Rept. 1986 296530 I
    Dept of Civil engng colo. State Univ. Fort Collins (August 1986)

(5) J.B. HORN - D.W. HENDRICKS - J.M. SCANLAN - L.T. ROZELLE - W.C. TRNKA
    Removing giardia cyst and other particles from low turbidity waters
    using dual - stage filtration
    JAWWA - Vol. 80, n° 2, pages 68 to 77 - February 1988

(6) Pascal MIGNOT - Etude fiabilité Ivry - January 1988
    Etude SAGEP - SAGEP document

(7) Ministry of Health - Draft decree on drinking water
    January 1988 version

(8) MONTIEL A.
    Les halométhanes dans l'eau. Mécanismes de formation et élimination
    CIFEC 1980

(9) MONTIEL A.
     Importance du choix du coagulant pour l'élimination des micropolluants
     minéraux
     CHA - AGHTM days at Clermont Ferrand - 1983

(10) MOUCHET J.
     Sanitary inspection of public water supply in Paris
     Rapporteur at the Departmental Hygiene Council of Paris : Mme MOUCHET
     CRECEP - March 1988 (29 pages).

(11) WHO - Guidelines concerning drinking water - 1984

(12) SCEVP - City of Paris
     Saint-Maur pilot study - phases 1 to 4
     Study financed by the "Agence de Bassin Seine Normandie" - 1981/1984

(13) V. UDOMRATANASILPA
     Application of horizontal flow prefiltration and slow sand filtration
     for small community water supply
     Asian Institute of Technology
     Bangkok Thaïland - Thesies n° EV.84.11 (1984)

1.5   MANAGEMENT OF SLOW SAND FILTERS IN RESPECT TO GROUND WATER QUALITY

Helmut Sommer — Institute for Water Research, Zum Kellerbach 46, D-5840 Schwerte – Geisecke, Federal Republic of Germany

ABSTRACT

In a catchment area with artificial ground water recharge an optimization model controls the delivery of infiltration rates upon the available slow sand filters in respect of different boundary conditions. This management conception affects an inprovement of the ground water quality. A revised utilization of natural cleaning processes during the underground passage of recharged ground water and an optimal mixing of different ground water types can be influenced by control of hydraulic gradients.

INTRODUCTION

In the Federal Republic of Germany the artificial ground water recharge is primarily used to replenish the ground water resources for municipal drinking water supply. Only in some cases an infiltration of surface water into the underground is applied to regulate the ground water situation for industrial water supplies or to avoid ecological problems.

The majority of installations for artificial ground water recharge are built nearby great rivers (figure 1). In 1986 the portion of artificially recharged ground water on the drinking water supply was nearly 10 percent.

As a result of special historical or regional features there are different techniques of infiltration into the aquifer. A common method of water works is the recharge of ground water by slow sand filtration.

QUALITY PROBLEMS IN CATCHMENT AREAS WITH ARTIFICIAL RECHARGE
OF GROUND WATER

Although the recharged ground water normally fulfills the quality standards for drinking water in Germany, in some catchment areas increasing concentrations of several substances in the ground water have been observed. These problems are caused by different chemical and hydraulic factors:

- One reason for quality interferences in catchment areas is an increasing pollution of the untreated infiltration water by substances, that cannot be eliminated. This group of chemical compounds passes the infiltration media and the underground sediments without an appreciable decrease of concentration.

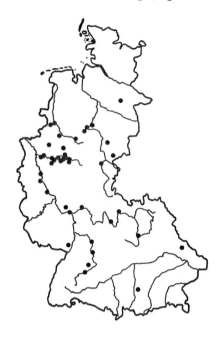

figure 1

Artificial Ground Water Recharge
in the Federal Republic of Germany
(BGW Wasser-Statistik 1987)

Normally water works have no influence on the quality of surface water used
for infiltration. So quality interferences in water catchment areas by substances, which cannot be eliminated, can only be avoided by using technical
treatments.

Other quality problems in drained ground water are affected by hydraulic conditions in the underground. An insufficient residence of infiltrated water
during the underground passage or an unfavourable mixture of recharged and
native ground water produces increasing concentrations in the ground water
in obtaining wells. In extreme situations the maximum permissible concentration of many substances may be exceeded.

In this case, however, controlled variations of the hydraulic conditions could
be a practical approach to avoid quality interferences of water catchment operations.

AN APPROACH TO REDUCE QUALITY PROBLEMS

A great number of microbiological and hydrochemical processes influence the purification capacity of an aquifer. These processes are directly related to the
residence time of artificially recharged ground water during the underground
passage.

An optimal management of local infiltration and discharge rates could improve
the residence time of recharged ground water in the aquifer by low variations
of hydraulic gradients. This could be a simple and cost-saving procedure to increase the purification capacity of an existing water work area.

In the same way mixtures of different ground water types in the underground of
a catchment area are affected by general hydraulic conditions.

A controlled infiltration of surface water could build a hydraulic barrier and routes out unwanted ground water from the drainage area. On the other hand, a strong subsidence of the ground water levels by high discharge rates could result in an uncontrolled inflow of ground water from the boundaries of catchment areas.

Therefore, consistent management of hydraulic gradients by optimal infiltration and discharge quantities offers the possibility of changing the rates of different ground water types in the pumping wells. It finally controls the chemical character of drinking water.

In the following, an optimized conception of groundwater management with its quantity and quality aspects is explained.

A MANAGEMENT CONCEPTION OF THE HENGSEN CATCHMENT AREA

The Hengsen water catchment area in the Ruhr valley was used to develop an exemplary operation conception to control the artificial ground water recharge and pumping rates with qualitative setting of objectives.

A key map shows the essential structures of the catchment area (figure 2). It is surrounded by the Ruhr river, and an impounded lake, which delivers the raw water for infiltration. The artificial ground water is recharged by 3 slow sand filters (S). A drain tile (D) between the infiltration basins is used for ground water pumping. The average thickness of the aquifer is 5 - 6 meters.

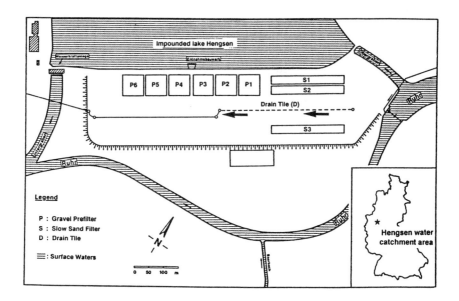

figure 2

Hengsen water catchment area with installations for artificial ground water recharge and drainage

THE GENERAL GROUND WATER FLOW SITUATIONS

The ground water flow in the catchment area is a result of different local infiltration and discharge rates. It can simply be demonstrated by using ground water simulation models. A comparison of various extreme situations shows the possibilities of this water work to influence the hydraulic conditions in the underground.

An extreme unfavourable operation with high discharge and low infiltration rates leads to an increasing inflow of bankfiltrate into the drain tile (figure 3).

In contrast to this, the flow situation resulting from regular water works operation is demonstrated in figure 4. It is based on a uniform distribution of the infiltration rates on the three slow sand filters. An optimal infiltration on both sides of the drain tile routes the bank filtrated ground water out off the contribution region. Only small sectors at the end of the drain tile are reached by bankfiltrate.

QUALITY PROBLEMS BY USE OF BANK FILTRATED GROUND WATER

An important quality problem of the Hengsen catchment area is caused by enhanced concentrations of iron, manganese and other heavy metals. The observed concentrations of some substances in the drain tile repeatedly exceed the permissable value of these parameters in Western Germany. Extensive investigations have shown that these concentration peaks are directly affected by increasing rates of bank filtrated ground water.

In the Hengsen catchment area occurs an anaerobe bank filtrated ground water. This ground water remobilizes several compounds of iron, manganese and other heavy metals, which are primarily bounded in the underground sediments. Iron and manganese were measured in the bank filtrate in concentrations of about 1 ppm. Both substances could be regarded as tracer elements for a group of chemical compounds in bank filtrated ground water. A decrease of manganese and iron concentrations in the catchment area by minimization of bank filtrated ground water in the drain tile will also diminish the concentrations of other critical substances. This clearly leads to the conclusion for minimizing the amount of bankfiltrate inflow to the drainage area.

INCREASE OF QUALITY BY OPTIMIZATION OF INFILTRATION AND DISCHARGE RATES

For this reason an optimal management conception to control the local infiltration and discharge rates was developed for the Hengsen catchment area. A minimization of bank filtrated ground water could be realized by a controlled variation of hydraulic gradients between the drain tile and the boundaries of the catchment area.

In the following, some basic elements of the management conception will be demonstrated by using the parameter manganese.

To develop an optimal management conception, the Hengsen catchment area was intensively investigated in a long time research programme (1). This recognized the consequences of different hydraulic factors on the quality situation in the aquifer.

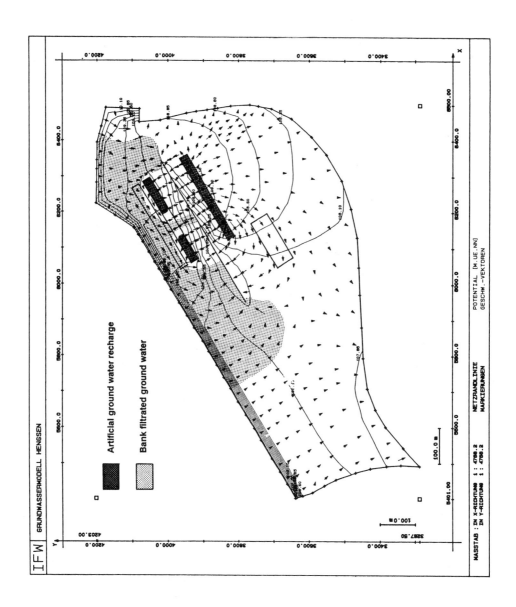

figure 3

Ground Water situation with high bankfiltrate inflow

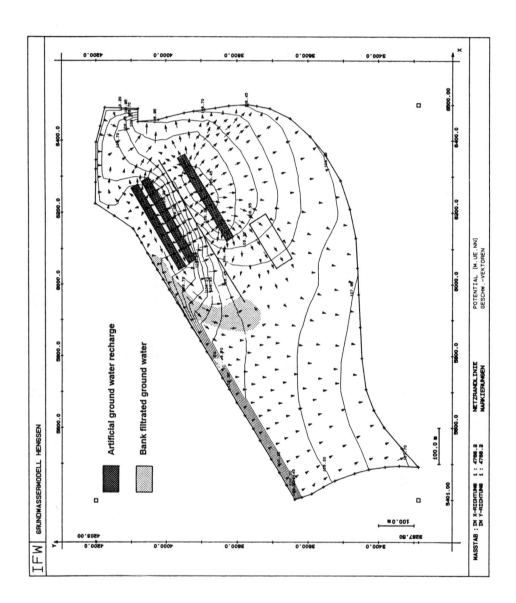

figure 4

Ground Water situation with low bankfiltrate inflow

At first, control lines between sectors of bank filtrated ground water and the drain tile were defined inside the water plant with the help of ground water simulation models. These control lines were used to estimate the quantities of inflowing bankfiltrate by hydraulic gradients of the ground water surface. The extreme points of these lines were placed in existing ground water observation wells. In some cases, additional wells were built. Some defined control lines are shown in figure 5.

Long term measurements of chemical parameters and water levels in the observation wells allow together with regression analysis the determination of mathematical relations between the ground water quality in the drain tile and the hydraulic gradients on representative control lines. It can be shown, that the magnitude of manganese concentrations in the drain tile is directly influenced by the hydraulic gradients on such control lines (figure 6).

Based on simple statistical relations between hydraulic and chemical data, the conception to control the local infiltration and discharge rates of the Hengsen catchment area was developed (figure 7).

In addition to the presented problems which result from enhanced bank filtrated ground water in the drain tile, the conception includes the residence time of artificially recharged ground water during the underground passage. The conception also considers the loads of ground water with heavy metals and some organic compounds.

The essential parts of the control model are:

- The effects of different infiltration and discharge rates and some other boundary conditions on hydraulic heads and gradients could be simulated by ground water models.

- An external ground water simulation was used to develop linear influence equations which describe the effect of one hydraulic factor upon the ground water situation at points of interest. The important hydraulic factors of this catchment area are the infiltration quantities of each basin and the total discharge rate of the drain tile.

- Next these singular equations are transformed to a complex response unit matrix. In this procedure the total ground water situation, resulting from several influencing factors could be determined by the multiple well equation of FORCHHEIMER (2).

- The linearized relationships between the hydraulic and hydrochemical situation could be included into the matrix.

- The optimization model is based on linear programming (simplex algorithm). The objective function is to minimize the total infiltration quantity because the cost of the used gravel prefilters and slow sand filters increase with the passage of water.

$$Q_{infiltr.} = Q_{basin1} + Q_{basin2} + Q_{basin3} \implies \text{Minimization!}$$

A first practical application of the optimization model already resulted in a significant decrease of iron and manganese concentrations in the drain tile. However the considerably simplified linear equations only provide a limited improvement of the ground water quality with respect to organic compounds. Therefore in future, a nonlinear optimization model will be developed to reduce the concentration of these substances.

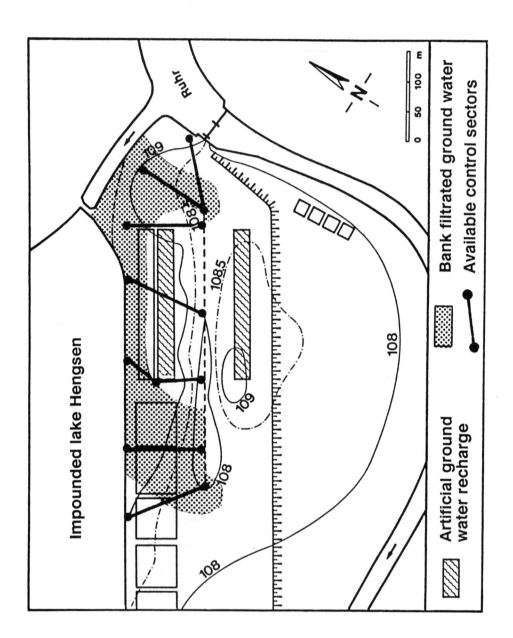

figure 5
Sectors to control bank filtrated ground water

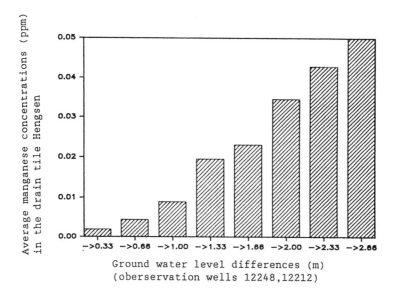

figure 6

Ground water quality in relation to hydraulic head
differences on control lines

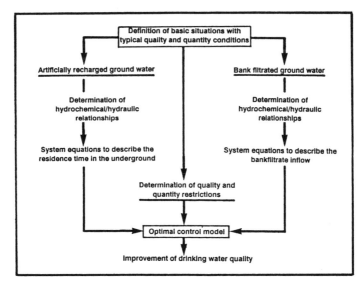

figure 7

Optimal management model

The consequences of different restrictions, which are used in the optimization model, for the results of management conceptions are exemplarily shown for two situations in figure 8. An optimal delivery of infiltration rates upon the three slow sand filters is calculated by the management model.

Both infiltration conceptions are developed in respect of a necessary pumping capacity which is defined by the water consumption. Additional two mostly observed surface water levels of the Ruhr river are considered, which influence the general ground water flow situation.

The two different optimal management conceptions are:

- A first management conception includes only technical restrictions (e.g. maximal practicable subsidence of the ground water levels at the drain tile, the maximal infiltration capacity of a slow sand filter, technical priorities by using the filter systems).

- The second management conception includes technical and qualitative restrictions. In this concept the quality factors clearly influence the delivery of infiltration rates upon the three slow sand filters. The two infiltration basins next to the inpounded lake are favouredly used. This infiltration conception affects a ground water situation, in which a bank filtrated ground water of the lake is routed out the contribution region of the drain tile.

However a comparision of these two different management conceptions suggests, that a control of a catchment area with artificial ground water rechage in respect of hydraulic situations in the underground and ground water qualities can influence the total pumping capacity.

CONCLUSION

The presented conception of controlling infiltration and discharge rates can be used to reduce the concentrations of several substances in a catchment area with artificial ground water recharge. Quality interferences can be avoided by using an optimized ground water management model, for which in most water works the necessary information about the hydraulic and chemical relations are available or can easily be obtained with small efforts. Therefore, in some cases the optimization conception should be considered as a real low cost alternative to technical concepts.

ACKNOWLEDGEMENT

This work was carried out with the support of the German Research Community (DFG) / Bonn and the Public Utility Company of Dortmund.

REFERENCES

1.   Sommer, H. (1983) Measurements to optimize the operation of a ground-water extraction area by control of infiltration and pumping, Proceedings of the International Symposium Ground Water in Water Recources Planning, Koblenz

2.   Forchheimer (1898) Grundwasserspiegel bei Brunnenanlagen, Zeitschrift des österr. Ingenieur-Vereins

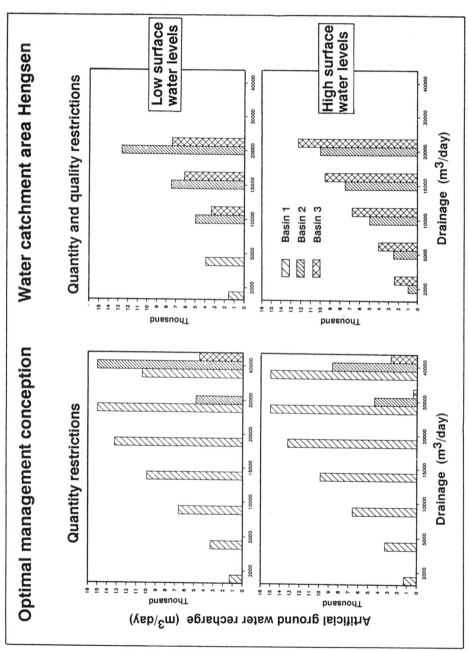

figure 8

Optimal management conception with different restrictions

2

Pretreatment methods

2.1 ROUGHING GRAVEL FILTERS FOR SUSPENDED SOLIDS REMOVAL

M. Wegelin — IRCWD/EAWAG*, CH-8600 Duebendorf, Switzerland

ABSTRACT

Reasonable operation of Slow Sand Filters (SSF) is only possible with raw water of low turbidity. Therefore, pretreatment of the generally turbid surface water is usually required. Prefiltration is an adequate process applied in the past but nowadays often replaced by chemical water treatment techniques. However, roughing gravel filters recently received considerable attention because of their simple design and reliable operation. Raw water characteristics, local physical conditions and available resources will determine the most appropriate prefilter type. Comparative field tests are planned to develop selection criteria. Prefilters and SSF present a potential treatment combination which will be gaining increasing importance in rural water treatment technology.

INTRODUCTION

No other single water treatment process can improve the physical, chemical and bacteriological water quality of surface water better than Slow Sand Filtration (SSF). The technology associated with SSF is simple requiring no chemicals, little or no machinery and is of reliable operation. The major drawback of SSF, however, is its sensitivity to turbidity. The suspended solids of a turbid water will rapidly clog a SSF, resulting in unacceptably short filter operation runs. Pretreatment of turbid surface water is therefore necessary in order to run SSF effectively.

Plain sedimentation and even prolonged storage are often unable to separate the suspended solids concentration to such level which would enable a successful SSF operation. Chemical destabilisation of the suspension is an unstable water treatment process difficult to operate and therefore often not an adequate pretreatment method for SSF. Furthermore, the use of chemicals will change the chemical water quality which might negatively affect the SSF biology.

---

* IRCWD - International Reference Centre for Waste Disposal
EAWAG - Swiss Federal Institute for Water Resources and Water Pollution Control

Prefiltration is an alternative option for suspended solids removal. Physical processes possibly supported by biological activities are the main mechanisms responsible for solid matter separation by prefilters. Hence, prefiltration hardly changes the chemical characteristics of the water and therefore does not impair the SSF activity. The layout of prefilters can be as simple and their operation as reliable as SSF. Therefore, prefiltration is an adequate pretreatment process for SSF. Prefilters combined with SSF represent a most appropriate water treatment process for rural water supply schemes in developing countries.

HISTORICAL REVIEW

The natural water treatment potential was used well before chemical water treatment methods were discovered and applied. Gravel and sand used as filter media are key components in natural treatment processes. While sand was able to maintain its important role since the development of the first SSF beginning of last century, the use of gravel filters was successively replaced by chemical water treatment processes. A few cases of early gravel filter application are presented hereafter.

Numerous castles and forts were constructed in Europe during the Middle Ages. The supply of water to the inhabitants of these structures was not always easy. This often

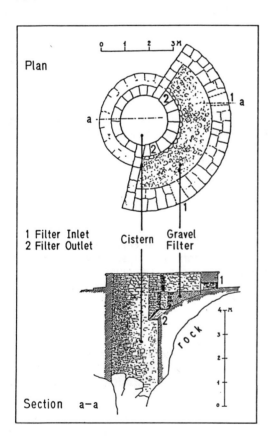

Fig. 1 Cistern with gravel filter at the castle Hohentrins, Switzerland (1)

Plan

1 Filter Inlet
2 Filter Outlet

Cistern   Gravel Filter

Section   a−a

led to ingenious solutions, a good example constitutes the former castle of Hohentrins which is located on top of a steep rocky reef in Switzerland. The people of this fort depended on rainwater collected from the yard and stored in a cistern. Water pollution caused by man and animal in this extensively used area was obvious. A gravel pack, as illustrated in Fig. 1, was installed around the inlet of the cistern and thereby a first kind of roughing gravel filter was born (1).

The first known filter used in a public water supply was constructed by John Gibb at Paisley, Scotland in 1804 (2). Fig. 2 illustrates the filter installation in which the water flowed from a ring-shaped settling chamber through two lateral-flow filters filled with gravel and sand to a central clear water basin. This is definitely an application of gravel filters prior to sand filters !

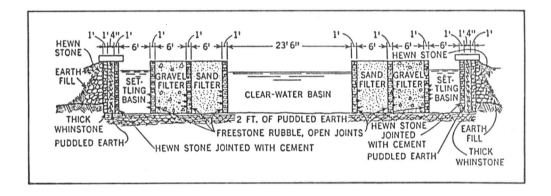

Fig. 2 Gravel and sand filter for the city water supply of Paisley, Scotland (2)

But John Gibb's contribution went even further as can be read in "The Quest for Pure Water" (2, page 78):

> "Water from the River Cart flowed to a pump well through a roughing filter about 75 feet long, composed of "chipped" freestone, of smaller size near the well than at the upper end. This stone was placed in a trench about 8 feet wide and 4 feet deep, covered with "Russian matts" over which the ground was leveled."

This is indeed a clear description of an intake gravel filter. Many other filter installations constructed in Great Britain followed these first examples.

In France, Puech-Chabal designed a system of multiple filters and aerators to treat surface water as illustrated in Fig. 3. The water passes through four down-flow roughing filters and one prefilter before it is treated by a finishing filter. The filter material decreases successively in size, and the filtration rate is also reduced from filter to filter. The first Puech-Chabal filter system was completed in 1899 to treat part of the water supply of Paris city. 125 of such plants were built in France by 1935, nearly 20 in Italy and a few in other countries (2).

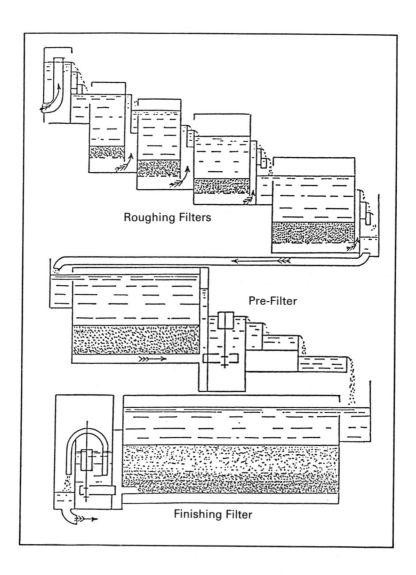

Fig. 3 Puech-Chabal filter system developed in France (2)

As time passed, the roughing filters were converted virtually into rapid or mechanical filters, and coagulation combined with sedimentation was introduced as pretreatment method. However, the roughing filter technology has been revived in recent years by the use of artificial groundwater recharge and simple water treatment plants.

In the early 1960's, the water works of Dortmund, Germany, constructed horizontal-flow roughing filters for an artificial groundwater recharging plant as illustrated in Fig. 4. The prefilters of 50-70 m length are operated at a filtration rate of approx. 10 m/h. The raw water inlet is located above the filter bed while the filter inlet zone is progressively impounded with increasing running time (3).

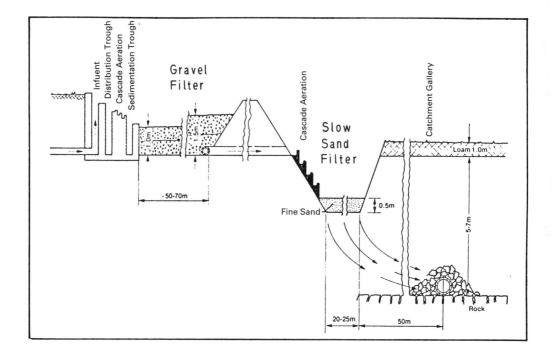

Fig. 4 Layout of artificial groundwater recharging plant of Dortmund, FRG (3)

Other water works in Switzerland and Austria followed the example of Dortmund with modified designs. Salient data of such plants are given in Table 1.

Table 1. Examples of Roughing Filters for Artificial Groundwater Recharge

Name of plant	Country	Name of river	Susp. solids conc. mg/l		filter length (m)	filtration rate (m/h)
			mean	max.		
Dortmund	W. Germany	Ruhr	8	20	50-70	10
Aesch	Switzerland	Birs	7	40	15	5-8
Graz	Austria	Andritzbach	5	20	10	4-14

The river water used for artificial groundwater recharge is normally of low turbidity. Filter operation is stopped at periods of high turbidity, i.e. during heavy rainfall. In such periods, the continuous supply of water to the consumer is maintained by the use of the aquifer's storage capacity. The prefilters used in Europe are consequently loaded with raw water of low suspended solids' concentration.

In tropical countries, roughing filters usually have to cope with raw water of permanent or seasonally high turbidity. In the absence of an aquifer, these water supply installations have to maintain operation throughout the year. Reliable operation is especially required during the rainy season when the risk of epidemic outbreaks of diarrhoeal diseases is greater as a result of washed off faeces not properly disposed. In need of reliable and simple water treatment processes, roughing gravel filtration has received considerable attention in recent years. Studies on the design and performance of prefilters to cope under tropical water quality conditions were conducted at different institutes. Some of these results will be presented hereafter.

FILTER CLASSIFICATION

Size of filter material and rate of filtration are two possible criteria for filter classification. Rapid and slow sand filters differ from intake and roughing filters by their smaller filter material size as seen in Table 2. Furthermore, their filtration rate encompasses the respective rates applied by intake and roughing filters. Since rapid and slow sand filtration are well-established conventional water treatment processes, they do not require additional explanation.

Table 2. Filtration Classification

Characteristics	intake filtration	roughing filtration	rapid sand filtration	slow sand filtration
filter material size (mm)	6 - 40	4 - 25	0.5 - 2	0.15 - 1
filtration rate (m/h)	2 - 5	0.3 - 1.5	5 - 15	0.1 - 0.2

On account of the coarse filter material and the low filtration rate applied in roughing filtration, filter resistance hardly increases during filter operation. Therefore, mechanical equipment is not required for flow control nor is such equipment used for filter cleaning. The roughing filters have a considereable silt storage capacity and, consequently, filter cleaning is required at intervals of weeks to months.

The characteristics and layout of the different roughing gravel filters are summarized in Fig. 5. Intake and dynamic filters located directly in the river or canal improve the water quality already at the point of abstraction. The roughing gravel filters require separate watertight structures. The flow direction in roughing filters can either be vertical as is the case for downflow and upflow filters, or horizontal as in the HRF.

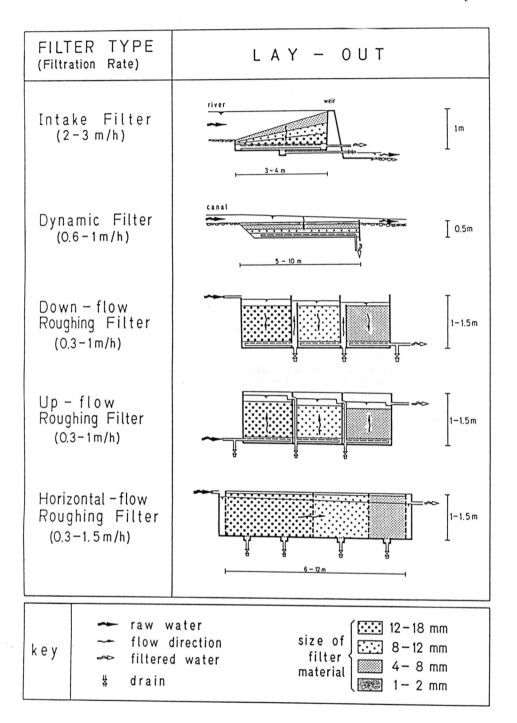

Fig. 5 Characteristics of prefilters

DESIGN AND APPLICATION OF ROUGHING GRAVEL FILTERS

Intake Filters

Intake filters are installed directly in small and narrow river beds. They consist of a small weir and different gravel layers installed upstream from the weir. For filter operation, the river is impounded at the weir. Part of the river water flows through the filter into the intake pipe, the remaining of the water falls over the weir into the downstream river bed. For filter cleaning, the entire river flow is discharged through the bottom outlet in the downstream river bed. The drained filter is cleaned manually. An example of an intake filter is illustrated in Fig. 6.

Intake filters might be appropriate for small rivers which can easily accommodate a weir. To avoid uplanding and sealing of the filter layers, sediment transport should be minimal. Its application is limited to rivers discharging turbid water only during a short time period.

Fig. 6 Intake filter in Columbia (6)

Dynamic Filters

Dynamic filters are installed in canal beds as exemplified in Fig. 7. During filter operation, part of the canal water is filtered through a series of sand and gravel layers, the remaining water is returned to the river. Filter cleaning must be carried out when the filter surface is sealed with deposited solids. The filter is cleaned by stirring the filter material. The accumulated solids are thereby resuspended and washed back to the river.

Dynamic filters are more easily accessible compared to intake filters. They require a certain hydraulic head and an adequate topography for the accommodation of the canal. Dynamic filters should also preferably be used with relatively clear river water of short turbidity peaks during rainfall.

Fig. 7 Dynamic filter of Chorro de Plata, Columbia

Downflow Roughing Filters

Downflow roughing filters are supplied at the filter top with raw water, which then flows through the filter material to the underdrain system. There, the prefiltered water is collected and sent to the second and thereafter to the third filter compartment containing finer filter material as shown in Fig. 8. Filter cleaning is carried out hydraulically. Each filter section can be drained separately. The accumulated solids are flushed out of the filter by high velocity drainage.

Downflow roughing filters can treat surface water of a moderate, but long-lasting turbidity. The filter efficiency is restored periodically by a hydraulic filter flush. After intervals of several years, the filter material has to be removed manually, washed and refilled into the filter boxes.

Fig. 8 Downflow roughing filter at Azpitia, Peru (7)

Upflow Roughing Filters

Upflow roughing filters are operated and cleaned similarly to downflow roughing filters. The main difference is the flow direction during operation. Upflow roughing filters are supplied with water at the filter bottom. The prefiltered water collected at the filter top is brought to the bottom of the next filter compartment. For filter cleaning, the raw water supply is stopped and each filter compartment undergoes an individual hydraulic filter flush.

The hydraulic cleaning of an upflow roughing filter is shown in Fig. 9. For hydraulic flush cleaning, a fast opening gate is required allowing the cleaning procedure to start immediately with a high water discharge as this should be the case for all roughing gravel filters. This shock drainage flushes the accumulated solids out of the filter, after which filter operation can be restarted.

Fig. 9 Hydraulic filter cleaning at El Retiro, Columbia (8)

Horizontal-flow Roughing Filters

Horizontal-flow roughing filters have the simplest hydraulic filter layout of roughing gravel filters. The water runs from the inlet compartment in horizontal direction through a series of differently graded filter material separated by perforated walls as shown in Fig. 10. Filter cleaning is also carried out with hydraulic filter flushes. Either a single drainage gate or several gates can be opened simultaneously to achieve a shock drainage of the entire filter. Depending on the efficiency of hydraulic filter cleaning, manual filter cleaning will also be necessary after a few years.

The top of horizontal-flow roughing filters is dry and the free water table remains under the gravel surface in order to prevent algal growth. Due to their considerable filter length and silt storage capacity, horizontal-flow roughing filters can handle raw water of high turbidity.

Fig. 10 Hozirontal-flow roughing filters in Mafi Kumase, Ghana

PRACTICAL EXPERIENCE WITH ROUGHING GRAVEL FILTERS

Intake and Dynamic Filters

Examples of intake and dynamic filters might be found all over the world. However, investigations and applications of these filters are centred in Latin America (4,5), e.g. in Argentina, Chile, Columbia, and Ecuador. Especially in Columbia, numerous intake filters have been installed over the past 15 years, and several dynamic filters have been set up recently. However, only limited information on the performance of these filters is available. Therefore, a study group of the University del Valle in Cali, Columbia is currently carrying out a corresponding monitoring programme sponsored by the city of Zurich and supervised by IRCWD. The results of this evaluation will be published in a field manual. Preliminary results (6) of the study reveal a cleaning interval of 2-3 months during the dry season, of 1-4 weeks during the rainy season. Filter efficiency with respect to turbidity reduction and bacteriological water quality improvement seems to be low during periods of minor river discharge. However, filter performance needs further assessment, especially also during the short intervals of high turbidity.

Roughing Filters

Development of the horizontal-flow roughing filter technology for the treatment of moderately to highly turbid surface water has started about 10 years ago. A demonstration project aimed at introducing and field testing this filter technology in different developing countries is currently under way and supervised by IRCWD. More information on horizontal-flow roughing filters is presented in the next chapter.

Investigations on the layout and performance of downflow and upflow roughing filters have more recently been conducted in Peru (7) and Columbia (8). A few full-scale plants applying these filtration processes are in operation. Layout and performance of 3 specific plants using one of the different roughing filter types are exemplified in Table 3. These prefilters do not only reduce peak turbidities by 80 to 90%, but they also improve the bacteriological water quality to a similar extent.

Filter efficiency of the downflow roughing filter at Azpitia is reduced by the relatively coarse filter material installed. Turbidity of the prefiltered water is therefore further decreased by filter fabrics placed on the slow sand filter bed. Further information on this project will be presented in an other paper of this seminar.

Operational problems were experienced with chemical flocculation and sedimentation at the treatment plant "El Retiro" in Cali, Columbia. The conventional pretreatment process was therefore replaced by 3 upflow roughing filters. Prefilters and slow sand filters are now running smoothly. Faecal coliforms are reduced to 89.4% by the prefilters and to 99.9% at the effluent of the slow sand filter (8).

Table 3. Examples and Practical Experience with Roughing Filters

Layoutand performance	Azpitia, Peru	El Retiro, Columbia	Blue Nile Health Project, Sudan
type of roughing filter	downflow	upflow (multi-layer filter)	horizontal-flow
filter length and size (mm) of material	60 cm, Ø 40-25 60 cm, Ø 25-12 60 cm, Ø 12-6	20 cm, Ø 18 15 cm, Ø 12 15 cm, Ø 6 15 cm, Ø 3	270 cm, Ø 25-50 85 cm, Ø 15-20 85 cm, Ø 5-10
filtration rate design capacity	0.30 m/h 35 m³/d	0.74 m/h 790 m³/d	0.30 m/h 5 m³/d
turbidity(NTU) raw water prefiltered water	50-200 15-40	10-150 5-15	40-500 5-50
faecal coliforms (/100ml) raw water prefiltered water	700 160	16,000 1,680	> 300* < 25*
reference	(7)	(8)	(9)

* as E. coli

DEVELOPMENT OF THE HORIZONTAL-FLOW ROUGHING FILTERS (HRF)

The **Asian Institute of Technology (AIT)** in Bangkok started laboratory and pilot-scale filtration tests in 1977. The investigators tested a prefilter of 5-m length containing 7 gravel layers. The average gravel size in these layers varied from approx. 18 mm to 5 mm. The finest filter medium was installed in the centre of the filter as illustrated by Fig. 11. In spite of the good results reported by AIT (10), no reasonable argument favours the installation of the smallest gravel fraction in the filter centre. The following gravel layers have by nature a smaller suspended solids removal efficiency and hence, further improvement of the respective water quality will only be minimal.

3 full-scale water treatment plants applying the AIT prefilter design and SSF were then constructed. The performance of the filters was monitored for approx. half a year after commissioning of the treatment plants. The available records (11) report a good performance of the prefilters which enable SSF runs of several months.

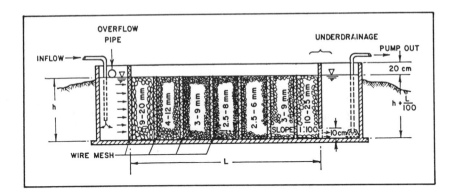

Fig. 11 Horizontal flow coarse media filter (10)

The **University of Dar es Salaam (UDSM)**, Tanzania embarked on filtration tests in 1980. Initial experiments with downflow coarse gravel filters revealed short running times and consequently high filter cleaning frequency. Subsequently, the design for a Horizontal-flow Roughing Filter (HRF) was developed. The schematic layout of an HRF is illustrated in Fig. 12. The filter is divided into three parts: the inlet structure, the filter bed and the outlet structure. In and outlet structures are flow control

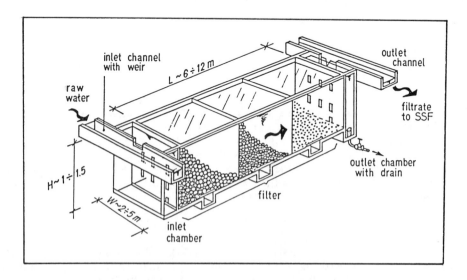

Fig. 12 Main features of a Horizontal-flow Roughing Filter (HRF)

installations required to maintain a certain water level in the filter as well as to establish an even flow distribution across the filter. The main part of an HRF consists of the filter bed composed of 3 to 4 gravel packs decreasing in size. The HRF efficiency in turbidity reduction was investigated by laboratory and field tests (12). These investigations indicated that HRF in combination with SSF could be a viable combination for turbid surface water treatment.

At the **International Reference Centre for Waste Disposal(IRCWD)** in Duebendorf, Switzerland, extensive filtration tests were conducted from 1982 to 84. A model suspension of kaolin was used to investigate the mechanisms taking place in HRF and to develop design guidelines for HRF. Filtration tests in the laboratory revealed that filter efficiency in suspended solids removal does not depend on the characteristics of the filter media surface for the examined filter size range (3-25 mm) and the applied filtration rates (0.5-2 m/h).

Another important phenomenon observed during the filtration tests is filter efficiency regeneration by means of drainage. The loose structure of the suspended solids accumulated on top of each filter grain will collapse and get flushed to the filter bottom when the water table in the filter is suddenly lowered. The laboratory test results were presented together with a semi-empirical filtration model in a recent publication (13). The more practical aspects for HRF implementation are compiled in a corresponding design, construction and operation manual (14). Financially supported by the Swiss Development Cooperation (SDC), IRCWD is coordinating a project on HRF promotion. The HRF process is introduced in form of full-scale demonstration plants in a number of developing countries. The process is being field tested under different technical and sociocultural conditions in order to establish its applicability and limitations. Some preliminary results of this HRF promotion project are presented hereafter.

The filter material originally used is gravel, however, according to the laboratory results (13), it can be replaced by any inert, clean, insoluble and mechanically resistant material. On the practical side, economic considerations besides availability of appropriate material are important factors in the selection of filter media. For instance, no gravel might be available in an alluvial plain as this is the case for the Gezira irrigation scheme in the Sudan. Therefore, the gravel of the HRF constructed by the Blue Nile Health Project has been partly replaced by broken burnt bricks (9). Field studies are in progress to explore the filter efficiency of HRF exclusively filled with differently-sized bricks. The first preliminary results of the monitoring programme indicate similar efficiencies for the different filter media. In another HRF project implemented in Java, Indonesia, the coarse gravel fraction has been replaced by injuk, a local palm fibre. Apparently, this fibre does not release taste nor odour to the water. This interesting fibre might be a potential filter material also due to its large specific surface area and high porosity (90-92%) which considerably increases retention time of the water in the filter and hence, most probably also filter efficiency.

Algal removal might be another objective of water pretreatment as large quantities of algae will affect SSF operation. A village community in Ghana recently constructed in a self-help project 2 HRFs and 2 SSFs to treat water from a small reservoir. These HRFs have to remove predominantly the plankton from the water. Remarkable in this case is the relatively small size of the algae. According to Table 4 in which preliminary microscopic surveys are compiled, the 7-m long HRF run at 0.75 m/h is able to reduce a variety of different algal species. High HRF efficiency in reducing the green algae colour of the raw water has also been reported from Burma (15). In this case, a 3-m long HRF pilot unit run at 0.6 m/h reduced the dry season raw water colour from approx. 400 Pt-Co units to an average of approx. 170 Pt-Co units, and the subsequent SSF further reduced it to an average of approx. 50 Pt-Co units.

Filter cleaning carried out regularly is required to maintain a good filter efficiency. Initially, HRF cleaning was intended to be carried out manually. This might be appropriate for small filter units and for situations where washwater disposal is difficult. The small HRF and SSF constructed by the Blue Nile Health Project in the Sudan are for example adequately cleaned twice a year by the community. However, manual cleaning of larger HRF units might become too cumbersome and pose an organisational problem. Hence, larger HRF units were provided with perforated pipes or troughs in order to achieve fast drainage necessary during hydraulic filter cleaning. The HRFs of 3 water treatment plants in Peru are equipped with troughs and fast opening gates for filter drainage. Field evaluations (16) revealed that the accumulated solid matter could not fully be removed by hydraulic filter cleaning. Perforated pipe systems might be more efficient provided the discharge capacity of the drainage system generates sufficient draging force to rinse the sludge into the pipe system. Additional field studies are required in this area. However, the installation of a drainage system is still recommended since regular filter drainage restores the filter efficiency and prolongs the running time before manual filter cleaning is required.

Table 4 Separation of algae by HRF

algae species	size (μm)	HRF inlet a	HRF inlet b	HRF outlet a	HRF outlet b
Cyclotella	10-15	++	+	+	0
Staurastrum	45-80	++	++	−	0
Chlorophyta	2-10	++	++	+	0
Tracholomonas	14	++	+	+	−
Mesismopedia	0.5	++	++	++	0
Gomphosphaeria in colonies	50-60 (ø colony)	++	++	++	−

Note:
a   sampling after   3 days of filter operation
b   sampling after 30 days of filter operation
++ very many, + many, − few, 0 nil

OUTLOOK AND CONCLUSION

The SSF process has been applied and improved since the beginning of last century. Therefore, this technology is well-established and new design modifications might only result in minor changes. However, the pretreatment technology of roughing gravel filters presently attracts much attention. Different ongoing projects aim at developing new prefilter designs and at assessing their applicability and limits under field conditions. Each of the presented prefilter types has its specific characteristics and field of application to be determined by practical field tests. In this context, a joint project with the University of del Valle in Cali, Columbia, IRC and IRCWD is planned to investigate the performance of different pretreatment methods by comparative tests carried out at one location under similar conditions. This project will help to develop selection criteria and design guidelines for the different pretreatment methods.

Prefilters together with Slow Sand Filters offer a simple but efficient treatment combination. The treatment processes are of physical and biological nature and thereby do not depend on the supply of chemicals. The reliability of a safe water quality is enhanced by the high stability of the treatment processes. The presented simple pretreatment methods might help to reduce the treatment problems presently experienced in rural water treatment plants.

REFERENCES

(1)   **Probst, E.,** Von Besonderheiten in der Wasserversorgung auf Burgen, *Nachrichten Burgenverein,* 6/1937

(2)   **Baker, M.N.,** The Quest for Pure Water, Vol. 1, AWWA, 1982

(3)   **Kuntschik, O. R,** Optimization of Surface Water Treatment by a Special Filtration Technique, *JournalAWWA,* Oct. 1976

(4)   **Perez Farras, L.E.,** Filtros dinámicos, Servicio Nacional de Agua Potable, Argentina, 1972

(5)   **Rodriguez, D.G.,** Análisis de la Eficiencia de Filtros Dinámicos para Tratamiento de Agua Potable, Universidad de Chile, Santiago, 1977

(6)   Proyecto de Evaluación del Sistema de Captación de Lecho Filtrante Dinámico Grueso en Capas, Progress Report, Universidad del Valle, Cali, Columbia, June 1988

(7)   **Pardon, M.,** Consideraciones, Desarrollo y Evaluación de un Sistema de Tratamiento que implementa la Filtración Gruesa de Flujo Vertical en Grava, CEPIS, Aug. 1987

(8)   **Galvis, G. and Visscher, J.T.,** Filtración Lenta en Arena y Pretratamiento, Proceedings of an International Seminar on Simple Water Treatment Technology, organized by ACODAL in Cali/Columbia, Aug. 1987

(9)   **Basit, S.E. and Brown, D.,** Slow Sand Filter for the Blue Nile Health Project, *Waterlines*, Vol. 5, No. 1/1986

(10) **Thanh, N.C.,** Horizontal-flow Coarse-Material Prefiltration, Research Report No. 70, AIT, 1977

(11) Monitoring and Evaluation of Village Demonstration Plants, Technical Report, Project Managing Committee and AIT, Oct. 1981

(12) **Wegelin, M. and Mbwette T.S.A.,** SSF Research Reports Nos. 2 & 3, University of Dar es Salaam, 1980/82

(13) **Wegelin, M., Boller, M. and Schertenleib, R.,** Particle Removal by Horizontal-flow Roughing Filtration, *AQUA* 2/1987

(14) **Wegelin, M.,** Horizontal-flow Roughing Filtration: A Design, Construction and Operation Manual, IRCWD Report No. 6/1986

(15) **Fraser, R.W., Kearton, R.D. and Randall, F.K.,** Designing a plant for Burma's need, *Asian Water & Sewage*, March 1988

(16) **Pardon, M.,** Evaluation of the Water Treatment Plants of Cocharcas, La Cuesta and Compin, DelAgua, May 1988

2.2 UPFLOW COARSE-GRAINED PREFILTER FOR SLOW SAND FILTRATION

Luiz Di Bernardo — Department of Hydraulics and Sanitation, School of Engineering of São Carlos, Av. Dr. Carlos Botelho, 1465 13.560 – São Carlos – Brazil

ABSTRACT

This work is a contribution to the study of upflow coarse sand and gravel prefilters as pretreatment methods for surface waters. Results of an experimental investigation using filtration rates varying from 4 to 36 m/day applied to different media composition are presented herein. It is shown that turbidity, apparent colour, total coliform and iron can reasonably be reduced when using this technology.

INTRODUCTION

Slow sand filtration has been recognised as an appropriate technology for treating water to human consumption in developing countries due to its simplicity of construction and the ease of operation and maintenance, even when people of the local community are involved. Although slow sand filters may be very efficient in removing particulate suspended matter, this material accumulates on the top of the filter bed, resulting in a rapid clogging of the media. Thus, a pretreatment process may be required when handling surface raw waters with a relatively high turbidity (> 10 TU).

Several pretreatment methods can reduce the level of particulate suspended matter in raw water and make it suitable for further treatment by slow sand filtration. During the last ten years, many investigations have been carried out to evaluate the performance of horizontal and downflow vertical gravel prefilters. Some studies

report on turbidity reduction from several hundreds to average values consistently lower than 25 TU allowing a feasible application
of slow sand filters in sequence, especially in rural areas and in
developing countries.

In Brazil, a tropical climate prevails in several regions through
out the year, where skilled technical supervision for construction, ope
ration and maintenance of plants requiring coagulation and   rapid
filtration is not available in the communities. Adoption of   slow
sand filtration, preceded by some sort of pretreatment, may   solve
the problems related to public health in those communities.   Based
on the large experience in the School of Engineering of São Carlos
(EESC-USP) related to the upflow direct filtration technology,   it
was proposed at the end of the 1970's by this author, to develop an
experimental programme to study upflow prefilter units using different
media as the pretreatment process.

In a previous study in 1980 (2) the perfomance of two units ope
rated at rates varying from 4 to 10 m/day, one upflow and the other
downflow, were compared. As shown in figure 1, the units could also
function in series. The downflow filter had a support layer consist
ing of three gravel sublayers, each one of 10 cm thickness (3,2   -
4,8 mm; 4,8 - 7,9 mm; 7,9 - 12,7 mm); the upflow filter had   also
three sublayers of the same gravel size as the downflow filter, but
with different thicknesses (1st: 10 cm; 2nd: 5 cm; 3rd: 5 cm). The me
dia was the same for both units, consisting of sand with grains varying in size from 0,15 to 1,00 mm, effective size of 0,25 mm   and
thickness of 80 cm. Table 1 presents maximum and minimum values   of
turbidity, apparent and true colour for the raw water and the   filter effluents for both filters operated in parallel and   the   run
lengths for the rates studied. Typical results are shown in figures
2, 3 and 4. Although good results were achieved in the upflow   filter, problems related to media clogging and the impossibility   of
performing effective cleaning of the media after some runs   indicated the necessity of using coarser sand at that time.

The use of horizontal flow units for prefiltration as described
by Than and Quano (6) in 1977, indicated the possibility of   using
coarse material for pretreatment. The authors presented results from
experimental investigations carried out on a pilot plant consisting
of four separate layers in series of crushed stone with sizes   varying from 25,4 to 2,4 mm. For a rate of 14,4 m/day, it was conclu-

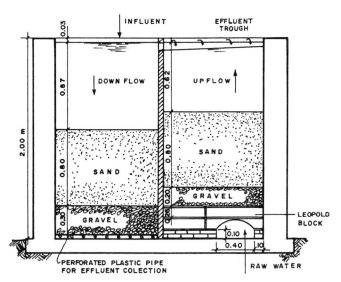

FIGURE 1 - UPFLOW AND DOWNFLOW FILTERS (2)

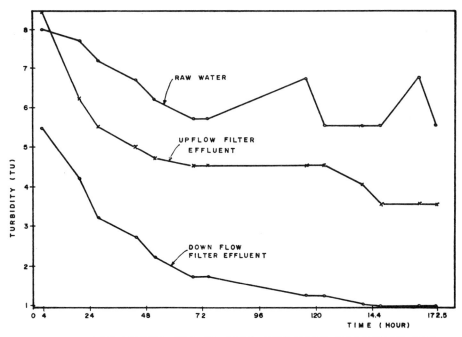

FIGURE 2- RAW WATER AND FILTERS EFFLUENT TURBIDITY FOR
FILTRATION RATE OF 8 m/day (2)

TABLE 1: Results of Operation of Downflow and Upflow Slow Sand Filters (2)

APPROACH VELOCITY (m/day)	TURBIDITY (TU)						COLOUR (mg.l^{-1} Pt - Co)												RUN LENGTH (Hour)	
	RAW WATER		DOWNFLOW FILTER EFFLUENT		UPFLOW FILTER EFFLUENT		RAW WATER				DOWNFLOW FILTER EFFLUENT				UPFLOW FILTER EFFLUENT				DOWNFLOW FILTER	UPFLOW FILTER
	Maximum	Minimum	Maximum	Minimum	Maximum	Minimum	Max App.	Max True	Min App.	Min True	Max App.	Max True	Min App.	Min True	Max App.	Max True	Min App.	Min True		
4	45	5.2	9.5	4.0	27	2.2	100	40	20	5	30	15	10	5	50	20	7	5	340	955
4	45	5.2	37.0	2.0	-	-	100	40	20	5	100	40	5	0	-	-	-	-	456	-
6	60	7.5	8.2	1.7	8.5	1.0	150	50	30	15	30	15	5	0	30	15	3	0	330	427
8	100	15	55	8.0	47	6.7	400	70	40	10	100	40	15	5	50	30	10	5	239	166
4	100	15	32	3.7	57	4.2	400	80	50	15	100	30	10	3	150	40	5	3	356	524
6	90	17	35	2.5	35	6.5	400	60	70	30	100	40	5	3	100	30	20	10	194	353
4	22	10	10	2.0	12	5.2	60	20	30	15	30	10	0	0	40	20	5	0	263	263
6	12	9	6.5	1.0	7	3	50	20	40	20	20	10	5	0	30	10	10	5	193	193
8	100	5.5	20	0.9	87	3.5	300	60	30	15	50	10	5	3	200	40	10	5	142	326
8	8	5.5	5.5	0.9	8.5	3.5	40	20	30	15	20	10	5	3	30	10	10	5	172	172
10	120	10	55	2.0	20	9	400	50	40	20	100	40	5	3	60	30	30	10	155	92
10	27	10	10	1.7	10	6	60	30	40	15	30	10	5	3	30	15	15	5	123	102
10	20	7.5	10	1.0	9.5	4.5	50	30	30	15	40	10	3	0	30	15	5	3	188	164
4	30	7.5	4.5	1.0	15	3	150	30	20	10	10	5	10	5	30	10	5	3	334	413
4	42	10	15	6.0	-	-	100	30	30	10	40	10	10	5	-	-	-	-	224	-
10	45	10	27	5.5	-	-	100	40	30	10	80	20	10	5	-	-	-	-	184	-
4	100	10	-	-	90	4.2	400	60	40	15	-	-	-	-	200	40	10	5	-	328

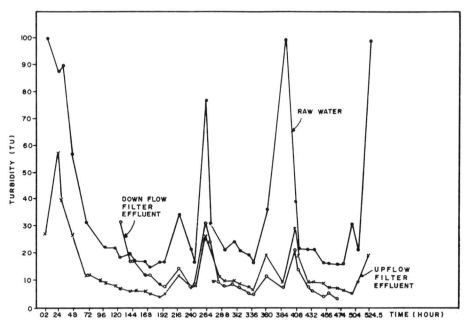

FIGURE 4 — RAW WATER AND FILTERS EFFLUENT TURBIDITY FOR
FILTRATION RATE OF 4 m/day (2)

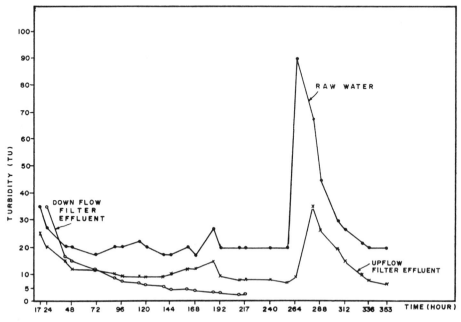

FIGURE 3 — RAW WATER AND FILTERS. EFFLUENT TURBIDITY FOR
FILTRATION RATE OF 6 m/day (2)

ded that turbidity removal of 60 - 70% could be achieved when   the
raw water had a turbidity varying from 40 to about 100 TU.

In 1984, Wegelin (8) presented the results of experiences in Eu
rope, Thailand and Tanzania, where horizontal-flow prefilter   units
were surveyed. He reported the use of filtration rates ranging from
3,6 to 14,4 m/day and the production of prefiltered water with turbidity varying from 10 to 20 TU when using raw water ranging   from
20 to 120 TU. Wegelin (8) also reported the results of studies concerning the influence of size and concentration of suspended solids
on the performance of prefilter units. Using Kaolin to simulate a sus
pension of fine particles (smaller than 20 µm), and transparent tubes and channels filled with gravel, it was observed that sedimenta
tion was the predominant mechanism for particle removal. In   addition, tests conducted to determine the influence of different parameters showed that particle size, type and size of filter   medium,
filter length and filtration rate played a very important role   on
the filter perfomance. Additionally, headloss recovery after   performing abrupt drainage was an observed phenomenon that could   increase run length, especially when rates as high as 24 - 48   m/day
were used.

In 1985, Perez and co-workers (5) presented a report to the Pan
American Health Organization containing results of pilot and proto-
type studies on gravel prefiltration. Their study included vertical
downflow and horizontal-flow prefilters at filtration rates ranging
from 2,4 to 48,0 m/day. Turbidity removal efficiency of almost   80%
was achieved under some conditions; in general, the higher the   raw
water turbidity, the higher the removal efficiency in terms of percentage. The authors concluded that prefiltration can   efficiently
be used to reduce the solids content of an influent   to slow   sand
filters and selected the main parameters that influenced the   process as being raw water quality, approach velocity, size of gravel,
size of suspended particles and whether the mode of operation   was
horizontal or vertical.

They recommended that a study was needed of an effective   way
for cleaning the media.

The results of these previous studies (5,6 and 8) supported the
idea of using coarser material in prefiltration as was   concluded
from previous studies at EESC-USP (2) at the end of the 1970's.   At
the end of 1984, a research programme was initiated concerning   the

use upflow prefiltration. Part of this work is the subject of this paper.

EXPERIMENTAL INVESTIGATION

Pilot Plant: Two pilot plants were used. The first consisted of a single unit with coarse sand supported on a gravel layer as shown in figure 5 (CSP). The sand had an effective size of 0,85 mm (0,59 - 2,4 mm), thickness of 1,25 m and uniformity coefficient of 1,8. The second plant consisted of another single unit with four gravel sublayers (GP) as shown in figure 6 (1st sublayer - size 50 - 75 mm, thickness 30 cm; 2nd sublayer - size 12,7 - 25,4 mm, thickness 30 cm; 3rd sublayer size 6,4 - 7,9 mm, thickness 80 cm; 4th sublayer - size 2,4 - 4,8 mm, thickness 45 cm).

Both units were constructed in two parts: a steel cylinder of 80 cm inside diameter, 7,9 mm thick and 1,75 m high; and a truncated inverted cone from 80 to 20 cm inside diameter made of steel, 7,9 mm thick and 60 cm high.

Raw water was abstracted from the local city water supply pipe that crosses the campus of EESC-USP, and pumped to a flow splitting box, from which the influent to the filters was conducted through individual vertical pipes provided with rotameters. A pipe arrangement at the inlet to each filter provided conditions to simultaneously waste the influent and perform an intermediate drainage of the media. A head of 1,2 m was provided for the CSP and 0,5 m for GP operation.

Prefilter Operation: Both pilot plants were operated in paralle and submitted to different approach velocities during the period of study, which comprised four phases, as follows:

- Phase 1: in this phase, the approach velocity was kept constant and equal to 12 m/day; the GP unit worked continuously during the 60 day period of operation, except when performing intermediate drainages (defined subsequently). Three runs were performed with the CSP: in the first run, the available water head of 1 m was consumed in 10 days and no drainage was performed except that after the end of the run; in the second run an intermediate drainage was performed after 10 days of operation, when the headloss in the sand layer was equal to 1 m, which provided a head recovery of 0,3 m and increased the total run length to 15 days; in the third run two intermediate drainages were performed, the first after 8 days of opera-

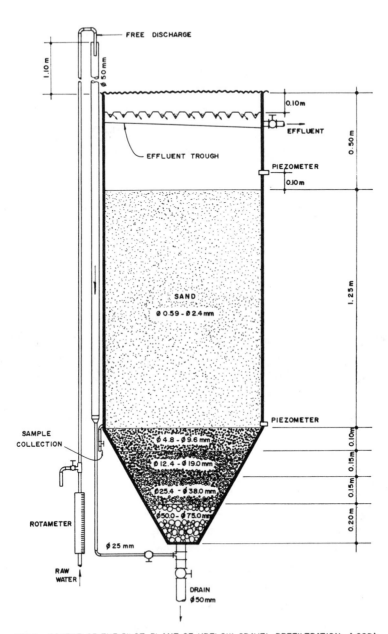

FIG. 5 - SCHEME OF THE PILOT PLANT OF UPFLOW GRAVEL PREFILTRATION (CSP).

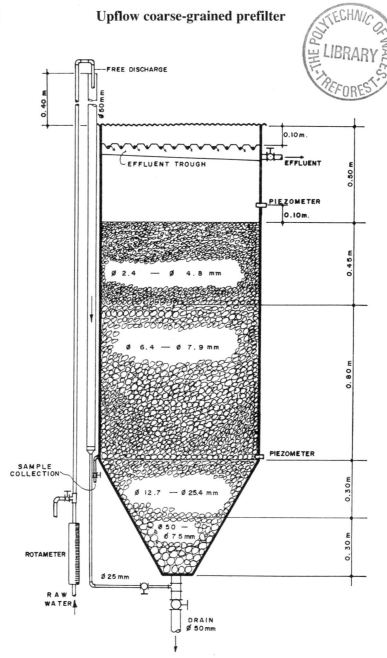

FIG. 6 - SCHEME OF THE PILOT PLANT OF UPFLOW PREFILTRATION (GP).

tion when the sand headloss was equal to 0,7 m, providing a head re
covery of 0,3 m, and the second after 15 days of operation, when the
sand headloss was equal to 1,0 m, providing a head recovery of 0,35
m. This procedure sustained the filter run for 20 days of operation
when it was terminated due to the consumption of the available head
of 1,2 m in the sand layer. The intermediate drainages in the GP u-
nit caused no significant variation in headloss, since it was very
low (around 0,02 m) at the end of this phase.

   After the end of one run with the CSP unit, the cleaning of the
sand was performed with several drainages (at least two); the upper
part of the unit was filled with stored, prefiltered water before exe-
cuting drainage.

   - Phase 2: in this phase both units were operated at 18 m/day for
20 days. However, for the CSP the approach velocity was   increased
incrementally at the beginning of the period from 12 to 18 m/day (at
2 m/day per 15 min period). Three intermediate drainages were performed in both plants; no significant variation in headloss   was
observed in the GP, in which it remained approximately constant   a-
round 0,025 m from the begining till the end of this phase; on   the
other hand, in the CSP, headloss recovery was equal to 0,2 m   (0,5
to 0,3 m) in the 1st drainage, 0,3 m (0,8 to 0,5 m) in the 2nd drai
nage and 0,3 m (1,0 to 0,7) in the 3rd drainage.

   - Phase 3: at the begining of this phase, a drainage was   performed only in the GP just before the approach velocity was   increased
gradually from 18 to 24 m/day. After cleaning the CSP as   described
earlier, the plant operated at 24 m/day for 30 days. Four intermedia
te drainages were performed in both plants; again, headloss   variation in the GP was not significant and it remained   approximately
constant and equal to 0,033 m. In the CSP, headloss recovery was e-
qual to 0,15 m (0,3 to 0,15 m) in the 1st drainage, 0,25 m (0,50 to
0,25 m) in the 2nd drainage, 0,35 m (0,70 to 0,35 m) in the 3rd drai
nage and 0,4 m (1,0 to 0,6 m) in the 4th drainage.

   - Phase 4: the same procedure as described in phase 2 was repeated
before the begining of this phase, in which the plants operated   at
36 m/day for 34 days. Six intermediate drainages were performed   in
both plants. Although headloss variation was not significant (around
0,04 m), a great content of material was visually removed from   the
GP when performing the intermediate drainages. In the CSP, headloss
recovery was equal to 0,2 m (0,5 to 0,3 m); 0,3 m (0,7 to 0,4   m);
0,2 m (0,7 to 0,5 m); 0,25 m (0,8 to 0,55 m); 0,15 m (0,95 to   0,80

m) and 0,15 m (1,0 to 0,85 m), respectively, after the 1st, 2nd, 3rd, 4th, 5th and 6th drainages.

<u>Evaluation</u>: Approach velocity, headloss and turbidity, apparent colour, iron and manganese of influent and effluent water were moni tored every two hours between 6 am and 6 pm and every six hours between 6 pm and 6 am during the experimental work. Total coliform counts of influent and effluent water were performed for some samples collected during the experiment, using the multiple tube fermentation method. Iron and manganese concentration was measured as a combined total using a colorimetric method; in general iron was the major component.

RESULTS

Figures 7, 8 and 9 show the results obtained during the experiment. All these figures show the four phases of study, but figure 7 is related to turbidity, figure 8 to apparent colour and figure 9 to iron and manganese of influent and effluent from both plants. Table 2 shows the values of MPN of total coliforms of influent and effluent water samples for different operating conditions.

DISCUSSION AND CONCLUSIONS

Figures 7, 8 and 9 show that the CSP produces an effluent much better than the GP, although the CSP had to be taken out of service for cleaning. These figures also show that intermediate drainage may increase the run length of the CSP due to head recovery, but no benefits could be noticed in the GP, because headloss in the media was very low, even for an approach velocity as high as 36 m/day. However, a significant amount of solids was observed to be present in the discharged water. All the drainages were performed by opening the drain valve and allowing the water level in the units to decrease to about 0,1 m above the top of the media. Thus, it is possible that the higher the water level in the units the more efficient was the drainage in removing the solids retained in the media, because of the higher intersticial and longer time of drainage.

Although intermediate drainages could increase the run length of the CSP, it caused disturbances in the media resulting in a deterioration of effluent quality in both plants, but the detrimental effects were more pronounced in the GP than in the CSP as can be seen in table 2. Also, it can be see that the higher the approach

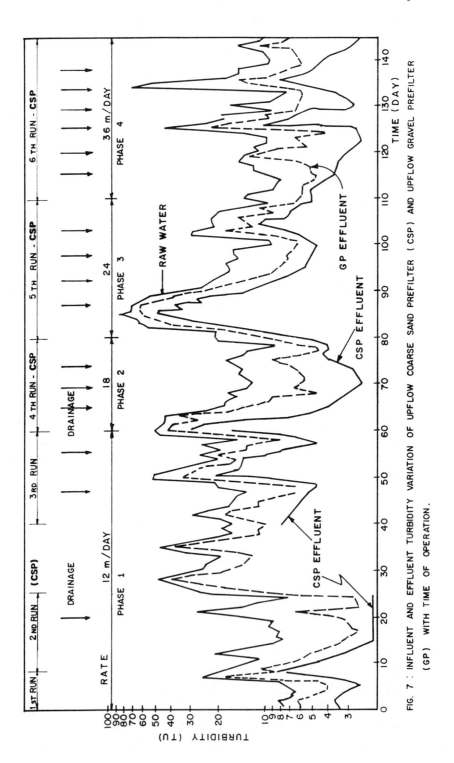

FIG. 7 : INFLUENT AND EFFLUENT TURBIDITY VARIATION OF UPFLOW COARSE SAND PREFILTER (CSP) AND UPFLOW GRAVEL PREFILTER (GP) WITH TIME OF OPERATION.

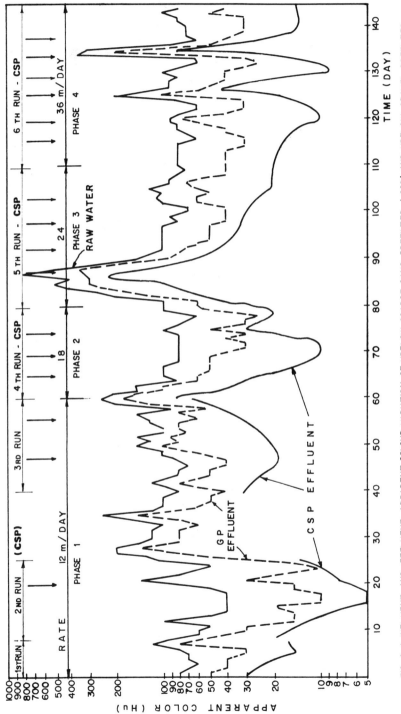

FIG. 8 : INFLUENT AND EFFLUENT APPARENT COLOUR VARIATION OF UPFLOW COARSE SAND PREFILTER (CSP) AND UPFLOW GRAVEL PREFILTER (GP) WITH TIME OF OPERATION.

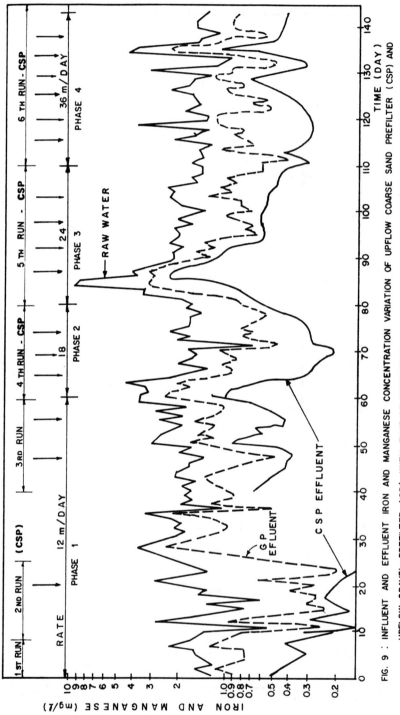

FIG. 9 : INFLUENT AND EFFLUENT IRON AND MANGANESE CONCENTRATION VARIATION OF UPFLOW COARSE SAND PREFILTER (CSP) AND
UPFLOW GRAVEL PREFILTER (GP) WITH TIME OF OPERATION.

TABLE 2: Results of MPN of Total Coliforms of Influent and Effluent
         Water.

OPERATING CONDITION	MPN of Total Coliform (coli/100 ml)		
	Influent	CSP Effluent	GP Effluent
- At 20th day of operation (2nd run of CSP)	93	36	36
- At 20th day of operation, just before the increase in approach velocity from 12 to 18 m/day	460	-	23
- At 60th day of operation, two hours after increasing the approach velocity	-	-	240
- during phase 2, just before performing 2nd drainage	240	9,1	43
- during phase 2, two hours after performing 2nd drainage	-	75	210
- At 80th day of operation, just before the increase in approach velocity from 18 to 24 m/day.	930	-	150
- At 80th day of operation, two hours after increasing the approach velocity	-	-	730
- during phase 3, before performing 1st drainage	1100	93	210
- during phase 3, two hours after performing 1st drainage	-	430	930
- At 110th day of operation, just before increasing the approach velocity to 36 m/day	210	-	93
- At 110th of operation, five hours after increasing the approach velocity to 36 m/day	-	-	240
- during phase 4, just before performing 5th drainage	1500	93	150
- during phase 4, five hours after performing 5th drainage	-	240	430

velocity the worse were the negative effects of intermediate draina
ge on bacteriological effluent quality of both pilot plants.

An interesting aspect of behaviour highlighted in figures 7, 8
and 9 concerns the variation in influent water quality. In general,
each time a peak in quality (turbidity, apparent colour, iron and
manganese) occurred in the influent, it was followed by a quality
peak in the effluents of both plants. This effect appeared to be in
dependent of the approach velocity. This behaviour was also observed by Perez and co-workers (5) using only turbidity as the performance parameter. It can be concluded that these prefiltration units
don't have a significant capacity for attenuating sudden changes in
influent water quality. In the particular case of turbidity and appa
rent colour (mainly turbidity), the trend of effluent peaks follow
ing influent peaks is consistent with physical filtration theory.
However, substantial differences in the magnitude of removal may re
flect more complex changes in the influent water quality, such as
particle size distribution. Some attempts were made to measure particle Zeta potential and dissolved solids concentration but the results were inconclusive. Future work on prefiltration at EESC-USP
will make use of particle counting equipment to monitor the changes in
the particulate content of the influent water.

Based on the experimental investigation carried out, the follow
ing conclusions can be drawn:
a) Since the plant start-up, effluents with better quality than the
   influent can be produced in the CSP and the GP units with significant reduction of turbidity, apparent colour and iron and manganese concentration;
b) Attenuation capacity of both units due to occurrence of influent
   water quality peaks is very low;
c) The higher the influent water peak of turbidity, apparent colour
   and iron and manganese, the higher the effluent peaks in the CSP
   and the GP;
d) Approach velocities in the range of 12 and 36 m/day appear to in
   fluence very little the quality of effluents produced in the CSP
   and GP units;
e) A better effluent quality can be produced in the CSP than that
   in the GP unit;
f) Intermediate drainages in the CSP can increase the run length be
   cause of headloss recovery, but it can cause detrimental effects

mainly on the bacteriological quality of the effluent;

g) For the GP, intermediate drainage caused a substantial removal
   of solids observed in the discharged water and resulted in a ne
   gligible variation in headloss, but a serious detrimental effect
   in bacteriological effluent quality occurred with a high increa
   se of MPN of total coliform after the plant was put in service.

ACKNOWLEDGEMENT

   This work described in this paper was supported through funds
provided by FINEP - Financiadora de Estudos e Projetos and   consists as a part of a wider project under study in the School of En
gineering of São Carlos - University of São Paulo concerning   the
technology of slow sand filtration.

REFERENCES

1 - CEPIS, Modular Plants for Water Treatment, Technical   Docu-
       ment nº 8, Panamerican Center of Sanitary Engineering   -
       WHO, Lima - Peru, March, 1982 (in spanish).

2 - COSTA, R.H.R., Experimental Comparison Between Downflow   and
       Upflow Slow Sand Filters, M.S. Thesis presented at the De
       partment of Hydraulic and Sanitation of the School of Engi
       neering of São Carlos - University of São Paulo, 1980   (in
       portuguese).

3 - DI BERNARDO, L.   Comparison Between Upflow Gravel and Upflow
       Coarse Sand Prefiltration for the Pretreatment of Surface
       Waters, Technical Report presented to FINEP, São Carlos,
       SP, Brazil, Dec. 1986 (in portuguese).

4 - DI BERNARDO, L. et al.   Considerations on the Use of   Gravel
       Prefiltration for the Pretreatment of Surface Waters, 14th
       Brazilian congress of Sanitary Engineering, São Paulo, Bra
       zil, Sept. 1987 (in portuguese).

5 - PERES, J.M.C. et al.   Prelimnary Report on the Gravel Prefil-
       tration Investigation, Panamerican Center of Sanitary Engi
       neering - WHO, Lima - Peru, May 1985 (in spanish).

6 - THAN, N.C. & QUANO, E.A.R.   Horizontal Flow-Coarse   Material Prefiltration, Asian Institute of Technology, Research Re port nº 70, Bangkok - Thailand, July 1977.

7 - UNIVALLE, Slow Sand Filtration and Gravel Prefiltration,   In ternational Seminar on Simplified Technologies for   Water Treatment, Universidad del Valle, Cali - Colombia,   Aug. 1987 (in spanish).

8 - WEGELIN, M.   Horizontal-Flow Roughing Filtration: An Appropria te Pretreatment for Slow Sand Filters in Developing   Coun tries, WHO - IRCWD NEWS Number 20, Aug. 1984.

2.3   PRETREATMENT WITH PEBBLE MATRIX FILTRATION

K. J. Ives and J. P. Rajapakse — Department of Civil & Municipal Engineering, University College London, WC1E 6BT

ABSTRACT

Pebble Matrix Filtration reduces suspended solid concentrations in highly turbid waters from values up to 5000 mgl^{-1} to below 25 mgl^{-1}.   This protects slow sand filters from excessive clogging by monsoon river waters. The Pebble Matrix Filter consists of a deep layer of pebbles, approximately 50 mm in size, infilled in its lower part by sand usually less than 1 mm in size.   Runs of about 20 h to head losses not exceeding 1.5 m , have been achieved with a flow rate of 0.72 m/h.   Cleaning is simply achieved with 2 drainage cycles followed by backwashing with raw water only.

INTRODUCTION

A Pebble Matrix Filter operating at 0.7 - 1.2 mh^{-1} appeared to meet the requirements for the pre-filtration of high turbidity flood waters in tropical conditions, as a pre-treatment technique for slow sand filtration.   Tests up to 5000 mgl^{-1} produced filtrates below 25 mgl^{-1}.   One particular advantage of this technique is that it does not require any chemicals for its operation, even at extremely high turbidities in the raw water.   The cleaning of the filter has been kept at a simple level (no air scouring) because of its intended application in semi-rural conditions in tropical countries.

This filter, first known as Skeleton-fill filter, has been developed in Tashkent, USSR, at an experimental scale for tertiary treatment of sewage, with a maximum suspended solids load of 20 mgl^{-1} (1).

In principle, the filters consist of a matrix of large pebbles about 50 mm in size (the "Skeleton") which is infilled for part of its depth with sand, as shown in fig. 1.

The flow direction is downward through the media. Therefore, the suspension approaching the filter first passes through a layer of large pebbles, and then through a layer of mixed pebbles and sand.   This creates a crude two layer filter where the pebbles alone have a pre-filtering effect (see fig. 2), and the filter can retain very large quantities of deposit without a great loss of permeability (as shown in fig. 3).

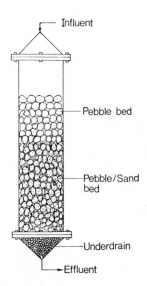

PEBBLE MATRIX FILTER

FIGURE 1

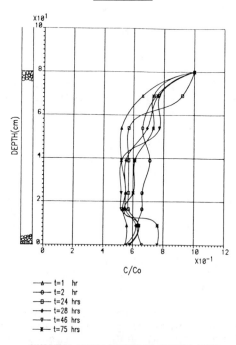

SUSPENDED SOLIDS REMOVAL IN PEBBLES ALONE
—RUN63

FIGURE 2

FIGURE 3

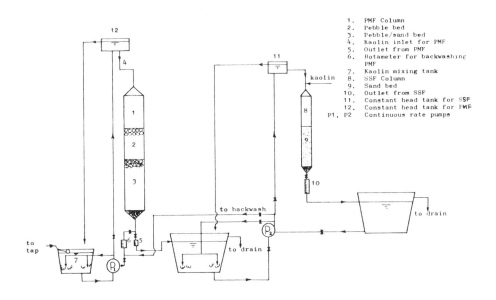

1.	PMF Column
2.	Pebble bed
3.	Pebble/sand bed
4.	Kaolin inlet for PMF
5.	Outlet from PMF
6.	Rotameter for backwashing
	PMF
7.	Kaolin mixing tank
8.	SSF Column
9.	Sand bed
10.	Outlet from SSF
11.	Constant head tank for SSF
12.	Constant head tank for PMF
P1, P2	Continuous rate pumps

FIGURE 4

LABORATORY INVESTIGATIONS

Filtration

The experimental investigations were carried out employing a model Pebble Matrix filter which was constructed for this study. The unit comprised a 245 mm ID, 1.30 m long perspex column with the top and bottom ends covered with two cones made of fibreglass. Filtration velocities of 0.5 mh^{-1} to 1.5 mh^{-1} with suspended solids in the raw water from 100 mgl^{-1} to 5000 mgl^{-1} were tested. The filtrate quality was monitored continuously using a HACH-Ratio turbidimeter and pen-recorder. This general arrangement is shown in fig. 4.

Various experiments, using different inlet concentrations, flow rates, sand sizes, and depths of media, were carried out to determine the best design for pretreating London tapwater containing kaolin clay. These are listed in Table 1.

At a filtration velocity of 0.72 mh^{-1} with fine sand (d_{10} = 0.38 mm) the filter produces an effluent of below 1 mgl^{-1} (suspended solids) even with as high as 1000-5000 mgl^{-1} concentrations at the inlet. At all filtration rates mentioned earlier an effluent quality of less than 25 mgl^{-1} can be obtained. A typical graph showing the variation of effluent quality with time (for sand d_{10} = 0.56 mm) is given in fig. 5.

Comparative studies with ordinary sand filters showed that the pebble matrix filter has a remarkably low pressure drop due to its high permeability caused by the presence of pebbles in sand (fig. 6). Consequently, the silt storage capacity of the bed is also considerably higher (30 - 70 kgm^{-2} per run) than of the conventional sand filters. The head loss development and the pressure distribution within the bed are shown graphically in figs 7 and 8 respectively.

Filter cleaning

For efficient operation the filter must be maintained in good condition by removing the deposited material from the bed at the end of a filter run. Several cleaning processes have been investigated satisfactorily.

1. By draining down the filter and refilling it with raw water, and draining down again, the majority of the deposit can be removed (> 70%). By reverse flow washing with clean water the sand was fluidised to occupy all the spaces between pebbles.

2. The draining procedure was the same as in (1) but washing was first accomplished with raw water followed by clean water. This reduces the clean water consumption during backwashing by about 50%.

Inlet clay conc$_n$ (mg/L)	Pebble/Sand depth (cm)	Total depth (cm)	approach velocity Va (m/h)	Run time T (hours) t_c or t_h	Headloss (cm) at T hours
		SAND 8/16			
500	61	77	0.72	14(t_c)	2.5
500	84	102	0.72	34(t_c)	6.5
500	84	102	1.17	10(t_c)	4.5
		SAND 16/30			
500	31	62	0.72	18(t_c)	14.5
500	75	102	0.72	60(t_c)	61.5
500	75	102	1.17	16(t_c)	32.0
1000	75	102	0.72	28(t_c)	63.5
1000	75	102	1.17	12(t_c)	33.0
2000	75	102	0.72	12(t_c)	34.0
2000	95	130	0.72	18(t_c)	41.5
500	95	130	0.72	116(t_c=t_h)	150.0
		SAND 22/44			
500	34	64	0.72	35(t_c)	62.8
1000	34	64	0.72	16(t_c)	56.2
500	75	102	0.72	44(t_h)	108.0
500	75	102	1.17	24.5(t_h)	102.0
500	75	102	1.56	19.0(t_h)	124.0
1000	75	102	0.72	27(t_h)	129.7
1000	75	102	1.17	14(t_h)	133.2
5000	75	102	0.72	8.5(t_h)	126.0

t_c = breakthrough by filtrate quality

t_h = breakthrough by headloss

Table 1

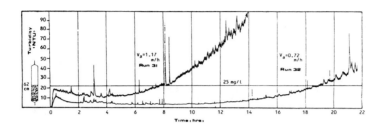

VARIATION OF EFFLUENT TURBIDITY vs TIME

FIGURE 5

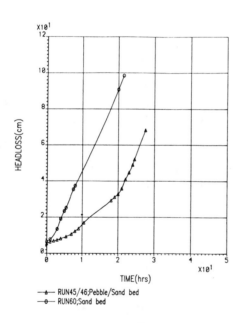

COMPARISON OF TOTAL HEADLOSS VARIATION WITH TIME
FOR PEBBLE/SAND AND SAND BED;
C=1000mg/l;Va(Sand)=2.0m/h;BED DEPTH=62.0cm

FIGURE 6

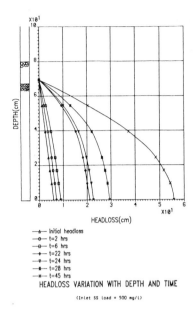

HEADLOSS VARIATION WITH DEPTH AND TIME

(Inlet SS load = 500 mg/L)

FIGURE 7

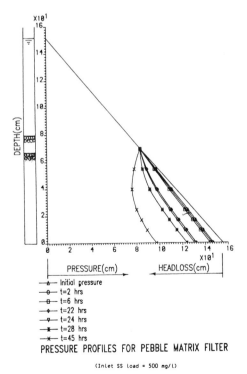

PRESSURE PROFILES FOR PEBBLE MATRIX FILTER

(Inlet SS load = 500 mg/L)

FIGURE 8

3.   After draining down the filter it was backwashed using
      raw water only.   This was sufficient to clean the filter
      of the accumulated clay so that when filtration was
      re-started the initial head loss was very similar to the
      one attained after washing with clean water (see fig. 9).
      Several consecutive runs incorporating this cleaning method
      produced similar head losses, effluent quality against time
      graphs and run times.   One advantage in this method is
      that raw water would be plentiful during a monsoon period.

ADDITIONAL EXPERIMENTS

Model Slow Sand Filter

     As a preliminary study a model slow sand filter (110 mm
diameter, sand depth 0.6 m) was operated with organically
contaminated (glucose) tapwater with a schmutzdecke seeded from
the Coppermills waterworks of Thames Water Authority.
This established normal run patterns, and was cleaned by scraping,
to the usual depth.   Subsequent runs with kaolin-clay laden
water established the limit of 25 mgl^{-1} suspended solids in the
influent.   Damage to the operation of the slow sand filter was
assessed by rate of rise of head loss, biological activity of the
schmutzdecke and penetration of clay into the sand.

Bacteria Removal by Pebble Matrix Filtration

     Although the slow sand filter carries the responsibility of
bacteriological improvement, the contribution to bacterial removal
by the Pebble Matrix Filter was assessed.   The inlet was seeded
with about 10^3 E. coli per ml.   The results showed that the
removal of E. coli was associated with the clay, indicating their
adsorption on the clay particles led to their removal by filtration.

Utilisation of Stones

     The principal experiments utilised smooth rounded pebbles from
the seashore.   As it is possible that such pebbles may not be
universally available, some tests were made with rough rounded
pebbles, and broken angular roadstone of the same general size
(50 mm).   These tests have not yet been evaluated (July 1988)
but some modification to the cleaning procedure may be necessary.

DESIGN CRITERIA (PRELIMINARY)

     When designing the Pebble Matrix Filter as in any pretreatment
process one important factor that has to be taken into
consideration is the maximum suspended solids concentration in
the river that would occur during a monsoon period.   Suspended
solids concentration would be expected to vary significantly with
river flow as shown in fig. 10 (2).   However, these monsoon
turbidity/suspended solids values for rivers are not readily
available in most parts of the developing countries and,
therefore, the appropriate authorities should be encouraged to
collect this data in general and more particularly during a
preliminary investigation of a pretreatment project.

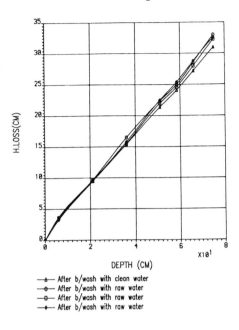

INITIAL H.LOSS AFTER B/WASH WITH RAW WATER ONLY
(SS concentration in raw water=500mg/l)

FIGURE 9

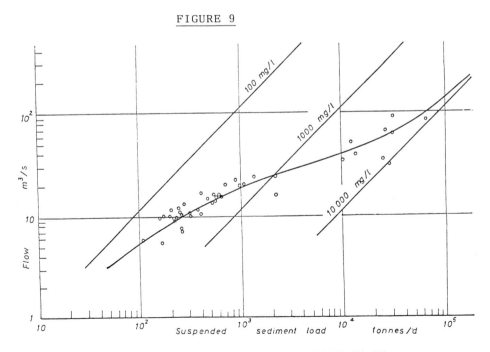

DEPENDENCE OF SUSPENDED SOLIDS CONCENTRATION ON
RIVER FLOW   (Ref. 2)

FIGURE 10

As in all sample analyses, such data should be obtained
over a period to allow for seasonal changes, alterations in
river flows, etc.   The raw water quality determines the
choice of treatment process.   However, it is the gross
approach velocity (V_a) that determines the plan area of the
filters, which is a major contribution to the capital cost
of the plant.   Therefore, choosing the right approach
velocity is also an important decision in the design stage.
A guideline to choose the gross approach velocity for the
Pebble Matrix Filtration is given in the diagram on fig. 11.

Filter Area and the number of units

Having decided the approach velocity V_a the area A can be
calculated to be equal to or greater than Q/V_a where Q is the
total hourly volume of water to be treated by the slow sand
filters.

The number of filter units should preferably be at least
2 or more   The formula n = ¼ $\sqrt{Q}$ suggested by Huisman & Wood (3)
can be used as a first approximation, when n is never less than
2 and Q is expressed in $m^3 h^{-1}$.

Constituent Parts

In addition to the above, it is necessary to give consideration
to the following constituent parts of the Pebble Matrix Filter
during its design stage:

1.   supernatant water reservoir
2.   free pebble bed
3.   pebble/sand bed
4.   underdrainage system
5.   flow control devices
6.   filter box

Since there is virtually no head loss within the pebble bed alone,
the depth of the supernatant water reservoir can be fixed so that
the head of water above the pebble/sand bed is sufficient to
overcome the resistance of the pebble/sand bed and allow a
downward flow through the bed.

Depths which are too short would create a negative head
within the bed and consequently shorter filter runs.   A value
of between 1.0 m and 1.3 m would be a suitable figure to give
a reasonable length of run time.   Two factors should be borne
in mind when determining the depth of the free pebble bed:

a)   suspended solid removal in pebbles alone (fig. 2)

b)   sufficient depth to allow the sand bed to expand
     up into the free pebble matrix during backwashing.

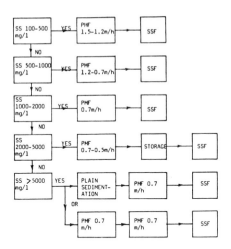

GUIDE FOR PROCESS SELECTION

  SS - Suspended Solids

PMF - Pebble Matrix Filter

SSF - Slow Sand Filter

     FIGURE 11

If rounded pebbles are not available some other media such as road materials or broken bricks may be used. For the sand bed both finer and coarser materials have been found to work satisfactorily under laboratory conditions, but the selection will actually have to depend on the locally available materials and pilot plant observations as in the case of pebble bed. Effective diameters of (d_{10}) between 0.35 mm and 1.00 mm have produced filtrates of extremely good quality as mentioned earlier. However, in addition to the removal efficiency other factors such as filter run time, maximum head loss, bed depth and approach velocity have to be considered as a whole.

Underdrains serve the purpose of collecting the filtered water and distributing the washwater during cleaning. The simplest form is a system of perforated lateral pipes packed round with gravel.

A floating effluent weir on the outlet is a suitable device to maintain the constant flow through the filter against the rising head loss due to clogging.

CONCLUSIONS AND RECOMMENDATIONS

1. Pebble Matrix Filter is an appropriate pretreatment method for slow sand filtration, suitable for semi-rural conditions in the developing countries where monsoon turbidities reach several thousand mg/l.

2. Final design, construction and operation criteria has to be decided upon completion of pilot plant studies.

3. It would be a worthwhile effort to encourage water authorities in the developing countries to prepare a record of turbidity/ suspended solids and particle size analysis for rivers during monsoon periods.

LITERATURE

1. KHABIROV, R.S., SLAVINSKI, A.S., KRAVTSOVA, N.V. (1983) Optimization of the technological and design parameters of skeleton-fill filters (in Russian). Khimiya i Tekhnologiya Vody, 5, (4), 352-353.

2. IVES, K.J. (1983) Treatability of water, J. Inst. Wat. Engrs. Sci., 37, (2), 151.

3. HUISMAN, L. and WOOD, W.E. (1974) Slow sand filtration, WHO Geneva.

ACKNOWLEDGEMENTS

Funds received by the U.K. Overseas Development Administration and University College London are gratefully acknowledged. Thanks are also due to Mr. I. Sturtevant for his expertise in constructing the experimental apparatus and Dr. A. Appan for additional information.

2.4 OZONATION AND SLOW SAND FILTRATION FOR THE TREATMENT OF COLOURED UPLAND WATERS – PILOT PLANT INVESTIGATIONS

G. F. Greaves*, P. G. Grundy** and G. S. Taylor** — *Ozotech Ltd., Burgess Hill, Sussex. **North West Water, Western Division.

ABSTRACT

North West Water Authority have substantial assets at their Llanforda Waterworks, Oswestry, which include land, civil structures associated with slow-sand filters of capacity 200 Ml.d^{-1} and mechanical plant. In order to comply with E.C. legislation final water colour needs to be reduced to 20° The use of coagulants is inappropriate and pre-ozonation is being investigated as a viable alternative.

Pilot plant work has indicated satisfactory colour removal but work is continuing to investigate other quality aspects and process economics.

INTRODUCTION

The E.E.C. Directive on the Quality of Water for Human Consumption has allowed exceptions and derogations to the maximum admissible concentration (M.A.C.) for colour of a natural origin in drinking water. Such derogations are however shortly to be repealed and there are certain cases where the conventional colour removal process of coagulation, flocculation and separation is inappropriate.

Slow sand filters are unable to be backwashed and depend, for effective treatment, on biological activity in the top few centimetres of the filter medium. The use of an inorganic coagulant prior to slow sand filtration would not only pose problems of rapid headloss development and the precipitation of dissolved residual coagulant deep in the filter bed, where it could remain for several filter cycles, but could also inhibit the proper development of the schmutzdecke. Consequently there is interest in the use of ozone to bleach coloured humic materials prior to treatment within slow sand filters, particularly where such filters represent a substantial asset to the water undertaking and where their replacement would be expensive or inconvenient. The major supplies to the cities of Liverpool and Aberdeen represent such a case

Lake Vyrnwy in North Wales is an artificial impoundment in an upland catchment area comprising moorland, rough pasture and both coniferous and deciduous woodland. The raw water is acidic (pH 6), low in dissolved solids and moderately coloured. Raw water colour varies between about 10 and 50°H and is typically 20-30°H. Up to 400ug l^{-1} Fe is also present in the water together with typically up to 100 ug l^{-1} Mn, although very occasionally this latter value is exceeded by 100% or more.

Prior to its supply to Liverpool, up to 200 Ml d^{-1} is treated at the Llanforda filters of North West Water Authority at Oswestry, Shropshire. The reservoir, filters and inter-connecting aqueducts were constructed about 100 years ago. Currently 11 of the 23 slow sand filters, each rate at 2 m.g.d. (9 Ml d^{-1}) are being remodelled to restore their design capacity and to remove hydraulic restrictions in the site pipework. A significant recent investment has also been made in the provision of machinery to mechanically skim and resand the filters and in an automatic sand washing plant. It is known however that slow sand filters remove only about 20% of the raw water colour, so given values of up to 50°H and an MAC of 20°H for the treated water, a colour removal stage will be necessary.

As North West Water have operating experience in the use of ozone at their Watchgate and Sunnyhurst works, it was a logical decision to carry out pilot trials into the use of ozone prior to slow sand filtration at Oswestry. Laboratory tests carried out by Ozotech Ltd. indicated that ozone would effectively reduce the raw water colour. The results of the tests, in terms of dose applied, together with the dimensions of existing pilot filters which had been built to investigate the effects of different sand granulometry and under-drain type on filter performance, determined the capacity of the ozone generator to be supplied by Ozotech Ltd. for these trials.

In 1866 the river abstraction works, filters and associated aqueducts providing a supply from the River Dee, at Banchory, to Aberdeen were complete. Successive extensions to this works, culminating in the addition of 3 filters in 1979, making a total of 14, provide a treatment capacity of 70 Ml d^{-1}. An interesting feature of the Invercannie Waterworks is the provision of two large, serpentine raw water storage reservoirs which both assist with maintaining the reliability of supply and facilitate the settlement of river solids. Water quality is not dissimilar to the Vyrnwy supply, being of upland origin, but despite a low alkalinity is close to being of neutral pH. Mean colours are similar, being between 20 and 30°H but the maxima and minima vary more widely from the mean, 1984 values being 85°H and 2°H respectively. Mean turbidity values are similar for both sources at about 1 NTU. but the maximum value is higher at Invercannie. There is less manganese in the river derived source at Invercannie than in the Vyrnwy water although the mean, maxima and minima for iron are of the same order of magnitude in the two supplies.

North West Water Authority at Oswestry and Grampian Regional Council at Invercannie have substantial assets, which although elderly in design and construction have been refurbished and extended. Neither, however, is able to continuously produce a water of the required colour, despite meeting most other water quality criteria with ease. The requirement to improve colour whilst utilising the existing assets, without the need to import chemicals to and export sludge from these substantial water treatment plants, persuaded the management within each Undertaking to explore the possibility of ozonation as an effective colour removal process. Consequently pilot plant trials were started independently by each Undertaking, but within a month of each other, in the Autumn of 1987.

DESCRIPTION OF PILOT PLANT

Air is compressed by a tank mounted Gast compressor to 5.5 - 7.0 bar before being fed to a Trailigaz 'Labo' ozoniser at an airflow of up to 500 l.h^{-1}. The 'Labo' ozoniser has been modified by Ozotech Ltd. to ensure that the high voltage discharge ceases on failure of the cooling water flow (401 h^{-1}) or loss of air pressure. These modifications allow the plant to be left unattended during operation at a production rate of up to 9gO$_3$ h^{-1}.

The ozonised air line is fed to the control board on the contactor assembly where it is split to supply each of two contact columns. Excess ozonised air can also be bled off via a by-pass valve to a thermal ozone destructor operating at about 350°C.

The supply to each contact column is metered through a 'Platon' control valve and airflow meter and passes through an anti-syphon chamber before being discharged through a sintered stainless steel diffuser under 4.3m static head of water. The raw water passes through a flow control valve and meter to flow downwards, counter-current through column 1. It ascends column 2, co-current with the ozonised air, before discharging through a cascade outlet where the gas is displaced from the liquid phase. Provision is made for venting off-gas from the top of the columns and also for the accommodation of any foam which might form at the water surfaces.

At Oswestry the ozoniser and compressor are housed in a small wooden shed whilst the contactor assembly is bolted to the outside wall of the micro-strainer house. A ply-wood and insulated weather-proof structure encloses the contact columns to protect them against the extremes of weather.

At Invercannie the contact columns are housed in a redundant micro-strainer chamber and the electrical and mechanical equipment on the decking above. Here the off-gas is vented outside the building and the atmosphere within the building is monitored for trace levels of ozone to comply with health and safety requirements. The monitor used is the Ozotech "Otrameter" which works on the amperometric principle.

At Llanforda a block of eight pilot slow sand filters, each of surface area $4m^2$ (2m x 2m) had previously been constructed in brickwork to evaluate various filter sands. Filters 7 and 8 and 1 and 2 were respectively selected as the experimental and control filters and each have 750mm washed sand and 270mm graded gravels (in 3 layers) over a drainage tile filter floor. Initial test have been conducted at a filtration velocity of 150mm h^{-1} with the intention of increasing the rate to 200mm h^{-1} later in the experimental programme. To cater for the maximum projected flow of 800 l h^{-1} to each experimental filter with a small surplus to overflow, the maximum flow through the ozonation plant was determined at 1800 l h^{-1}. At this flow rate the maximum available ozone dose is 5 mg O_3 l^{-1} and in order to achieve an overall contact time of 5 minutes, the contact columns were sized at 150mm dia.

At Invercannie existing filter shells were not available so 2 circular units, each of area $15m^2$ were constructed in G.R.P. one as an experimental filter, the other as a control. Filter medium is as per the main plant. A large header tank was provided upstream of the ozone contact columns and a balancing tank, controlled by ball-valves, was provided upstream of each filter. The flow through the filters is regulated to give a filtration velocity of 4 in h^{-1} (100mm h^{-1}).

EXPERIMENTAL PROGRAMME

At Oswestry the principal physical and chemical operating parameters recorded on a weekly basis are as follows:

    pH,
    colour, °H spectrophotometric method
    colour, °H long tube method

turbidity, NTU

Iron, ug Fe 1^{-1} (total)

Iron, ug Fe 1^{-1} (sol.)

Manganese, ug Mn 1^{-1} (total)

Manganese, ug Mn 1^{-1} (sol.)

Coliform bacteria, n.dl^{-1}

E.Coli   n.dl^{-1}

Aluminium, ug Al.1^{-1} (total

Aluminium, ug Al.1^{-1} (sol.)

particle content, um.m^{-1}

   In addition to the above, it is intended that during the Summer period May
- September 1988, when raw water quality might be expected to vary only a
little in terms of colour, turbidity and metals concentration, variations in
ozone dose will be made to determine the effect of dose on filtered water
quality. Ozone dose is to be varied through four steps from 2.5mg O_3.1^{-1} to
5.5mg O_3.1^{-1}, with each dose to be applied over a 4 week period. During this
series of tests the ozone in air concentration is to be maintained at 18g.$m^3$$^{-1}$
and the ozonised air applied to the contact columns in the proportion 2:1 to
the first and second columns respectively. Overall contact time will be 6
minutes.
   As well as monitoring basic quality parameters during this period it is
also intended to investigate the effect of dose on such phenomena as colour
return, chlorine demand, lime usage to maintain the target pH, THM
formation and organic quality. This latter is to be measured by GC/MS
analysis, in terms of products of ozonation, including the effect on THM
precursors.
   The initial operating conditions of the plant, set at commissioning, were
as follows, water flow = $1m^3$ h^{-1}, ozone dose = 3.5mg O_3.1^{-1}. The ozone was
applied to the first column only giving a nominal contact time of 4.6 min.
Subsequently the ozone dose was reduced to 2.5mg.1^{-1} whilst being proportioned
at 2:1 to colums 1 and 2 respectively. With the waterflow maintained at
1.0m^3.h^{-1} this gave a contact time of about 9 min. A subsequent increase in
dose to 3.5mg.O_3.1^{-1}, whilst keeping the other conditions constant, was made
in early April, in response to an increase in raw water colour, as determined
by the Water Authority's own 'long tube method'.
   These early tests provided a period of familiarisation with the pilot
plant and allowed a thorough scientific programme of sampling and analysis to
be developed whilst at the same time generating an abundance of data on the
treatability of Vyrnwy water with ozone. On completion of the Summer trials a
further evaluation of the progress to date will be carried out. This will be
followed, if the evaluation so indicates, by a programme of process
optimisation. This programme will proceed throughout the period when water
quality is expected to be at its worst and should allow the investigation of
the relationships between raw water quality, ozonation conditions and final
water quality. Any correlations indicated will be important in specifying the
size of plant to be installed and in defining the operating parameters on

commissioning.

At Invercannie daily samples are taken for measurement of the following parameters:

pH,

colour, °H

turbidity, NTU

total inorganic carbon, mg TIC 1^{-1}

total organic carbon, mg TOC 1^{-1}

At weekly intervals the samples are also analysed for the following:

nitrite, $ug.No_2.1^{-1}$

nitrate, $mg.No_3.1^{-1}$

ammonia, $mg\ NH_3.1^{-1}$

Iron (total), $mg\ Fe.\ 1^{-1}$

Manganese (total) $mg\ Mn.\ 1^{-1}$

Aluminium (total), $mg\ Al.\ 1^{-1}$

Frequent comparative tests are also made to compare filter performance, with and without pre-ozone, in terms of the removal of Coliform and E.coli bacteria. Grampian Regional Council are also investigating colour return, ozone demand, the development of filter headloss, THM formation potential and microbial growth potential.

The ozone plant at Invercannie was initially operated to ensure a small residual concentration of $0.1 - 0.2$ mg $O_3\ 1^{-1}$. However, during periods of increasing raw water colour the residual soon disappears and treatment is incomplete. The operating dose has therefore been raised to that required to produce an ozone residual of $0.4 - 0.5$ mg $O_3.1^{-1}$. This higher residual effectively increases the ozone/water contact time from the nominal 5 min. (water flow rate to plant = 30 l min^{-1}) in the columns because treatment will continue in the balancing tank whilst the residual persists.

As previously mentioned the filters at Invercannie are operated at a lower filtration rate than at Oswestry and filtration is terminated at a lower headloss of 0.6m rather than 0.7m.

Grampian Regional Council protect the residual biomass at the sand surface after skimming off the top 2-3cm following the termination of a run by re-filling the filter with raw water before ozonation is recommenced. This procedure prevents residual, disinfecting ozone reaching the sand surface. North West Water have re-seeded a newly skimmed sand-bed, taking particular care to include aquatic worms in the seed, which are believed to be important in keeping the filter surface 'open'.

INTERIM RESULTS

(a) Colour removal - During the 7 month period 23rd November to 22nd June 1988 the mean true and apparent colour of the raw water at Oswestry have been 19.3°H and 29.5°H respectively. Slow sand filtration brings about an average 30% reduction in apparent colour to give a mean value of 20.5°H whilst ozonation followed by filtration reduces the colour by 74% to a mean value of 7.7°H. True mean colour values are reduced on average   from 19.3°H to 15.4°H (20% reduction) by filtration and to 6.6°H (74% reduction) by ozonation and filtration. (See fig. 1)

(b) Turbidity - Over the same 30 week period ozone could not be seen to enhance the degree of turbidity removal in the filters.   Raw water turbidity ranged from 0.6 NTU to 2.6 NTU with a mean of 1.2 NTU. Filtrate quality from the experimental and control plant were as follows:

Turbidity (NTU)	Raw Water	Non-ozonated filtrate	Ozonated filtrate
Max	2.6	0.93	0.54
Min	0.6	0.21	0.23
Mean	1.2	0.35	0.36
% removal	-	71	70

Ozonation may therefore have some value in reducing peak turbidity values but not the average filtrate concentration.

(c) Iron - From the following table it can been seen that   ozonation improves the average iron removal performance of the filters by 15%.   (See also fig. 2)

Iron (total) (ug Fe.1^{-1})	Raw water	Non-ozonated filtrate	Ozonated filtrate
Max	315	150	117
Min	160	65	52
Mean	210	98	81
% removal	-	53	61

(d) Manganese - Manganese levels in the slow sand filtrate are invariably of the order of 10 ug Mn.1^{-1} and no effect of pre-ozonation can be determined.

(e) pH - Raw water pH of 5.8 - 6.6 is normally enhanced by filtration to 6.5 - 6.8 and depressed by ozonation to 4.9 - 5.4.   Ozonation followed by filtration however leads to a marginally higher filtrate pH of 6.9 - 7.1.

(f) Filter run length - During the experimental period up to May 6th reference filter No. 2 had gone through three filter cycles averaging 9.2 weeks. On the other hand experimental filters 7 and 8 had completed three filter cycles between them at an average length of 12.3 weeks.   There is therefore some indication that within the Winter and Spring period, filter run length may be increased by about one third. In general terms the development of headloss in the experimental (pre-ozonated) filters does not exhibit an exponential phase as in the control filters where this phenomenon is typical. An exception to this rule can however be seen in Graph 3 during the period when the ozoniser was shut down.

The work carried out at Invercannie confirmed the degree of colour removal to be expected by ozonation and filtration by reducing a mean raw water colour of 33°H to 8°H, i.e. a reduction of 77%. Work on colour return indicated that after a period of 7 to 10 days a degree of colour return occured which was

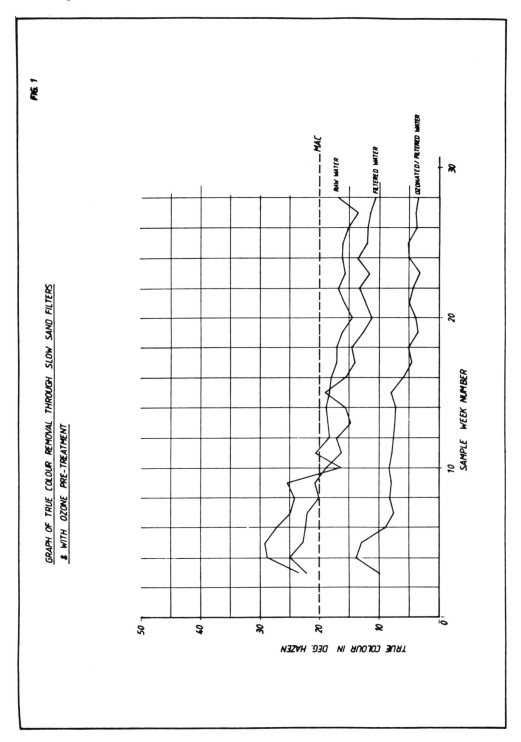

FIG 1

GRAPH OF TRUE COLOUR REMOVAL THROUGH SLOW SAND FILTERS
& WITH OZONE PRE-TREATMENT

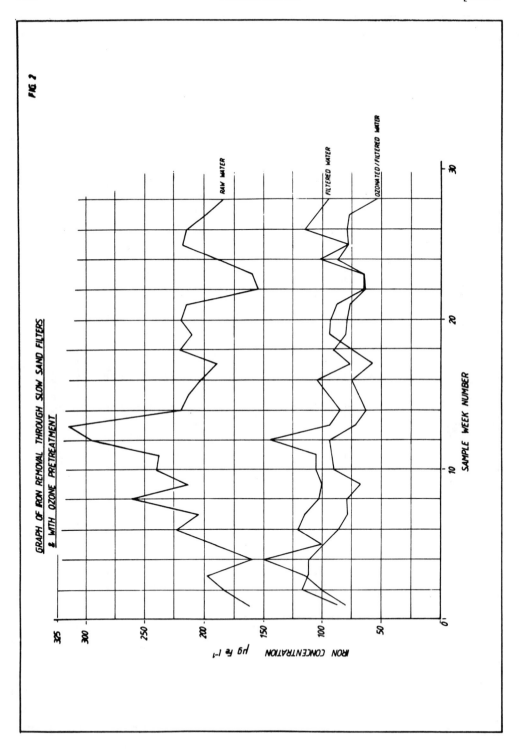

FIG 2

GRAPH OF IRON REMOVAL THROUGH SLOW SAND FILTERS
& WITH OZONE PRETREATMENT

RAW WATER

FILTERED WATER

OZONIZED / FILTERED WATER

IRON CONCENTRATION µg Fe l⁻¹

SAMPLE WEEK NUMBER

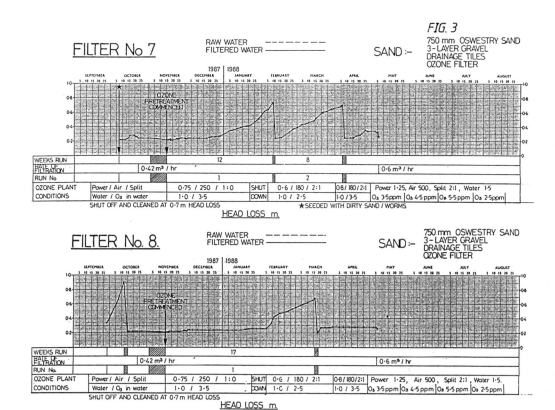

FIG. 3

FILTER No 7

RAW WATER – – – – – – –
FILTERED WATER ————

SAND :–   750 mm OSWESTRY SAND
3-LAYER GRAVEL
DRAINAGE TILES
OZONE FILTER

HEAD LOSS m.
★ SEEDED WITH DIRTY SAND / WORMS

SHUT OFF AND CLEANED AT 0·7 m HEAD LOSS

FILTER No. 8.

RAW WATER – – – – – – –
FILTERED WATER ————

SAND :–   750 mm OSWESTRY SAND
3-LAYER GRAVEL
DRAINAGE TILES
OZONE FILTER

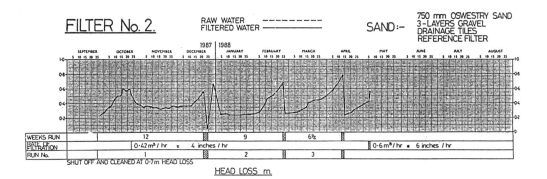

SHUT OFF AND CLEANED AT 0·7 m HEAD LOSS
HEAD LOSS m.

FILTER No. 2.

RAW WATER – – – – – – –
FILTERED WATER ————

SAND :–   750 mm OSWESTRY SAND
3-LAYERS GRAVEL
DRAINAGE TILES
REFERENCE FILTER

SHUT OFF AND CLEANED AT 0·7m HEAD LOSS
HEAD LOSS m.

independent of the original raw water colour. The test was carried out at pH9 and colour returned to a level of about 13°H. Also, as at Oswestry, ozonation brought about no enhancement in the removal of turbidity or manganese.

For the period reported 18th November 1987 to 6th April 1988 iron levels in the raw water were approximately half of those reported at Oswestry. At this level ozonation was not seen to make any improvement in the degree of iron removal.

Total organic carbon levels at Invercannie are almost entirely dependant on the raw water colour and are not significantly reduced by ozonation alone. However, it has been reported (1) that even at low water temperatures of between 1.5°C and 8.4°C pre-ozonation and slow-sand filtration will reduce TOC levels by 25-35% whereas the reference filter, without pre-ozonation removed only 8 - 15%.

The work carried out by Grampian Regional Council to date indicates that the conversion of TOC to AOC (assimilable organic carbon) by ozonation and its subsequent removal on the filter is unlikely to present quality problems in distribution. Work is, however continuing in this field, both in relation to raw water quality and to filter operation. As the results could form the basis of a future paper they will not be discussed further here.

CONCLUSIONS

Colour can be effectively removed by pre-ozonation although the limits in terms of practical and economic application have yet to be determined. It would seem that ozone residual and contact time might be more important than initially assumed. A small degree of colour may return on enhancing the pH and applying lengthy storage.

Pre-ozonation has not enhanced turbidity removal in general except that it appears to have reduced peak but not mean values at Oswestry.

Filter headloss develops progressively after pre-ozonation and not exponentially towards the end of the run as in the control filter without pre-ozonation. At Oswestry this led to extended filter runs through the Winter and Spring period.

ACKNOWLEDGEMENT

The authors wish to thank North West Water Authority, Grampian Regional Council and Ozotech Ltd. for permission to publish this paper, the contents of which are acknowledged to be the views of the Authors and not necessarily those of the Authorities named above.

REFERENCES

1. Maclean, C., <u>Personal Communication</u>.

3

Biological aspects

3.1 THE ECOLOGY OF SLOW SAND FILTERS

Annie Duncan — Department of Biology, Royal Holloway & Bedford New College, Egham, Surrey, TW20 OEX, U.K.

ABSTRACT

Changes in the abundance and vertical distribution of the interstitial sand fauna and flora (ciliates, worms, bacteria and chlorophyll a) as well as the sand particulate organic matter is described throughout a run of an operational slow sand filter bed and is related to its head loss and input carbon loading.

In contrast to the wealth of published information on the biology of used-water treatment systems (Curds & Hawkes, 1975, 1983a,b), the literature on the biology and ecology of slow sand filter beds is rather sparse. In Britain, this literature can be found in (1) the Bacteriological and Biological Sections of the Annual Reports of the Metropolitan Water Board which were written by Drs N Burman and J E Ridley (41st 1963-64; 42nd 1965-66; 43rd 1967-68; 44th 1969-70; 45th 1971-73) or (2) in a series of theses produced for the PhD Higher Degree (Brooks, 1954; Bellinger, 1968; Lloyd, 1974; Lodge, 1979; Goddard, 1980; Galal, in preparation) or for the G.I. Biology Degree (Bayley, 1985) or (3) in some published papers arising out of these (Bellinger, 1967, 1968, 1979; Brooks, 1952, 1953, 1954; Burman, 1961; Burman & Lewin, 1961; Lloyd, 1973; Richards, 1974; Ridley, 1971). Most of this work is sited in the Walton, Hampton and Ashford Common Treatment Works of the Thames Water Authority and concerns the bacteria, algae, protozoans and meiofauna (oligochaete and nematode worms, rotifers, microturbellarian flatworms and harpacticoid copepods) inhabiting the interstitial sand as well as touching on the more complex biological community of the surface 'schmutzedecke'. European published research on the biology of slow sand filtration is mainly German from the Water Treatment Works of Bremen (Husmann, 1958, 1968, 1974), Berlin (Ritterbusch, 1974, 1976) and Dortmund (Schmidt, 1963; Frank & Schmidt, 1965) and deals with meiofauna and bacteria. There is one paper on the flora and fauna of Indian sand filter beds (Anuradha et al., 1983).

This paper deals with the unpublished results of Lodge (1979), Goddard (1980) and Galal (in preparation), all of whose work was supervised by the author. Lodge worked on Bed 45 at the Hampton Treatment Works when it was run at the slow rate of o.2 m per hour and studied the meiofauna and population of the largest species, <u>Enchytraeus buchholzi</u> (Oligochaeta). Goddard worked on the same bed

but when it was run at the faster rate of 0.4 m per hour and investigated its
ciliate and flagellate fauna. For one run, she was able to make a comparison
with a slower rate Bed 44. The ciliates and bacterial flora of three runs at
0.4 m per hour at the Ashford Common Treatment Works were examined by Galal
during 1985 and 1986 when Thames Water Authority was conducting detailed
investigations into their performances, thus making available more detailed
information on flows, head losses, particulate organic carbons and chlorophyll
a concentrations in the sand than is normally available. There is some difference in the sand grain composition of the beds studied at Hampton and Ashford
Common which is illustrated in Fig. 1.

TECHNIQUES

     Lodge (1979) developed the core-sampler, illustrated in Plate 1, which was
subsequently used by the other two workers. After filling the inner core (10 cm
by 31 cm) with washed sand, these were inserted into their outer cores set at
random on the central area of the sand bed after it had been cleaned. They were
then retrieved by boat during the course of a run and at known run-ages in days.
Thus, sand samples at 1 cm depth intervals down to 30 cm depth were obtainable,
as shown in Plate 1. One whole hemisphere of sand (40 cm^3) at a known depth was
used for extracting the meiofauna by washing and decanting whereas the other
sand hemisphere was used to determine the particulate organic carbon content,
after shaking, using   wet combustion and a titration with an amperometric end-
point. For sampling the protozoans and sand bacteria, sub-samples of 1 cm^3 were
taken within the sand hemispheres.

     Goddard (1980) developed the technique for determining total bacterial
densities which was subsequently used more intensively by Galal for his work in
Ashford Common. The sand sub-sample was shaken for an optimal period of 30
minutes in a known dilution of sterile water in order to release the bacterial
cells into suspension. The suspension was subsequently suitably diluted, stained
with a fluorochrome Acridine Orange and filtered onto a black 0.22 μm membrane
for counting by epifluorescence microscopy, as described by Jones & Simons (1975).
The effect of 30 minutes shaking on the bacterial flora of sand grains taken from
1 cm depth after a 16 day run is illustrated in Plate 2, which is an EM micrograph
taken by M I Galal. He estimates that about 11% of the cells are not removed
whereas Goddard (1980) gives a much higher value of 66% but this may reflect
seasonal changes in the nature of the bacterial matrix.

     Goddard also developed the method adopted for counting the ciliates and
flagellates whilst still alive so that they could be both identified and sized.
A cm^3 of sand was washed with 100 cm^3 of protozoan-free water and the protozoans
were allowed to sediment for 5 hours at 5°C in a 100 cm^3 Utermohl sedimentation
chamber and onto a cover-slip. Subsequent counting involved the use of an
inverted microscope at 235 times magnification and involved about 2 hours' work.
With the lower ciliate densities found in the Hampton Bed 45, Goddard was able
to carry out total counts but Galal had to modify the counting technique because
of the higher ciliate densities in the Ashford Common Bed 14. The washed sand
grains left behind were subsequently examined microscopically in order to count
the attached ciliates such as <u>Vorticella</u> spp.

THE INTERSTITIAL SAND COMMUNITY

     The commoner organisms recorded by Lodge (1979), Goddard (1980), James (pers.
comm.) and Galal (in prep.) in the sand interstices of slow sand filter beds of
Hampton and Ashford Common are listed in Table 1. Algal species are also present
but were not determined microscopically. Bellinger (1979) has demonstrated that
motile diatoms such as <u>Navicula</u> spp, <u>Nitzschia linearis</u> and <u>N. acicularis</u> can

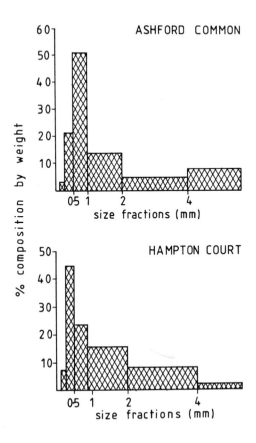

Fig. 1 The sand composition by weight in the slow sand filter
beds at the Hampton and Ashford Common Treatment Works of Thames
Water Authority (data of Lodge (1979) and Galal, pers.comm).

Plate 1   The core-sampler used to sample the interstitial sand organisms
          in slow sand filter beds and the procedures involved in inserting,
          retrieving and cutting sand samples from different depths.

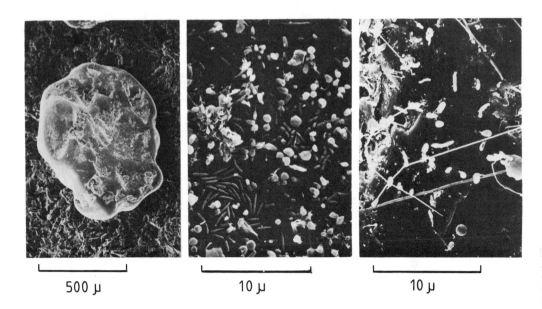

500 μ   10 μ   10 μ

Plate 2   Electron microscope micrographs of surfaces of sand grains collected
           from 1 cm depth in Bed 14 at Ashford Common Works in July 1985,
           after it had run at 0.4 m per hour for 16 days (Galal's data).

           The sand grain on the left shows the distribution of bacterial cells
           on an unshaken grain, with concentrations within the grain crevices.

           The effect of 30 minutes shaking upon the bacterial cover of sand
           grains can be seen by comparing the right micrograph (shaken) with
           the one in the centre (not shaken).

Table 1

The commoner living organisms recorded in the sand interstices of the secondary filter beds at Hampton (H) and Ashford Common (AS) treatment works of Thames Water Authority. Groups or species which occurred only rarely have been omitted.

Major Groups	Genera or Species	Treatment Works	
		H	AS
Aerobic bacteria		H	AS
Flagellates		H	AS
Ciliates	✗ *Lionotus lamella*	H	AS
	✗ *Lionotus fasciola*	H	AS
	✗ *Hemiophrys bivacuota*	H	AS
	✗ *Hemiophrys polysabrobica*	H	AS
	• *Cinetochilum margaritaceum*	H	AS
	○• *Chilodonella* spp	H	AS
	• *Cyclidium* spp	H	AS
	✗○• *Tachysoma pellionella*	H	AS
	✗○• *Stylonichia mytilus*	H	AS
	• *Vorticella convalaria*	H	AS
	• *Vorticella companula*	H	AS
	✗○• *Oxytrichia falax*	H	AS
Rotifers	•		
Flatworms (Microturbellaria)	✗ *Catenula lemmna*	H	
	✗ *Stenosoma* spp		AS
Gastrotrichs			
Nematoda (Round worms)	○ *Trilobus gracilis*	H	
	○ *Chromadorita leukarti*	H	
	• *Monhystera vulgaris*	H	
	Panagrolaimus rigidus	H	
Anellida (Segmented worms)	○ *Aelosoma hemprichi*	H	AS
	Pristina idrensis	H	
	Pristina foreli	H	
	Stylaria fossularis		AS
	Nais elinguis	H	
	Nais communis		AS
	• *Enchytraeus buchholzi*	H	
Arthropoda (Harpacticoids)	✗ *Canthocamptus staphylinus*	H	AS

KEY

- • Filter-feeding on suspended particles or bacteria
- ●○ Grazing by engulfing or grasping individual bacterial (●) or algal (○) cells
- ● Grazing by engulfing organic material (Detritivores)
- ✗ Predatory seizing of other organisms (Carnivores)
- •● Bacterivores;   ○ Herbivores;   ✗○● Omnivores.

penetrate down to 30 depth of sand and are even detected in the outflowing water. These can produce substantial total densities of several thousands per cm^3 of sand within the top 2 cms of sand and of about 100 cells per cm^3 down to 10 cm depth and represent measureable amounts of chlorophyll a.

Galal's EM micrographs (Plate 2) show that the bacterial cells of the Ashford Common Bed 14 consisted of different shapes, of which two-thirds were cocci (about 0.34 μm^3 volume) and one-third were rods (about 0.19 μm^3) and fusiforms (about 0.22 μm^3). A depth series of sand grains from this 16-day old bed in July showed a decreasing density per μm^2 of sand grain surface which ranged from 0.36 cells per μm^2 at 1 cm depth to 0.12 cells per μm^2 at 30 cm depth. Goddard (1980), who produced similar EM micrographs for sand grains taken from a 28-day old bed at Hampton, found filamentous forms which increased with depth. The total bacterial densities recorded by Galal using the epifluoresecent technique range from 10^9 to 10^{10} cells per cm^3 of sand, which are similar in order of magnitude recorded by Burman (1967-68, 43rd Annual Report MWB) for sand from Ashford Common beds, using the $22^{\circ}C$ colony count technique with dilute nutrient agar. Burman records other bacterial forms in the sand although much less abundantly: coliforms, E.coli, $37^{\circ}C$ colony counts, aerobic spores, anaerobic Clostridial spores and Clostridium perfringens (an obligate anaerobe). He also records the presence of Bdellovibrios which parasitise pseudomonad and enteric bacteria and Myxobacteria capable of killing and digesting other bacteria.

The interstitial sand protozoans, which number in thousands of cells per cm^3, consists of flagellates, ciliates and am oebas (43rd Report 1967-68). Table 1 lists the commoner species recorded in the Hampton and Ashford Common beds, with some omissions (Euplotes in Ashford Common and Aspidisca costata and Glaucoma spp in Hampton). There were differences in the protozoan faunas of the two studied beds, which may be related to the differences in their sand grain composition (Fig. 1). The ciliates were ten times more numerous in Ashford Common and were, in general, larger in cell size. There were also differences in the relative proportions of the various kinds of ciliate groups, which are normally distinguished by the relative simplicity or complexity of their mode of feeding (Laybourn-Parry, 1984). Holotrichs such as Litonotus, Hemiophrys and Chilodonella) possess a cell mouth with no oral cilia and feed by engulfing larger 'particles' such as flagellates and ciliates or, in the case of Chilodonella, algal cells. Other holotrichs (Cinetochilum, Cyclidium)possess in addition a few ciliated membranella which enable them to filter clear water of suspended small 'particles' such as bacterial cells. According to Fenchel (1980), such holotrichs can clear a volume of water that is 3.10^3 to 3.10^4 their own cell volume per hour. The spirotrichs (Tachysoma, Stylonichia, Oxytricha) with their many more membranelles which extend over the cell surface appear to be omnivorous, capable of capturing a wider range of sizes and kinds of 'particles: ciliates, flagellates, algae and bacteria. The peritrichs, represented hereby species of Vorticella, possess the most complex oral ciliature which provides them with the greatest filtering area and these attached ciliates are probably the most efficient filter-feeders of the suspended particles. In the Hampton slow sand filter bed, 45% of all the ciliate cells belonged to the filtering bactivorous holotrichs and 35% to the omnivorous spirotrichs. On the other hand, in Ashford Common, 56% of the total belonged to the large-particle engulfing holotrichs (half carnivorous and half the algal-feeding Chilodonella). The vorticellids formed 13% of the ciliates in Hampton and 9% in Ashford Common.

The rest of the fauna consist of metazoan groups of animals which are capable of inhabiting the sand interstices because of their small sizes, which range from 0.01 to 5 mm (Lodge, 1980). Enchytraeus buchholzi is the largest organism (4-5 mm) as well as the most numerous of the meiofaunal species. The nematodes and other worms were less than 3 mm in length and the predatory flat-

worms and harpacticoid copepods were small at less than 1 mm. Lloyd (1974)
records a similar species composition in the sand meiofauna of the beds at
Ashford Common but with differences that may be accounted for by the sand
compositions. Similar taxonomic groups of animals are listed by Husmann
(1958) and Ritterbusch (1974) for the filter beds of Bremen and Berlin but
with differences in the species. In general, the meiofauna of slow sand
filter beds is characteristic of that found in natural sandy habitats. The
most numerically abundant species was the worm E. buchholzi which was also the
only worm actively reproducing sexually with the production of cocoons containing several ova; the other worms reproduced asexually by budding. Lodge (1980)
suggests that the success of E. buchholzi in this man-managed environment may
be due to its possession of a resistant cocoon which can both survive the
washing and storage of sand better than the adults and other worms and can
form the new inoculum when new beds are re-sanded. Another reason is that it
is a detritivore, feeding by engulfing the organic masses which cumulate in
the beds.   The other meiofaunal species feed by grasping individual bacterial
or algal cells or by filter-feeding (the rotifers) or are predatory on the
other metazoans and larger protozans.

DENSITIES AND DEPTH DISTRIBUTIONS

        The use of a core sampler permitted the retrieval of sand samples from
different depths during the course of a filter run. It was thus possible to
follow the changes in the numerical densities of the commoner organisms with
depth and age of run.

        Fig. 2 illustrates the density of E. buchholzi at different depths in
Bed 45 run at 0.2 m per hour at the Hampton Works during a run which lasted
from January to March 1976. Substantial worm populations did not develop until
day 23 of the run when the surface particulate carbon content (POC) was about
1 mgC per cm^3 and in the sand below 2-3 cm much lower values. Already by day
23 and more markedly by days 37 and 51, the worms were avoiding the surface high
concentrations of POC and were establishing peak densities in the depths below.
Lodge (1980) suggests that the ever-increasing cumulation of POC in the surface
sand layers makes them less penetrable by this relatively large species, thus
confining its feeding impact to the depths below 3 cms. Fig 2 provides some
values of the population biomass of this species, as a mean for the sampled
column of sand. These values represent 1 to 2 % of the mean sand column POC
mass.

        The time course of the depth distributions of POC, chlorophyll a, bacterial
and total protozoan densities for Bed 14 at Ashford Common is illustrated in
Fig. 3 for run 4 in July 1985 (Galal, pers. comm.). Already by day 5 in this
bed run at the fast rate of 0.4 m per hour, substantial concentrations of POC and
chlorophyll a had accumulated in the surface 2 cms of sand but not in the lower
depths where there were already high densities of bacteria and large numbers of
protozoans. By days 12 and 20, the surface and deeper concentrations of POC were
greater than seen in the slower-rate Bed 45 and the development of large populations
of bacteria in the upper 3-4 cms of sand reflected this. The depth distribution
patterns of the total protozoan fauna is more complex, as might be expected from
the diversity of species present, but there is a marked avoidance of the upper
sand layers towards the end on the run on day 20.

COMPARATIVE BIOMASSES FOR BED 14 RUN 4

        The relative importance of the different living and non-living components
of the slow sand filter ecosystem can be demonstrated by their percentage

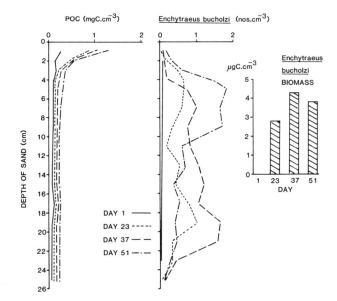

Fig. 2 The depth dist-
ribution of Enchytraeus
buchholzi (Oligochaeta)
and particulate organic
carbon in the sand on
different days in Bed
45 at Hampton Works
(TWA) when run at 0.2 m
per hour from January
to March 1976 (data of
Lodge, 1979).

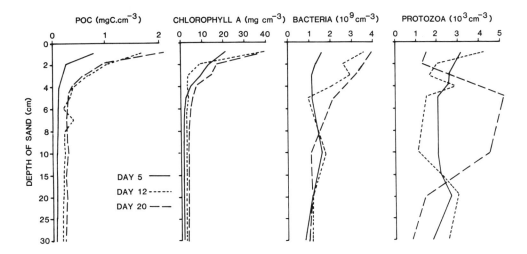

Fig. 3 The depth distributions of the particulate organic carbon, chlorophyll
a, bacteria and total protozoans in the sand on different days in Bed 45 at
Ashford Common Works (TWA) when run at 0.4 m per hour during July 1985 (data
of Thames Water Authority and Galal, pers.comm).

weight in carbon. This has been attempted for the July run 4 of Bed 14 at
Ashford Common and is illustrated in Fig. 4.

Of the elements involved, only the particulate organic carbon, which
includes all the living and non-living components, was measured directly in
carbon and this weight represents 100%. The numerical density of the bacterial
cells was converted to biovolume using a mean volume of 0.293 μm^3 per cell,
calculated using the most appropriate geometric shape from dimensions measured from the EM micrographs by Galal (pers.comm.) and bearing in mind the
relative proportions of the various shapes of cells. Assuming a specific gravity
of 1.06 (Zviaginzev & Rogachevsky, 1973), a dry weight to wet weight ratio of
0.2 and that carbon weight was 0.5 of dry weight, a final factor of 0.0311 pgC
per cell was used to convert cell density to carbon biomass per cm^3 of sand.
Both Goddard (1980) and Galal (pers. comm.) measured the dimensions of live
ciliate species whilst counting and calculated their cell volumes using
appropriate geometric shapes. These were converted to carbon weight using
0.106 pgC per um^3, which is on the low end of a series of measurements for marine
protozoans (Fuller, pers. comm.). Multiplication of the cell carbon by the
species' density provided an estimate of their biomass. The algal biomass in
the sand was measured directly by the Thames Water Authority's laboratory at
Wraysbury, using a spectrophotometric determination on methanol extracted
solutions. These were converted to carbon weight by using a factor of 25 ugC
per ug chlorophyll a, which is a C:a ratio determined on phytoplanktonic species
inhabiting the storage reservoirs (Steel, pers. comm.). The non-living fraction
of the particulate organic matter was determined by difference between the
total POC and the sum of the other components.

Fig. 4 illustrates the percentage composition of the carbon biomasses at
five depths on each of three run-days of Bed 14 during July 1985. The 100%
values for the total POC are shown in Fig. 3. The living biomass represents a
high proportion of the total POC on day 5 and particularly at the depths below
3 cm. The bacterial biomass predominates over the algal biomass and the ciliate
biomass is present rather more at 3 cm depth. By days 12 and 25, the non-living
carbon mass forms more than 50% of the total at all depths and the algal component is present in approximately equal proportions with the bacterial biomass,
which is a surprising result considering the continuous input of dissolved and
particulate nutrient material into the bed but may reflect biological interactions
in the form of bacteriovore versus algal ciliate grazing. By day 25, the main
protozoan biomass has shifted down to 5 cm depth.

Fig. 5 provides some information about the depth distributions and time
changes in Bed 14 during the July run 4 of the biomass of ciliates grouped into
different feeding types: bacteriovores, algal-feeders, carnivores and omnivores,
as described earlier. The algal-feeding Chilodonella and the omnivorous spirotrichs are the main contributors to the ciliate biomass peak that developed at
3cm depth on Day 5. By Day 12, the main ciliate biomass occupied the upper 5
cms of sand and consisted of bacteriovores such as the filtering holotrichs and
peritrichs together with substantial biomasses of omnivores, herbivores and
carnivores. In contrast to the situation in Day 12, the main ciliate biomass
had shifted down to 5 and 10 cms depth by Day 20 and consisted of herbivores,
onmivores and carnivores. Clearly, this figure is only indicative of the kind
of impact that ciliate feeding can exert upon the bacterial populations of sand
filter beds, although this impact, when eventually actually measured, is likely
to be proportional to the biomass or size of the species. The relatively high
presence of algal-feeders and omnivores in relation to bacteriovorous ciliates
is surprising and again may be related to the cell size of the species involved.

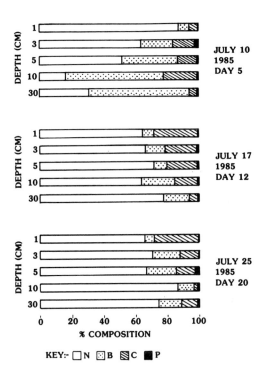

Fig. 4   The percentage composition by carbon weight
of the bacteria, algae and protozoans at different
depths and on different days in Bed 14 at Ashford
Common Works when run at 0.4 m per hour during July
1985.

KEY: N - non-living material; B - bacterial carbon
C - algal carbon derived from chlorophyll a;
P - protozoan carbon.

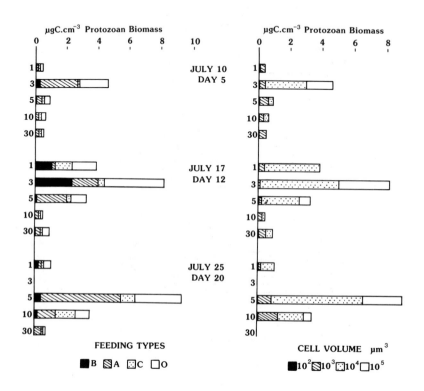

Fig. 5   The protozoan biomasses at different depths and on different days
in Bed 14 at Ashford Common Works when run at 0.4 m per hour in July 1985.

On the left, the biomass is analysed into ciliates of different feeding
types. On the right, the biomass is grouped into ciliates of different
cell volumes.

KEY: B - bacteriovores; A - algal-feeders; C - carnivores; O - omnivores

In Fig. 5 there is also provided some information on the ciliate biomass of Bed 14 during its July run 4, analysed by cell size based upon cell volumes calculated by Galal (pers. comm.): 10^2, 10^3, 10^4 and 10^5 um^3. It is immediately clear that the peak biomasses at the various depths on different days are due to the presence of the larger ciliate species in substantial numbers. The reasons for this is not known at present.

HEAD LOSSES, INPUT POC LOADINGS AND BIOMASSES

The results of Lodge (1979) and Goddard (1980) showed that the development of the populations of animals inhabiting the slow sand filter beds of Hampton were in some way linked with the changes occurring within the beds, as measured by head loss (m) and vertical flow rate (m per hour). A similar relationship is shown for Bed 14 during its July run 4, as can be seen in Fig. 6. Here, the POC content of the sand and the bacterial, algal and protozoan biomasses (all means for the sand column down to 30 cm) show an initial faster increase from days 1-12 and a subsequent slower increase from days 12-22. These changes parallel the periods of slow and fast changes in head loss and the periods of high and low vertical flow rates. The two periods are also distinguished by high and low input POC loadings (mgC per cm^3 per day), obtained from the daily water flows through the bed and the concentration of particulate organic matter in the inflowing water (data provided by Thames Water Authority).

It seems that the living organisms might be responding to the input POC loading to the bed and that the their biomasses growing within the bed was a reflection of the input carbon loading culated day by day, as a kind of cumulated food supply convertible into growth. Fig. 7 is an attempt to illustrate this idea. In the lower graph, daily water flows to the bed (m^3 per m^2 per day) multiplied by the input water POC concentration provided the daily carbon loading which was then cumulated daily to provide the Y-axis on a log scale and plotted against head loss. The output carbon 'loadings' was treated similarly for interest. The initial very sharp increase in input carbon loading (with a slope of 10 or 11) is associated with small head loss changes of 0.38-0.64 m and took place between days 1-12. A marked change occurred when the head loss became greater than 0.64 m which is reflected in the reduced quantities of water passing through the bed (Fig. 5), and the input carbon loading increases with a slope of only 0.15 but with very large changes in head loss from 0.64-3.0 m. The cumulated output carbon 'loading' reflect similar changes but at lower concentrations. The living biomasses of bacteria, algae and protozoans also increase at a faster rate at head losses up to 0.64 m and slow down a higher head losses, reflecting the food supply situation. The bacteria and protozoans appear to be able to maintain their population biomasses during the later phase described above whereas the algal populations decline. According to Bellinger (1979), the motile diatom species which penetrate the sand can remain viable but probably do not multiply so that their biomasses are the resultant of accumulation from the input water and removal by protozoan grazing rather than the result of in situ growth as is the case with the bacteria and protozoans.

The relationship between cumulated input carbon loading and head loss just demonstrated for the July run 4 of Bed 14 at Ashford Common is not unique. Fig. 8 shows that it also applies to the bed's May run 2 and the September run 7, spanning rather different climatic conditions and some differences in the top water POC levels. The changeover occurs, as before, at about 0.6 m head loss in all cases. The changes in the sand POC reflects the input carbon loading as before, although there are differences in the highest concentrations obtained which were greater in May than in September. It may be that the input carbon levels are too low when calculated only from the top water concentrations of

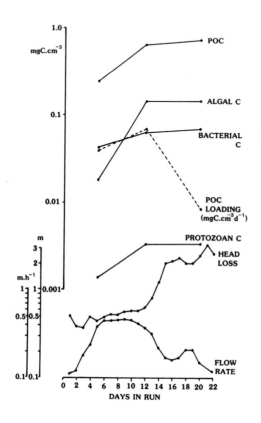

Fig. 6   The time course of Bed 14 at Ashford Common
Works during July 1985, showing the changes in head
loss, flow rates, sand POC and the biomasses of the
bacteria, algae and protozoans.   The input POC
loading is also shown.

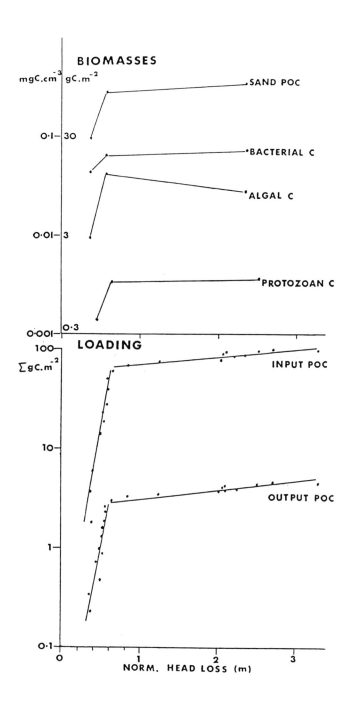

Fig. 7   Bed 14 at Ashford Common Works in July 1985. The changes in input
and output POC cumulated loadings and the changes in the biomasses of the
bacteria, algae, protozoans and sand POC related to changes in head loss.

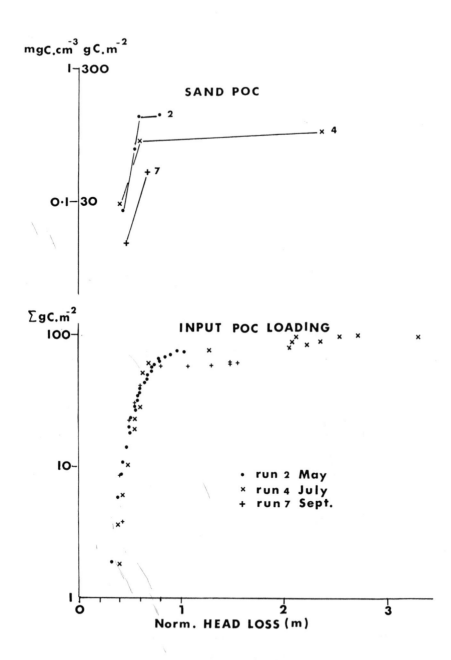

Fig. 8   Bed 14 at Ashford Common Works during runs in May, July and September 1985. The changes in the input POC cumulated loadings and sand POC masses related to changes in head loss.

particulate organic matter. Bellinger (1979) points out that, during photosynthesis, algal populations of the supernatant water and surface of the sand
release up to 30% of their photosynthetic products as extra-cellular organic
substances which will pass into the filter bed and which will form a readily
metabolizable food source for the interstitial organisms, the bacteria in
particular. The times of the year when this forms a substantial carbon input
will be climatically determined during periods of intense sunlight and may
account for the differences just noted in the sand POC levels during the three
runs of Bed 14.

Fig. 8 seems to suggest that this complex sand ecosystem of interacting
organisms responding promptly to environmental change behaves like a dynamically
stable system, similar to a continuous flow chemostat, in a way not fully understood as yet.

ACKNOWLEDGEMENTS

I am grateful to Thames Water Authority, Drs D V Lodge and Dr M R Goddard
and Mr M I Galal for use of their data without which this paper would not have
been possible. I thank Royal Holloway & Bedford New College for technical
assistance with the production of the paper.

REFERENCES

Anurdadhan, S., R. Gadkari, R. Paramasivam & S.K. Gadkari. Indian J. Environ.,
     Hlth., 25, 241-260 (1983).
Bayley, R.G.W. The ecology of meiofauna of slow sand filters in relation to
     operational conditions. G.I.Biol. thesis. N.E. Surrey Coll. Tech. (1985)
Bellinger, E.G. Br.phycol.Bull. 3, 410, (1967)
Bellinger, E.G. J.Soc.Wat.Treat. Exam., 17, 60-66, (1968)
Bellinger, E.G. Ecological and taxonomical studies of the algae of slow sand
     filter beds. PhD thesis. Royal Holloway College (U.L.), 240pp, (1968)
Bellinger, E.G. J.Instn.Wat.Engrs. Sci., 33, 19-29, (1979)
Brooks, A.J. Hydrobiologia, 4, 281-93, (1952)
Brooks, A.J. Hydrobiologia, 6, 333-51, (1953)
Brooks, A.J. Hydrobiologia, 7, 103-17, (1954)
Burman, N. Effl.Wat.Treat.J., 2. 674-77, (1962)
Burman, N. & J. Lewin. J.Inst.Wat.Eng., 15, 355, (1961)
Curds, C.R. & H.A. Hawkes. Ecological Aspects of Used-water Treatment. A.P.
     3 volumes, (1975, 1983a,b).
Frank, W.H. & Schmidt, K. Wasser u. Abwasser, 106, 565-69, (1965).
Fenchel, T. Microbial Ecology, 6, 13-25, (1980)
Goddard, M.R. The ecology of protozoan populations of slow sand filters, with
     particular reference to the ciliates. PhD thesis. Royal Holloway College
     (U.L.), 388pp, (1980)
Husmann, S. Abm. Brauschweg. Wiss. Ges., 10, 93-116, (1958)
Husmann, S. Gewasser u. Abwasser, 46, 20-49, (1968)
Husmann, S. Kunstliche Grundwasseranreicherung am Rhein. Wiss.Ber.Unt u. Plan.
     Stadwerke Wiesbaden, 3, 173-83, (1974).
Jones, J.G. & Simons, B.M. J.Appl.Bact., 39, 317-39, (1975)
Laybourn-Parry, J. A functional biology of free-living Protozoa. Croom Helm
     218 pp, (1984)
Lloyd, B. Water Research, 7, 963-73, (1973)
Lloyd, B. The functional microbial ecology of slow sand filters, PhD thesis.
     University of Surrey. (1974)

Lodge, D.V. An ecological study of the meiofauna of slow sand filters, with particular reference to the oligochaetes. PhD thesis. Royal Holloway College (U.L.), 333 pp, (1979)
Metropolitan Water Board, 41st Report for 1963-64, 90-91.
Metropolitan Water Board, 42nd Report for 1965-66, 54-58.
Metropolitan Water Board, 43rd Report for 1967-68, 29-32, 61-65.
Metropolitan Water Board, 44th Report for 1969-70, 16-25, 52-55, 83-86.
Metropolitan Water Board, 45th Report for 1971-73, 27-53
Richards, A.D. J. Protozoology, 21, 451-2, (1974).
Ridley, J.E. Proc. Soc.Wat.Treat.Exam., 16, 170-91, (1971)
Ritterbusch, B. Z. Angew.Zool., 61, 301-46, (1974).
Ritterbusch, B. Int. J. Speleol., 8, 185-93, (1976).
Schmidt, K. Ver. Hydrol. Forsch. Dortmunder Stadwerke, 1-51, (1963)
Zviagintzev, D.G. & Rogachevsky, L.M. Microbiologia, XLII, 892-98, (1973).Rus.)

3.2 SOME ASPECTS OF THE FILTRATION OF WATER CONTAINING CENTRIC DIATOMS

K. B. Clarke — University of East Anglia, School of Environmental Sciences, Norwich NR4 7TJ

ABSTRACT

          Filtration of water containing centric diatoms can be a major challenge to water treatment plants. This is well known to plant operators but accounts of the problems experienced are rare and inconsistent.   The paper briefly reviews our knowledge of the behaviour of centric diatoms in filters. These diatoms may possess chitin fibrils and the effect of these on the filtration process is considered. The account is illustrated by experiences with centric diatoms at two waterworks during 1988.

INTRODUCTION

          Difficulties experienced in the filtration of water containing discoid centric diatoms are well known but accounts of these difficulties are hard to reconcile. The first edition of the Manual of British Water Supply Practice (1) gave a table (p 638) which listed organisms reported to have caused difficulties in British Waterworks. Some organisms were listed as having caused acute difficulties both by blocking filters and by penetrating them. There seemed to be something rather contradictory in this behaviour. The organisms so listed were the centric diatom Stephanodiscus and the genus Mallomonas, a member of the Cryptophyceae which is characterised by its body being covered with hair-like processes. ROUND (2) elaborated on the problem of the behaviour of centric diatoms and removed some of the contradiction by his statement that " the larger forms block the fast filters but the smaller species (e.g. Stephanodiscus and Cyclotella) pass these and block the slow sand filters."
          The answer to this strange behaviour appears to lie in the posession of fibrils by some species at certain times. The fibrils alter the properties of these discoid diatoms very substantially. These fibrils had been known since the last century (3) but it is only recently that their nature and origin has been understood.   The advent of the scanning electron microscope suggested that their points of emergence from the cell might be identified as strutted processes on the frustule and McLachlan et al (4) were able to describe the fibres of Thalassiosira, a discoid diatom, in detail. They showed that the fibrils were pure crystalline chitin (poly-N-acetyl-D-glucosamine).   HERTH & BARTHLOTT (5) produced SEM photographs showing the fibrils and their emergence from the strutted processes in species of Cyclotella and Thalassiosira.

Having located the point of origin of the fibrils HERTH (6) went
on to examine the pores in serial sectioning and found that the
fibrils originated from special conical invaginations of the
plasma membrane.
It.seems that the fibrils are not always developed. There are no
reports about the conditions which cause development of the
fibrils but judging from the large amount of nitrogen which is
locked up in the glucosamine they would tend to be present at
times of high nutrient availability in the water.
MCLACHLAN ET AL(4) calculated that approximately 18% of the
nitrogen added to culture medium was recovered in the chitin which
they isolated.
            In view of these advances in our knowledge it
seemed opportune to reexamine the behaviour of discoid diatoms in
filtration. An occasion presented itself when, during normal
operation, the Ormesby Works of East Anglian Water Company changed
the source of water for Ormesby Waterworks from the River Bure,
which contained only pennate diatoms to Ormesby Broad which
contained substantial numbers of fibrilate discoid diatoms.
          A further occasion arose at the Lound Waterworks
of the Company when a species of Cyclotella gave rise to a "green
filter". Here the centric diatoms, had originated not by passing
the prefilters in considerable numbers, as was the case at Ormesby
but by growing in the water above the slow sand filter which had
been newly put to work.

                          METHODS

            Fibrils were examined by drying a drop of the water
sample on a 19mm cover slip gradually on the laboratory bench at
normal temperatures (19º C). The slip was inverted over a ring
of aluminium foil with the ring glued both to the slip and a
slide. This air mount allowed examination under oil immersion with
a magnification of x 1500 using phase contrast. Further details
of the cleaned diatom cells were studied on a scanning electron
microscope of the University of East Anglia (Hitachi S450).
Cores of damp sand from the filter beds were taken with a
half-round guttering corer, the sliding stainless steel cover for
the pvc commercial 75 mm guttering was retained during sampling
but removed to allow the core to be cut up. Some compression of
the core occured with the rapid filter sand but no compression on
the slow sand filters. The cores were cut into lengths of 25 or
50 mm and stored in plastic bags. About 50 ml of sand were added
to 200 ml of water in a measuring cylinder. Readings gave the
total volume occupied by the sand as well as the volume of the
actual grains. The number of diatom "disks" per ml, whether live
or dead was counted as "valves" using a haemocytometer type
cell. The number per ml was multiplied by the volume of water
used (200 ml) and divided by the total volume of sand to give the
number per ml of sand.
            Resistance to flow is defined as the time taken for the
water to fall a distance equal to the loss of head. This gives a
figure which is independent of the size or speed of operation of
the bed and independent of the system of units employed.

FIG. 1

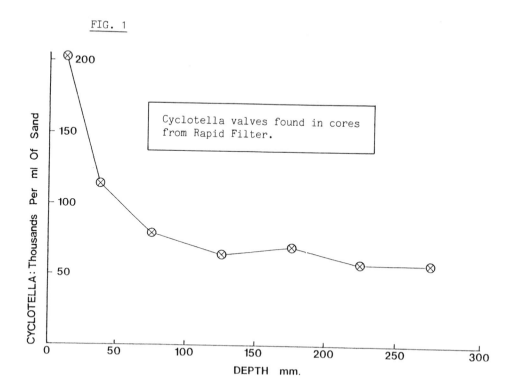

Cyclotella valves found in cores
from Rapid Filter.

BEHAVIOUR OF RAPID FILTERS

Accounts of the behaviour of centric diatoms in
rapid filters are almost totally contradictory. While we were
observing as little as 20% retention of Cyclotella comta by the
rapid filters at Ormesby, Watson (7) was reporting about 90%
removal efficiency of Stephanodiscus "astraea" by various
experimental rapid filters at Grafham. The size and shape of the
diatoms was not dissimilar.

Cores were taken of the rapid filters at Ormesby to
give a picture of the penetration (Fig 1). It can be seen that
there is an exponential decrease in numbers for the top 100 mm or
so, after which numbers per ml of sand remain fairly constant.
This suggested that the organisms were not being completely
washed out of the bed before the wash cycle ended. Confirmation
of incomplete washing was obtained by sampling before beds were
put into service again. The trend of the exponential curve was to
somewhere around 50,000 ml^{-1} of sand and the washed sand was found
to contain 42,300 ml^{-1} of sand. Between 1000 ml^{-1} and 2000 ml^{-1}
Cyclotella were leaving the bed in the filtrate. Voids represent
about 1/3 of the sand volume so that the mean concentration in the
interstitial water was 42,300 X 3 = 126,900 ml^{-1} of which only
around 1% was being mobilised into the final filtrate.

The mechanisms which cause removal of material by
a filter have been summarised by Huisman & Wood (8). At the
surface of the bed there is a straining action caused by the
limiting size of the apertures. Consideration of the effect of
fibrils in the past has always assumed that because they were

rigid extensions of the diatom they would increase its likelyhood
of being strained out at the surface. This does not seem to be the
case. The fibrils are flexible and will break rather than offer
resistance to a lateral force. They certainly permit diatoms with
a diameter of 30 um and a fibril length of up to 112 um to pass
right through a bed composed of uniform sand with a mean diameter
of 1.0 mm which would be expected to strain out material above a
limiting size of 140 um. Between fibril tips the Cyclotella must
measure up to 250 um.
            Within the bed the mechanisms fall into two
groups, those which get a particle into a position at rest close
to a sand grain and those which hold the particle in position once
it is closely adjacent to the sand grain. Posession of fibrils
means that the diatom is held clear of the sand grains in the area
of maximum velocity and will thus tend to aid the diatom in its
penetration of the bed by negating both types of filter mechanism.
            The distinction which Round made (2) between the
behaviour of larger and smaller forms of discoid diatoms in
filters has some foundation in experience although the distinction
may not be purely one of size. It may be connected with the
ability to form fibrils. There seems to be a greater tendency
among the smaller forms to form fibrils which would give them
greater ease in passing filters.
            A possible solution to the mystery of why Anne
Watson's filters were so successful would be if her centric
diatoms were devoid of fibrils. I put this possibility to her and
in March and April 1988, a year after her trials began, she
supplied me with material from the same source containing the same
centric diatom, Stephanodiscus "astraea" (possibly Stephanodiscus
neoastraea Hakansson and Hickel (1986)). As anticipated, this
diatom did not possess fibrils. This fact would seem to provide a
good explanation for the success of the Grafham experimental
filters in removing it.
            Much more difficult to explain is the fact that
the rapid filters were failing to wash clean under these
circumstances. It may be that the very properties which allowed
them to move easily under the viscous flow regime of the working
filter were a hinderance to being washed out in the highly
turbulent flow conditions of a rapid filter wash.
            The behaviour of Scenedesmus which was present at the
same time and in roughly the same numbers,was compared with that
of Cyclotella comta. Removal through the rapid plant varied from
80% to 95.5%. Wash efficiency was calculated by summating the
organisms present in the bed before and after washing. While
51.2% of Cyclotella comta remained in the bed after washing, this
figure was only 20.9% for Scenedesmus.
There was no penetration of the slow sand filters by Scenedesmus.

                BLOCKING OF SLOW SAND FILTERS

            In March 1988 No 6 Slow Sand Filter at Ormesby
Works was examined after a run of 63 days receiving centric
diatoms. It was found that the skin consisted of the Centric
diatom Cyclotella comta fo praetermissa which is known to secrete
a gelatinous coat besides posessing fibrils (9). Indian ink added
to the suspension showed that the Cyclotella did not have any
surrounding mucillage at all at that time. The dry skin was

almost white and capable of being picked up with tweezers although
in pieces of very limited size. A piece was gently agitated in
distilled water and the resulting material was diluted and slowly
dried in air. The centric diatoms had a higher than usual average
of attached fibrils but the chief feature of the slide was the
very large number of detached fibrils. Examination of the water
coming from the primary filters and the raw water also gave a
considerable quantity of detached fibrils. (Fig 2)

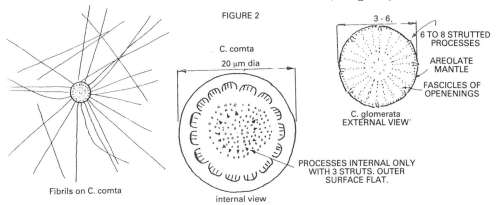

FIGURE 2

C. comta
20 µm dia

3 - 6.
6 TO 8 STRUTTED PROCESSES
AREOLATE MANTLE
FASCICLES OF OPENENINGS
C. glomerata
EXTERNAL VIEW'

PROCESSES INTERNAL ONLY
WITH 3 STRUTS. OUTER
SURFACE FLAT.

Fibrils on C. comta

internal view

                              They seem to not be losing their
fibrils while lying in the non-turbulent conditions of the filter
surface but acting as a sieve for suspended fibrils reaching the
surface. The large numbers of glass-like fibrils affected the
appearance of the dried filter skin.
                         Appreciable areas of the
filter bed surface did not dry but retained pools of a quite
viscous brown liquid.   McLachlan et al (4) cultured another
centric diatom Thalassiosira fluviatilis and noted the medium in
which Thalassiosira had been grown developed an apparently high
viscosity even at relatively low cell densities. This was due to
copious production of long fibres attached to the valve surface of
the cell and occuring free in the medium.

SLOW SAND FILTER BED PENETRATION

                    To give a picture of the rate at which
Cyclotella could penetrate slow sand filters, cores were taken
from three filter beds and examined for numbers of cells of
Cyclotella comta present at various depths (Fig 3).   It will be
seen that numbers which had penetrated in 8 days were much lower
than those which had penetrated the rapid filters in 1 day
although numbers retained at the surface were about the same.
                         The beds had been cleaned at different
intervals of time after the source had been changed from river
water (mainly pennate diatoms) to Broad water (mostly Cyclotella
comta around 3000 cells ml^{-1} of which about 1000 cells ml^{-1}
passing the rapid filters).   The result appeared to show that
the longer a bed ran with a loading of Cyclotella and the fewer
Cyclotella were present at depth in the sand.
Although all the slow sand filters at Ormesby were receiving the
same water their response was very different.   Those which had

been cleaned at the start of the period of <u>Cyclotella</u> fared worse.
Some of them sealed quickly, were cleaned and sealed again, while
those filters which had not been cleaned for several months
showed no sign of sealing. The best of these, No 5 filter, had
been cleaned in October. It was found to have a well developed
community of invertebrates associated with the skin and by
mid-March enough chironomid exuviae were present to form a black
scum in one corner of the bed.
        The presence of fibrils, both on the diatoms and at
the filter surface must interfere with the normal mechanisms which
keep the filter surface community stable. Above all they prevent
the discoid diatoms being ingested by grazers but they also appear
to hamper the sweeping action of worms which is important in
keeping the schmutzdecke material from clogging the surface pores
of the filter.

FIG. 3

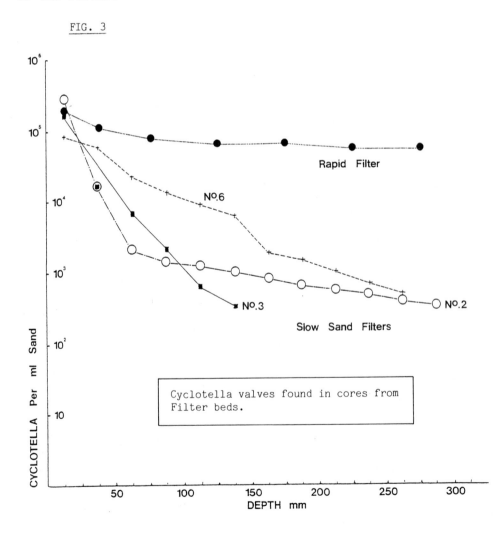

Cyclotella valves found in cores from
Filter beds.

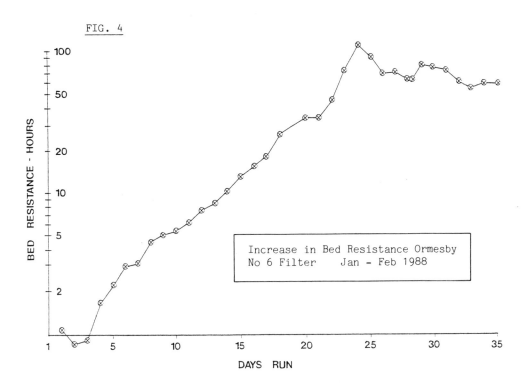

FIG. 4

Increase in Bed Resistance Ormesby
No 6 Filter   Jan - Feb 1988

The observations above, however, indicate that a well-established
filter surface community can deal with the arrival of the
fibrilate diatoms without the bed choking.
Also, providing enough time is allowed, even a cleaned filter will
develop a community which will inhibit further choking of the bed.
No 6 Filter took about 24 days to establish such a community (Fig
4) The community, which formed in mid-February,consisted of
Nematode and Nais worms as well as a few rotifers. There were
about 0.4 worms per sq.mm.of filter surface at that time. Numbers
of Cyclotella on the surface of the bed were almost halved but
this did not reduce the head loss which remained stable at a high
figure for about a fortnight.

                          GREEN FILTERS

          When a slow sand filter is put to work the water above the
filter occasionally becomes turbid and greenish. This is usually
due to the presence of microscopic motile green unicellular algae
which develop well in the absence of their usual predators.
Normally the population reaches its peak in a few days and then
quickly declines. While the population is high these algae may
pass through the filter and go into supply but otherwise the
occurence has little operational significance (10).(11).
          At the end of June 1988 there was a similar occurence on G
slow sand filter bed at the Lound Works of East Anglian Water
Company. It was due to a centric diatom rather than a green
unicell. G Slow Sand Filter measures 120 ft X 80 feet and
receives lake water which has been prechlorinated, sedimented

without coagulants and pre-filtered. It had been out of use for
twelve weeks, having been cleaned after a run of 154 days in mid
April and then left empty.
            On 20 June the bed was refilled by admitting filtered
water from below.   The bed was operated at a very conservative
rate of flow (31.5 mm hr^{-1}). Later this rate was trebled.
When a routine sand surface sample was taken after a fortnight it
was noticed that the bed had a heavy plankton above it, the water
having a turbidity of 2.6 units while the other beds and the water
coming onto the bed had turbidities around 0.6.   Rather
surprisingly the turbidity was found to be due to a fibrilate
centric diatom. It was not unlike Cyclotella glomerata Bachmann.
There was no tendency to form chains or bundles of cells.   Carter
(12) has pointed out the problem of allocating small centric
diatoms to species like Cyclotella glomerata, due to the
considerable morphological instability among these small forms.
The valves averaged 5.1 um diameter with eight strutted processes
and with 20 fascicles of areolae in 10 um. There were just less
than four fibrils per cell although the valves had provision for
12 to 16.

            THE PLANKTON POPULATION

            Although 220,000 cells per ml were present in the
water above the filter only 22 cell ml^{-1} were detected in the
water leaving the filter.   There were only around 1000 cells ml^{-1}
arriving at the bed in the prefiltered water.
            The population did not sediment very quickly. A
Winchester Quart bottle standing on the laboratory bench showed a
very slow decline in numbers at 5 mm below the surface of the
water for the first 48 hours. The count at 5 mm fell from 220,000
cells ml^{-1} on day 15 to 197,400 in 48 hours but the sample then
suddenly cleared completely.   On the filter bed numbers fell to
4300 cells ml^{-1}·on day 19 and on day 20 only about 10 cells ml^{-1}
remained in the plankton.
            The most important factor in the decline of the
population was infection by a chytrid fungus.   The chytrid had a
uniflagellate zoospore about 1.5 um diameter which settled on the
surface of the diatom and began very quickly to penetrate the cell
and ingest the contents.
The zoospore increased in size to 5 to 6 microns diameter and
appeared to produce new zoospores within it. This sporangium
contained a number of refractile bodies which may have
corresponded to the new zoospores. Dehiscence or dehisced
sporangia were not observed but the fungus was provisionally
identified as Rhizophydium cyclotellae Zopf which has been
recorded from other species of Cyclotella.
            During the third week of the life of the
bed zooplankton began to increase. Most prominent was the rotifer
Synchaeta pectinata E. The stomach contents could not be
identified in detail.
            Numbers were however small in logistical terms, the
most abundant, Synchaeta pectinata, was present in only about 10
per litre when the diatoms had cleared.

THE FILTER SURFACE

At two weeks resistance to flow through the bed was actually much less than would have been expected from a normal bed of this age.

During the third week, as the water above the bed began to clear, resistance to flow began to increase and at 20 days overall resistance was 4.28 hours. Very little of this resistance, however, was associated with the surface skin but appeared to be in the lower part of the sand.

When the water cleared a sample of the skin including the top 36 mm of sand contained 41.235 million cells cm^{-2}. This just accounts for the diatoms in the water column on the highest day and fails to account for the diatoms in the water which passed through the bed by several orders of magnitude.

Numbers of this tiny Cyclotella penetrating the bed completely were very small. Counts ranged between 4 and 22 per ml in the water leaving the bed.

DISCUSSION

We have seen from the above:
a) Rapid Filters were relatively ineffective against fibrilate centric diatoms. Part of the problem appears to be in the washing of the beds.
b) Slow Sand Filters, if mature can cope with a considerable load of fibrilate centric diatoms quite well.
c) Slow Sand Filters which have just been cleaned seal quickly and in doing so allow the diatoms to enter the bed in greater numbers than mature beds.
d) While fibrilate centric diatoms penetrated the rapid filters with ease they did not penetrate slow sand filters in any significant numbers even when the supernatant water contained over 200 thousand cells per ml of one of the smallest fibrilate centric diatoms.
e) Green Filters can be caused by centric diatoms. It may be that the practice of back-filling beds from below is a contributary cause of green filters and this is to be investigated.
f) The clearing of the green filter was due to a chytrid infection. This is yet another instance of dramatic reduction of algal numbers by chytrids.

ACKNOWLEDGEMENTS

The work was carried out in the laboratory of East Anglian Water Company and I am grateful to the Directors for these facilities and for permission to publish this work. I must record my special thanks to Dr. Ann Smith of Goldsmiths College, University of London and to John Carter of Hawick for their help and advice on fibrils. I am indebted to Anne Watson of University College London for discussions on the Grafham results.

REFERENCES

1.   " MANUAL OF BRITISH WATER SUPPLY PRACTICE " 1095 1st edn.
     Heffer, Cambridge.

2.   ROUND F.E. (1965) The Biology of the Algae. Arnold. London.
     (p 221)

3.   GRENFELL J.G. (1891) On the occurence of pseudopodia in the
     diatomaceous genera Melosira and Cyclotella. Quart. J. Micr.
     Sci. 32. 615-623

4.   McLACHLAN J., McINNES A.G., FALK M. (1965) Studies on   the
     Chitan (Chitin: Poly-N-acetylglucosamine) Fibers of   the
     Diatom Thalassiosira fluviatilis HUSTEDT. Canadian Journal of
     Botany 43. 707-713

5.   HERTH W. & BARTHLOTT W. (1979) The site of chitin fibril
     formation in centric diatoms: 1. Pores and fibril formation.
     Journal of Ultrastructure Research 68. 6-15

6.   HERTH W. (1979) The site of chitin fibril formation in centric
     diatoms: 2. The chitin forming cytoplasmic structures. Journal
     of Ultrastructure Research 68. 16-27

7.   WATSON A. (1988) The penetration of rapid gravity filters by
     plankton. Paper presented to Scientific Section IWEM. In press

8.   HUISMAN L. & WOOD W.E. (1974) Slow Sand Filtration. World
     Health Organisation Geneva.

9.   LUND J.W.G. (1951) Contributions to our knowledge of British
     algae XII A new planktonic Cyclotella. Hydrobiologia III (1),
     93-100

10. HEUSDEN G.P.H. van, (1944) Ervaringen bij de Bestrijding van
     "Groene" Filters.   Water 28. 37-41

11. HEYMANN J.A. (1920) Een lastig organisme in het Amsterdamsche
     waterleidingbedrijf.   Water 4 70-79

12. CARTER J.R. & BAILEY- WATTS A.E. (1981) A Taxonimic Study of
     Diatoms from Standing Freshwaters in Shetland. NOVA HEDWIGIA
     XXXIII 513-630

3.3 DEVELOPMENT OF A SLOW SAND FILTER MODEL AS A BIOASSAY

Chr. Schmidt — Institute for Water Research, Dortmund

ABSTRACT

Computer aided slow sand filter models at a laboratory scale are under development to simulate the biological, chemical and physical conditions of a real filter. Modern fermenter technologie (cultures of algae and microorganisms) in connection with the filter models can help to improve the knowledge of the limits of biological water purification, behaviour of pollutants during the filtration process, hazardous effects on filter organisms due to a high pollution load and toxic or subtoxic effects of water components on a biological system (system-bioassay).

INTRODUCTION

During the last years the interest in biological water treatment methods of artificial groundwater recharge rised due to the problems of a direct conditioning of river water in filters with activated carbon or denitrification problems in groundwater. In addition, unnoumerous attempts of contaminated soil and groundwater purification with biological methods should be mentioned. Such processes need a strict control of the biological processes during the treatment. But the application of special control parameters require a detailed knowlegde of the system.

A description or prognosis of the actual biological reactions during artificial groundwater recharge is very difficult, because the whole process is a combination of quite a few of single processes which depend on the river water quality or metereological conditions (1).

Generally, the fields of research with the aim of a possible improvement of biological water purification processes are:

- the biogenic processes in the raw water (e.g. algal born substances)
- the biological processes during bank filtration (e.g. reductive
conditions)
- the biology of the slow sand filtration.

Biological processes during slow sand filtration and underground passage
depend on special boundary conditions. Beside more general water quality
requirements the following paramters are important:

- Oxygen
- Nutrients
- Temperature
- Structure of population
- Composition of the filter material
- Infiltration velocity
- Infiltration amount
- Light
- pH-value
- CO_2
- Degradable DOC

These parameters are the main gains for biological decomposition processes
in water purification systems like a slow sand filter.

To optimize such a system, at least four general remarks have to be made:

1. Keep the oxygen content high
2. Avoid algal mass growth on the filters
3. Keep the operating rate of a filter as low and continuous as
possible
4. The undergound passage after a slow sand filtration should be as
long as possible.

These are the very complex factors and preliminary conditions for an attempt to simulate a slow sand filter process in a laboratory model.

Model-systems are sections or reductions of natural systems. A slow sand
filter can be seen as such a section, where surface water percolates
through a biologically active sand filter. In this filter, purification
processes act similar to a natural groundwater recharge. For this section
of a water-ecosystem we have a lot of measurements from practical work and
results of many years research. This knowlegde and the possibilities of re-

gulating such a system makes a filter model suitable for a complex test system for the behaviour of pollutants.
The model should be computer aided and the biological, physical and chemical conditions should be as natural as possible. On the other hand, the system should be reproducible and capable of mantaining special condition over a certain time to enable comparisons between different assays.

SYSTEM DESCRIPTION

Main part of the system are 4 slow sand filter models, which consist of industrial made glass-colums of SCHOTT (Fig. 1). There are two colums of 45 cm diameter and 50 cm height with a volume of 79 ltr, filled with 135 kg sand. Two other colums have 20 cm diameter and 100 cm height and a volume of 31 ltr, they are filled with 55 kg sand. The supporting stainless steel plates are perforated and covered with a thin gauze. To reduce the dead volume of the outlet, glass-pearls are used. After filling with the sand, the colums were totally sterilized.

The outlet is connected to a measuring chamber, where temperature, pH-value, oxygen, redox-potential and conductivity can be measured continuously.

The second part of the system constsits of three biofermenters. In two of these fermenters, algae and a mixed microbial population are cultivated. Bacterial free groundwater is connected with a third fermenter. Fromout this fermenter groundwater is dosed onto the columns. Several water paramters can be supervised and changed in this fermenter e.g. oxygen, pH, conductivity, temperature. Nutrients or substances of interest can be dosed into the fermenter.
Biological activities in the slow sand filter models can be induced by dosing algae or bacteria on the columms, which is normally done continuously. We run each of the four filter models under different conditions: with algae, with bacteria, with algae and bacteria, and one sterile.
The whole system is governed by a host computer, which realizes a complete supervision of the actual state of the system and enables us to measure and control the system.

With this device of modern fermenter technologie in connection with the filter models we can try to improve our kowledge of the

- limits of biological water purification,
- behaviour of pollutants during the filtration process,
- hazardous effects on filter organisms due to a high pollution load,
- toxic or subtoxic effects of water components on a biological system (system- bioassay).

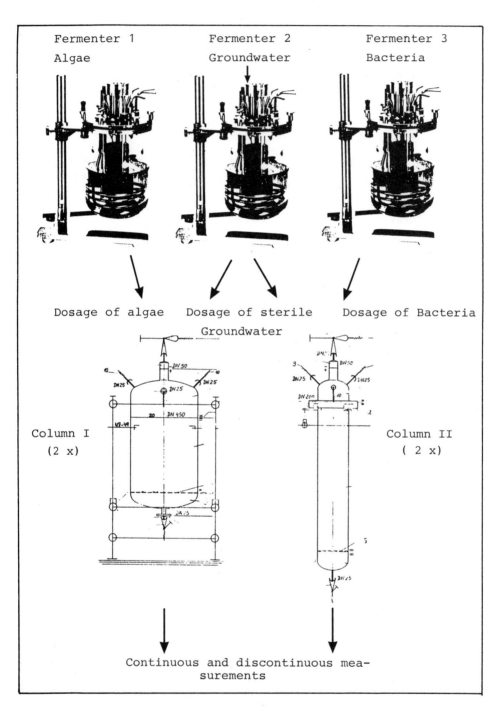

Figure 1: Priciple scheme of slow sand filter model

RESULTS AND DISCUSSION

Before discussing some results with our model, a short remark on the different aims of biotechnology and drinking water purification should be made.
Every technical or biological process tries to reach a high efficiency.
Biotechnology means to perform this process with a minimum of space and
time. In addition to the parameters temperature, pH, redox-potential and
oxygen the growth potential of organisms depending on the surrounding substrate concentration is essential. For a high yield it is therefore important to keep the substrate concentration very high near a saturation value
(2).

During a slow sand filtration, nothing shall be produced and all substances
should decompose into biologically inactive final products. The obligation
to maintain a low residual contamination has to result in a low substrate
concentration of the used surface water.

This difference is of fundamental importance for the control parameters and
the turnover velocity of these procedures. From that point of view, slow
sand filtration is a biotechnological process, which has to be clearly
distinguished from the scopes of biotechnology depending on its special
boundary conditions. However, the results of biotechnology must be integrated in the studies on the biology of a slow sand filter and they have confirmed at least three possibilities:

> 1. The number of microorganisms per unit filter body has to be in-
> creased (key-word: development of new filter materials).
>
> 2. The structure of the microorganisms biocoenosis has to be changed
> due to the water quality (key-word: development of bacterial specia-
> lists).
>
> 3. Increasing of the turn-over by increasing the substrate concentra-
> tion (key-word: Avoiding high output concentrations by running diffe-
> rent prefilters).

Thus, if keeping the biological control mechanisms at a level of optimal
reproducibility and adaptability to the requirements of a slow sand filter,
we can optimize such a system step by step.

After the installation of the test-system two columns run since July 1987,
two other columns since May 1988. The results discussed in this paper refer
to this phase which was a kind of breaking-in period. The effects of a dosage of several pollutants on the columns will not be presented in this paper; we began with the dosage of NTA and the results will be presented on
the seminar.

Germ Number

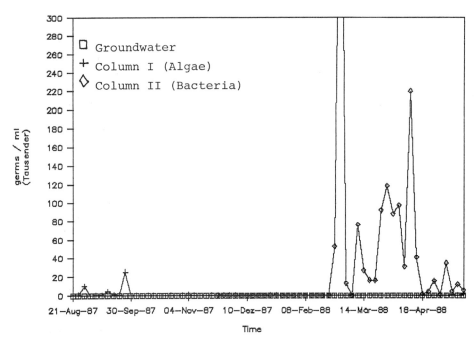

Figure 2: Germ number

Number of Germs / g Sand (FW)

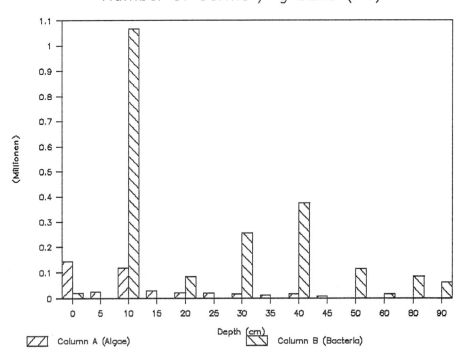

Figure 3: Distribution of germs in the filter material

All columns were run with <u>sterile groundwater</u> for a certain time, followed
by a continuous <u>dosage of algae</u> on column I <u>and bacteria</u> on column II. This
dosage was calculated referring to the normal concentration range of algae
in the River Ruhr surface water, e.g. 5 g Chlorophyll l^{-1} and 5.000-
10.000- germs ml^{-1}.

<u>Germ development on</u>
p-Agar $2,9 * 10^8$
German Standard of
Drink.-Water $1,2 * 10^8$
Endo-Agar $4 * 10^6$
E.coli $2,4 * 10^4$

<u>Percentage</u>

E.coli   5
Aeromonas hydrophilia   40
Enterobacter cloaceae   5
Klebsiella pneumonia   3,3
not determined   43

<u>Morphological Differentiation of 60 Isolates (p-Agar)</u>
<u>Percentage of Bacteria</u>

	Water of River Ruhr	Fermenter
Gram	75	76,7
pleomorphic	20	23,3
movable	46,7	43,3
color-building	18,3	21,7
swarming	30	23,3

Table I: Structure of the fermenter biocoenosis compared to the
         River Ruhr

One of the main questions was the maintaining of a mixed microbial population in the bacteria fermenter which should have a similar structure as the
River Ruhr bacterial community. In Table I the <u>community structure</u> of the
fermenter is compared to the community structure of River Ruhr water which

Conductivity

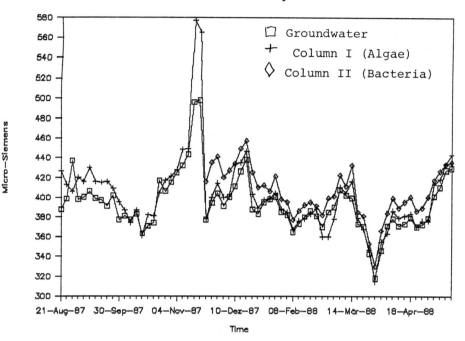

Figure 4: Conductivity of groundwater and columns I and II

pH–Value

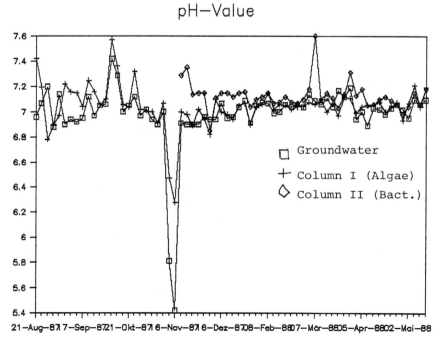

Figure 5: pH-value in the groundwater and the columns
outlets

was used to start the fermentation process. This structure was analyzed after 1 month of fermentation. It is one of 4 measurements, which have been
done by Prof. Nehrkorn, University of Bremen. The results show that under
morphological aspects the composition in the fermenter is very similar to
the composition in the water of River Ruhr. After 3 month, we determined a
sharpe change in this structure. Over 80% of the bacteria were E. coli.
Thus, the fermentation process was restarted and a new structure analysis
after 2 1/2 month gave similar values like the values shown in Table I.

The dosage of algae and bacteria led to a thick algal layer on the sand in
column I and a layer of mucous bacteria on column II. Fig 2 shows the germ
number at the outlets of the two columns. The germ number was determined
daily, the values in the dosed groundwater normally did not exceed 0, in
some cases they reached 50 to 100 germs ml^{-1}. These bacteria did not lead
to a constantly growing bacterial flora in the columns, because the normal
DOC-content of the groundwater (2 mg l^{-1}) is quite low.

The outlet of column I (algae) showed no significant rise in the germ number. The column II (bacteria) shows a discontinuous break-through of bacteria after the beginning of the dosage with bacteria from the fermenter.

The resulting distribution of germs in the columns is shown in Fig. 3. In
comparison to column I (algae) in the column II (bacteria) much more bacteria up to 1 Mio. germs g^{-1} sand (fresh-weight) can be counted. In this
column, the maximum of germs appear in 10 cm depth. In 90 cm depth still
some 80.000 germs g^{-1} sand can be measured. A differentiation of these
germs has not been made until now.

The conductivity of the dosed groundwater reaches values between 360 and
420 S cm^{-1} (Fig. 4). Conductivity normally does not change after passing
the columns. In the first time, the conductivity at the outlets of the columns was a bit higher. It seems as if substances contributing to a high
conductivity are washed out of the filter material. A decrease in the input
pH-values results in a certain increase in conductivity which then exceeds
the input values by more then 100 S cm^{-1}. This can be explained by a solution of carbonates and hydroxydes.

Natural groundwater in the River Ruhr aerea has a pH-value around 6.8 to
7.0. After passing the columns the pH-values raise slightly. A decrease of
the input pH-values is buffered by the filter material. At a decrease down
to 5.4 pH the outlet values never underceed 6.0 pH. A comparison between
the two columns (Fig. 5) shows lower pH-values at the outlet of the column
II (bacteria), but they do not differ significantly from column I (algae).

The oxygen content of the water normally does not change after passing the
sand as long as there is no biological activity in the columns (Fig. 6). A
decrease of the oxygen input down to values of 50% oxygen saturation results in the same values at the columns outlet. The sharp decrease of the
oxygen concentration is caused by a dosage of DOC (see below).

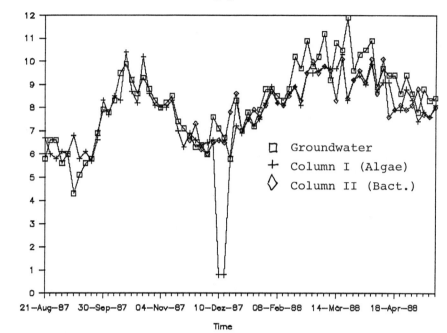

Figure 6: Oxygen concentration

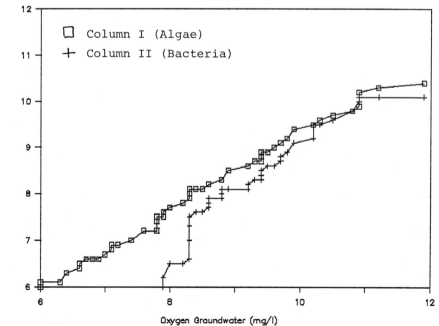

Figure 7: Comparison of oxygen concentration between
Columns I and II

CO2

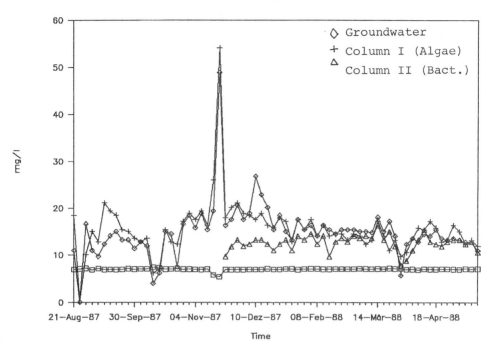

Figure 8: CO_2-concentration in the groundwater and the columns outlets

Turbidity

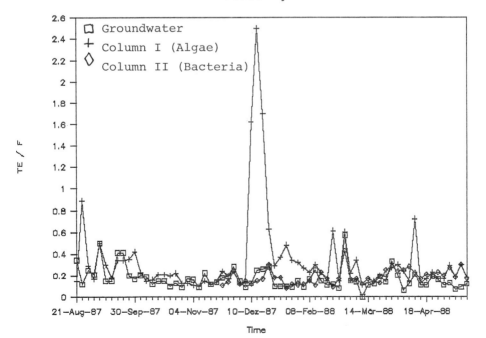

Figure 9: Turbidity of groundwater and columns I and II

A comparison between the two columns (Fig. 7) shows, that the activity of the algae normally does not alter the oxygen content. The dosed groundwater was adjusted at 80% oxygen saturation corresponding to 8 to 10 mg l^{-1} O_2. The greater number of bacteria in the column II (bacteria) led to a higher oxygen consumption in this column.

Carbon dioxyde measurements of the groundwater gave values between 10 and 20 mg l^{-1} CO_2. The outlet values normally follow the input values (Fig. 8). A pH-decrease produces up to 50 mg l^{-1} CO_2. No significant difference can be measured between the two columns.

Differences in the turbidity can only be detected in the first period Fig. 9). After the beginning of the dosage of bacteria, a sharpe increase can be measured during a time of high DOC input values (see below). This is an indication for a break-through of the dosed bacteria.

The N-components pass the columns as long as no biological activity can be measured (Fig. 10). Nitrate has normally a concentration between 16 and 20 mg l^{-1}, ammonia and nitrite are not measurable. During a 7 days dosage of 4 mg l^{-1} quickly decomposable DOC (glucose), we can measure a sudden decrease of the oxygen and nitrate content which is accompanied by an appearance of Nitrite (0,6 mg l^{-1}) and ammonia (1,1 mg l^{-1}). The end of the DOC-dosage results in a reestablishment of the former conditions: normal oxygen concentration, no ammonia and nitrite, full break-through of nitrate. This experiment shows that nitrifying bacteria have been settled in or on the model filter.

Fig. 11 shows the o-phosphate concentration in the groundwater and after the filter passage. During the passsage through column I (algae), the phospate does not decrease. This was not expected because a content of around 0.3 mg l^{-1} o-phosphate normally stimulates algal growth. A few weeks after the beginning of the dosage of algae, there is a remarkable increase in the phosphate concentration of column I (marked on the Figure). Parallel measurements showed an increase of DOC-values. Algae on the filter have either excreted a high amount of phosphate or were partly washed into the filter and died. The passage through column II (bacteria) leads to no rise of o-phosphate, which could be expected due to degradation processes.

Dissolved Manganese and iron (Fig. 12 and 13) is washed out continuously out of the filter material during the first period. The concentration of manganese in the columns outlets reaches values up to 70 g l^{-1}. A high output of mangenese can be measured during periods of a low pH and periods of low oxygen concentration. A high biological activity causes reductive conditions with low oxygen concentrations and manganese can thus be mobilized. During this period a certain amount of dissolved iron is also washed out of the column.

Striking a balance of the difference between input and output on the columns the following calculations can be made (Table II):

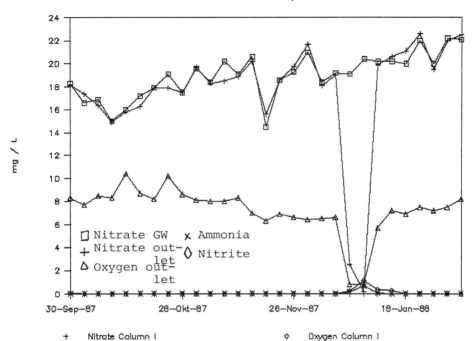

FIGURE 10: Column I (algae) Dosage of DOC and change of
N-Components

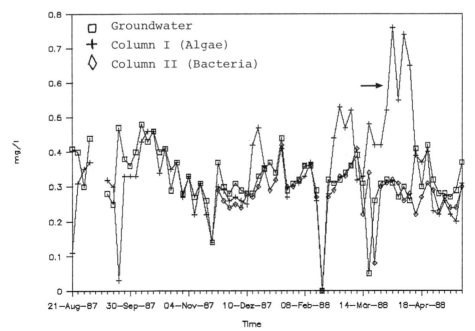

Figure 11: o-Phosphate output

Manganese

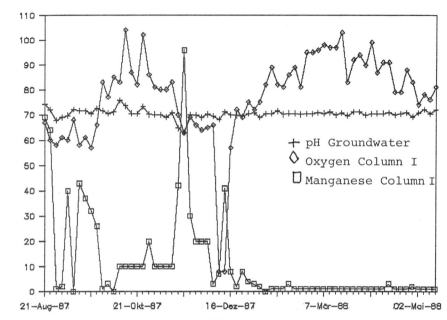

Figure 12: Manganese output Column I (Algae)

Total Iron

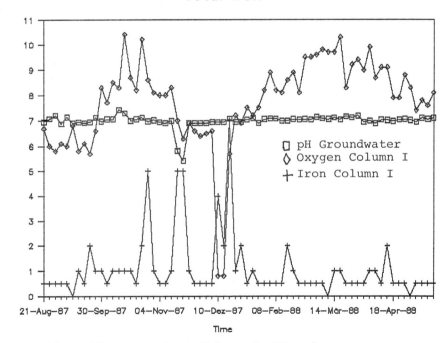

Figure 13: Iron output Column I (Algae)

<u>Column I (algae); 6 month run; 35,04 m^3 dosed groundwater:</u>

<u>Ammonia:</u>	Input:	651 mg
	Output:	4861 mg
<u>Nitrate:</u>	Input:	1.319 g
	Output:	1.268 g
<u>Phosphate:</u>	Input:	25,7 g
	Output:	23.5 g
<u>Manganese:</u>	Input:	0,15 g
	Output:	1,82 g

Table II: Balance of some constituents

Concerning the N-constituents, it is obvious, that the differences in the balance are caused by biological processes. A very small quantity of phosphate has been fixed in the sand. Manganese is washed out at the beginning until a new equilibrium has been established and manganese hydroxides are dissoluted under special conditions.

CONCLUSIONS

The experiments show, that in the filter model a stabilization of biological and chemical parameters can be established after a certain time. These conditions are similar to the conditions in a natural filter. It seems, that the integration of the above mentioned parameters in a control system is leading to a steady state in the model. Furtheron, the selection of these paramters seems to be sufficient. The system does not oscillate in an uncontrollable way, it follows the input paramters after a period of establishing at a new equilibrium.

The next phase of the experiments will show, in which way this system reacts on certain loads like the dosage of pollutants. A settlement of a third biological component like protozoa may help to adapt the system to the conditions of a natural slow sand filter.

REFERENCES

1. SCHMIDT, Kh (1985): <u>Künstliche Grundwasseranreicherung als</u>
<u>biologischer Verfahrensschritt bei der Wassergewinnung.</u>
DVGW-Schriftenreihe Wasser Nr. 45, pp 95-104
2. MUDRACK, K. (1988): <u>Biotechnologie und biologische Abwasserreinigung.</u>
Korrespondenz Abwasser 3, pp 205-210

3.4   THE REMOVAL OF VIRUSES BY FILTRATION THROUGH SAND

D. Wheeler, J. Bartram and B. J. Lloyd — Robens Institute of Industrial and
Environmental Health and Safety, University of Surrey, Guildford,
GU2 5XH. U.K.

ABSTRACT

A recently proposed dose-response relationship for the
incidence of diarrhoeal disease in children implies that for
agents with low infectious dose, small improvements in the
microbiological quality of drinking water may not result in
measurable health benefits (1). This hypothesis may have
fundamental implications for water supply engineering in
developing countries where waterborne transmission of microbial
pathogens with low infective doses eg viruses and protozoa, may
be responsible for a significant proportion of diarrhoeal
disease. The removal of protozoal cysts by filtration through
sand is generally assumed to be very efficient if not absolute.
In this sense, slow sand filtration may adequately compensate
for the comparatively low efficiency of disinfection against
encysted protozoa. However, data relating to the removal of
enteric viruses through slow sand filtration are relatively
scarce. Investigations by the Thames Water Authority (2,3) and
more recently by McConnell and co-workers (4) suggest that the
elimination of virus indicators such as poliovirus and reovirus
may be at least as efficient as the removal of faecal bacteria
through slow sand filtration. However, no studies have been
undertaken which place these observations in their
epidemiological context. It is the purpose of this paper to
describe recent research on the adsorption and elimination of
simian rotavirus and other organisms to surfaces in slow sand
filters. Such observations may assist in the assessment of the
likely efficiency of such filters in small community water
supplies.

INTRODUCTION

A comprehensive review of the impact of a number of water treatments on the removal of indicator bacteria, pathogenic bacteria, viruses and protozoa from natural waters has been presented by Kool (5). Unlike many reviews of this type, due recognition was given to processes such as dune infiltration and conventinal slow sand filtration in the removal of bacteriophage and some human enteric viruses (enteroviruses). Kool summarised data from a number of sources and concluded that infiltration through sand dunes was capable of reducing titres of bacteriophage and enteric viruses by four $\log_{10}$ units, this compared with a reduction through slow sand filters of 1-2 $\log_{10}$ units and through rapid sand filtration of 0 - < 1 $\log_{10}$ units;

It was noted that dune infiltration through 10 m of sand and a detention time of 8 days would simultaneously reduce thermotolerant coliform densities by > 2 $\log_{10}$ units and $37^{\circ}C$ and 22 $^{\circ}C$ colony counts by 4 $\log_{10}$ units. In comparison with sand filtration by this method, other processes proved reasonably effective. Under well controlled circumstances, bacteria, bacteriophage and enteroviruses are removed by chemical coagulation and flocculation by 1 - 3 $\log_{10}$ units, by ozonisation: 3 - >5 $\log_{10}$ units, and by reservoir storage : 1 - 3 $\log_{10}$ units.

In their review of the impact of water treatments on viruses, Lloyd and Morris (6) considered coagulation to be the single most effective chemical procedure (with the exception of disinfection) for the removal of enteric viruses. Percentage reductions using aluminium-based coagulants were reported; they were 99.99 per cent for poliovirus 1, 93.3 per cent for simian rotavirus (SA 11), 79 - 85 per cent for echovirus 7, and 57 - 99.9 per cent for T2 and MS2 coliphages. Reported reductions in virus titre using ferric salts were 99.9 per cent for poliovirus 1 and 92 - 99.4 per cent for f2 coliphage. In comparison with the potential of well controlled chemical coagulation, the removal of viruses by rapid sand filtration was noted to be poor.

According to this review, improvement of virus attenuation by rapid sand filtration may be achieved by the addition of alum floc or the presence of calcium cations. Removal of T4 coliphage by a rapid sand filtration plant was 0 - 87 per cent. However slow sand filtration was observed to be substantially more efficient, with reductions of 95 - 100 per cent in poliovirus 1 and 99.75 - 99.996 per cent in MS2 coliphage being achieved in some studies.

The usual microbiological criteria applied to slow sand filter performance relate to bacteriological parameters - in particular those of the faecal indicator bacteria eg thermotolerant coliforms (or Escherichia coli) and faecal streptococci. However, some work has been undertaken on the removal of specific micro-organisms of public health

significance, including protozoa (7) and helminths (8). Studies
specific to the removal of enteric viruses by sand filtration are
relatively few in number.

Work conducted by Robeck et al (9) confirmed the ability of
slow sand filters to reduce titres of poliovirus 1 by 83 - 98 per
cent. Lower flow rates were found to provide greater
efficiencies of poliovirus removal. However, removals were still
relatively low (69 per cent) at moderate conventional flow
velocities ie 0.15 mh^{-1}.

Using perspex cylinders of dimensions 200 x 40 mm, Lefler and
Kott (10) were able to simulate the transport of coliphage and
poliovirus through dune sand. Although these later experiments
were not continuous flow in nature, and thus not strictly
comparable with slow sand filtration, some interesting
observations were made. For example, it was noted that bivalent
calcium and magnesium cations at concentrations of 0.001 M and
0.01 M significantly increased adsorption whereas monovalent
sodium and trivalent iron cations had no such effect. Elimination
was greatest in the upper portions of the sand columns. Elution
of viruses from sand cloumns was not totally effective even after
10 washings with pH 10.5 buffer.

Detailed work was conducted by Poynter and Slade over a
period of years using perspex slow sand filter columns of surface
area 0.09 m^2 (2). At flow rates of 0.2 - 0.5 mh^{-1} and water
temperatures of 5 - 18°C, reductions in bacteria and viruses were
as follows: poliovirus 1 : 98.25 - 99.99 per cent, Escherichia
coli: > 88.0 - > 98.6 per cent, 37°C colony count : 81.2 - 93.7
per cent, and 22°C colony count : 94.5 - 98.4 per cent.

The effect of flow rate and sand depth on efficiency
suggested that to a certain extent, a reduction in efficiency
caused byhigher flow rates might be mitigated by increasing
sand depth. The observation that reductions in poliovirus 1
titres were consistently similar to or greater than reductions in
Escherichia coli densities led to the conclusion that the
mechanism of removal was essentially similar, and was
predominantly biological.

Subsequent observations by the Thames Water Authority (3)
showed that full scale filters (0.337 hectare) reduced
enterovirus levels in impounded river water with a consistently
higher efficiency than Escherichia coli. Sixteen samples taken
from two filters over a ten week period (water temperatures 6 -
9°C) showed that reductions in enterovirus titres in conventional
slow sand filtration were high (97.1 - 99.8 per cent) but not
necessarily equivalent to those observed in pilot scale
experiments.

However, from the perspective of this paper the most
relevant experiments on virus removal by slow sand filtration
have been conducted in the United States using reovirus as a
model (4). Reoviruses are closely related to rotaviruses and

they are readily isolated from raw sewage at comparable
concentrations ie 10^2 - 10^5 viruses per litre. Thus it is at
least arguable that reoviruses provide a relevant comparative
index for rotavirus removal in water treatment. Using
radioactively labelled reovirus and glass columns of dimensions
3 m x 150 mm, a number of important observations were made.

At a flow rate of 0.2 mh^{-1}, greatest removal (more than 65
per cent) occurred in the top 350 mm of the sand bed. At no time
was reovirus detected deeper than 1.2 m in the sand bed.
However, between depths of 0.3 and 1.0 m, reovirus continued to
be removed as a relatively constant proportion of the influent
reovirus titre.

In contrast to the studies of Poynter and Slade (2), there
was no significant difference in removal between clean and aged
(mature) sands. Furthermore, there was no difference in the
pattern of removal with depth for clean or aged sands, and graded
sands were no more efficient than ungraded sands in removing
reovirus. Due to the high overall removal efficiencies obtained
(more than 4 log_{10} units), reovirus densities in effluents did
not allow absolute comparisons to be made between clean and aged
(mature) sands. A summary of results with respect to sand type
is illustrated below.

SAND TYPE	PER CENT REOVIRUS RECOVERED	
	In Effluent	In Sand
Graded - Clean	3.3	96.7
Ungraded - Clean	5.2	94.8
Ungraded - Aged (with schmutzdecke)	5.9	94.1
ungraded - Aged (without schmutzdecke)	4.9	95.1

Table 1. Percentage of added ^{125}I-labelled reovirus detected in
effluent and sand samples collected from sand filter columns
containing various sand types (4).

Thus, it was concluded that biological activity within sand
horizons was not primarily responsible for elimination of
viruses. Indeed,because infectious virus was undetectable in
sand samples extracted at the end of filter runs from non-mature
columns, it was clear that either elution was ineffective or some
other factor had inactivated the viruses.

To examine the effect of related biological factors on the efficiency of virus removal during filtration through sand, a series of three experiments was undertaken. These were:

I   The removal of indicator bacteria and viruses through a small scale protected slow sand filtration system designed for use in less developed countries (11,12);
II   The removal of indigenous rotavirus through sand and anaerobic biomass beneath a wastewater lagoon and irrigation site in Peru (13); and
III   The adsorption and attenuation of simian rotavirus SA11 with various substrates extracted from both protected and conventional slow sand filters.

MATERIALS AND METHODS

Bacteriological Parameters

        Faecal indicator bacteria in water samples were assayed according to standard procedures (14). The faecal indicator groups examined for attenuation with depth in sand were thermotolerant coliforms and faecal streptococci. Plate count bacteria ($37^{o}C$ and $22^{o}C$) were enumerated by the spiral plate technique. Thermotolerant coliforms and faecal streptococci in sand smaples were assayed by a modified most probable number (MPN) technique. Sand was weighed in three series of five wet weights (5 x 1.0 g, 5 x 0.1 g and 5 x 0.01 g) and placed in sterile 25 compartment replidishes (Sterilin). To these sub-samples was added Minerals Modified Glutamate Broth (Oxoid) or Kanamycin Aesculin Azide Broth (Oxoid) for thermotolerant coliforms and faecal streptococci respectively. Dishes were incubated at $44^{o}C$ for 24 hours. Compartments showing growth were subcultured into confirmatory media : Brilliant Green Bile Broth for thermotolerant coliforms and Kanamycin Aesculin Axide Agar (Oxoid) for faecal streptococci and incubated at $44^{o}C$ for a further 24 hours. Quantitative MPN results were computed from standard statistical tables (14).

Bacteriophage

        Five bacteriophage (bacterial viruses) were employed in the investigation. They were selected primarily on the basis of their size, each having a head with cross-sectional area within the range $\pm$ 50 per cent of the cross-sectional area of rotavirus particles. Bacteriophage were drawn from three groups of the Tikhonenko classification (15), and possessed tails which varied in size between <20 nm and 150 nm. The source of the bacteriophage has been described (16) and their characteristics are summarised in Table 2.

Bacteriophage	Classification	Dimensions (nm) Head	Tail
Serratia marcescens	III	50 x 50	< 20
Erwinia carotovora	III	75 x 75	30
Escherichia coli K12	IV	50 x 50	150
Enterobacter cloacae	V	100 x 70	85
Bacillus licheniformis	V	70 x 70	150

Table 2   Morphological characteristics of bacteriophage used
in Experiment I. Classification according to Tikhonenko, (15).

Bacteriophage were assayed using a modification of the soft agar
overlay technique of Adams (17).   All bacteriological products
were obatained from Oxoid and chemicals were obtained form BDH
(Analar Grade). The basal medium was Blood Agar Base and the
soft agar overlay was a simple formulation of Nutrient Broth (1.2
gl^{-1}), Purified Agar (5.5 gl^{-1}) and sodium chloride (7.0gl^{-1})

    Sub-sample volumes of 0.1 ml were pipetted onto 90 mm petri
dishes of basal medium. To each sample was added 0.5 ml of an
overnight culture of host bacterium grown in a rich nutrient
broth comprising Brain Heart Infusion (20 gl^{-1}), Casein
Hydrolysate (20 gl^{-1}), potassium dihydrogen orthophosphate (5.0
gl^{-1}) and glycerol (2 per cent in distilled water). To the
mixture of sample and host was added 3.0 ml molten soft agar at a
temperature of 45°C. The plate was swirled vigorously to ensure
thorough mixing of the overlay. The petri dish was then placed
in a clean environment with the lid ajar for a period of five
minutes in order to promote rapid setting and avoid condensation
forming on the underside of the lid (drops of condensaton can
interfere with the assay). Finally, dishes were reassembled,
inverted and incubated at the optimum growth temperature for the
host. Plaque formation was usually within 6 - 12 hours;
following enumeration of plaques, results were expressed as
plaque forming units (pfu) per ml.

Rotaviruses

    Rotaviruses in aqueous samples were enumerated by visual
counts of fluorescent foci in multiwells.

    MA104 cells in 80 ml bottles (Nunc) were washed three times
in PBS before being stripped in versene-trypsin solution. Cells
were suspended in growth medium (18) to yield a concentration of
2 x 10^5 cells per ml. 2.0ml of suspension were added to each
compartment of a six well multiwell plate (Nunc). After allowing
30 minutes for cell attachment in an atmosphere of 5per cent

carbon dioxide and > 95 per cent relative humidity ($37^{\circ}C$), the medium was discarding and replaced by 3.0 ml of fresh growth medium. Cells were grown to confluence over a period of 24 - 48 hours and then sheets were bathed in serum-free growth medium for three periods of thirty minutes. Cell sheets were subsequently incubated for a further 12 hours in 3.0 ml serum-free medium to eliminate the influence of any residual anti-rotavirus antibodies in the foetal calf serum.

Before application of samples to the cell sheets, medium was drained off and replaced by 0.2 m fresh serum-free medium containing 10 ug per ml trypsin. To this was added 0.2 ml of sample (or sample dilution), in order to provide a shallow but complete immersion of the cell sheet. Failure to observe extreme care in this procedure was shown to result in fraying or desstuction of the cell sheet. The sample was incubated for a period of 60 minutes in an atmosphere of 5 per cent carbon dioxide and > 95 per cent humidity ($37^{\circ}C$) to allow attachment of virus to susceptible cells. Following this period, cells were washed carefully with one change of serum-free growth medium, and incubated for a further 24 hours at $37^{\circ}C$. No advantage was conferred by centrifuging samples to encourage virus attachment to cell sheets.

Cell sheets were then washed gently in PBS before being fixed for 15 minutes in approximately 2.0 ml of methanol at $4^{\circ}C$. After pouring off the methanol, cell sheets were air dried at ambient temperature. In this state, the fixed cells could be stored indefinitely, but in practice staining was invariably undertaken within two hours.

The staining procedure comprised two steps. Following washing in PBS, 0.2 ml of a 1:60 dilution of bovine anti-rotavirus serum was added to each cell sheet. The plate was incubated for 30 minutes in an atmosphere of > 95 per cent humidity ($37^{\circ}C$) and each cell sheet was then washed for ten minutes in three changes of PBS. The second step required the addition of 0.2 ml of a 1:60 dilution of lapine anti-bovine serum conjugated with fluorescein isothiocyanate (Wellcome MF-13) followed by a further incubation period of 30 minutes. Finally, each well was bathed in three further ten minute changes of PBS to remove all traces of excess reagent before being air-dried.

Cell sheets were scanned using incident UV light by Leitz microscope. Replication of rotavirus in individual cells was readily visualised by characteristic cytoplasmic staining, invariably observed as unmistakeable apple-green foci formng horseshoe or annulus shapes around the cell nucleus. The use of positive and negative contols together with careful contrasting of observed staining with cell morphologies under normal transmitted light ensured that false positives were eliminated. Counts were related to the volume of concentrate and subsequently to the volume of original sample and quoted as fluorescent focus forming units (fffu) per litre.

Rotavirus was eluted from sand smples using 3 per cent Beef Extract solution (Oxoid) at pH 9.5. Elution from filters was achieved using a volume of eluent in the range 200 - 400 ml. Elution from sand was accomplished by placing 100 g sand sample in the base of a sterile 1 litre flask and adding 500 ml beef extract eluent. A sterile magnetic stirrer was introduced, and the sand and eluent agitated for a period of sixty minutes. Eluents from sand were decanted.

Because of reported variations in the efficiency of beef extract flocculation (18), eluted virus was precipitated by a process of enhanced organic flocculation. Eluents were placed in sterile 500 ml or 1 litre flasks and sterile Skim Milk concentrate (Oxoid) was added and dispersed to provide for a final concentration of 1.5 per cent by weight. The pH of the flask contents was then adjusted to pH 3.5 by the addition of a predetermined volume of hydrochloric acid (1.0 M). A sterile magnetic stirrer was introduced and the flocculating mixture agitated slowly for a period of 30 minutes. The mixture was then centrifuged in 200 ml or 500 ml polypropylene bottles at 3000 g for 30 minutes at $4^{\circ}C$. Supernatants were discarded and precipitated pellets dissolved in the minimum volume of a 0.15 M solution of sodium dihydrogen orthophosphate (pH 9.2). This volume was typically 10 ml per 200 ml original eluent.

Shake Flask Adsorption

Untreated water was collected from the Hampton Water Treatment Works (Thames Water Authority) and sterilised by autoclaving at $121^{\circ}C$ for 20 minutes. A volume of 22 ml SA11 virus stock solution was added to 375 ml sterilised raw water and agitated slowly for a period of 60 minutes on a rotary shaker in an ambient temperature of $18^{\circ}C$. An aliquot of 5 ml of this stabilised virus suspension was removed and stored at $4^{\circ}C$ for subsequent assay.

100g samples of sand and 5g samples of fabric (dry weight) were obtained. Except in the case of controls, substrates were taken from functional protected slow sand filters and conventional slow sand filters. 75ml volumes of virus suspension were added to 250 ml flat-bottomed glass flasks containing pre-weighed quantities of substrate. Flasks and contents were vigorously mixed and supernatants immediated sampled. Thereafter, flasks were agitated gently on a rotary shaker and sampled on six further occasions over a period of 20 hours.

Simian rotavirus (SA11) densities were assayed by immunofluorescence, with counts of fluorescent foci in ten representative microscope fields being extrapolated to obtain the original density in supernatants.

RESULTS

Experiment I

To investigate the attenuation of indicator bacteria and viruses
with respect to depth, parallel protected slow sand filters were
operated for a period of more than 12 months at a flow rate of
0.28 mh^{-1}. Samples of supernatant and interstitial water were
taken in duplicate from a mature filter protected by six layers
of polypropylene fabric. Sampling ports enabled depth samples to
be taken at 0 mm (the fabric-sand interface), 20 mm, 100 mm,
200 mm, and 500 mm (the filter outlet). Samples were assayed
within four hours of collection and the mean results of a large
number of sampling occasions are summarised in Figures 1-4.

Figure 1 illustrates the attenuation of thermotolerant coliforms
with depth. Marginally more than 50 per cent of thermoltolerant
coliforms were eliminated through the six layers of polypropylene
fabric. The removal rate increased to 88 per cent at a depth of
100 mm in the sand and > 98 per cent at the filter outlet
(500 mm).

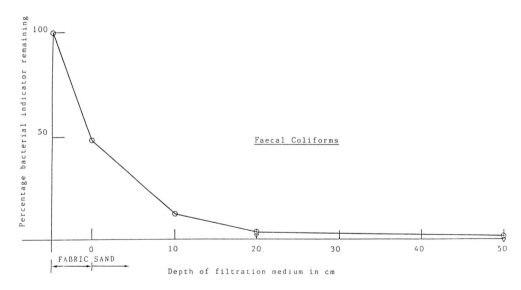

FIGURE 1. Percentage removal of bacterial indicator in a protected slow sand filter
operated at a filtration velocity of 0.28mh^{-1} ○

NB ◉ represents result 'less than' illustrated

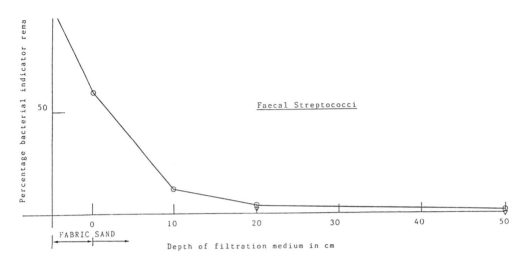

FIGURE 2.   Percentage removal of bacterial indicator in a protected slow sand filter
            operated at a filtration velocity of $0.28mh^{-1}$ ○

        NB   ♀ represents result 'less than' illustrated

Similar results were obtained for faecal streptococci (Figure 2).
However, in the case of the relatively larger, gram positive
organisms, more attenuation occurred in the first 100 mm of sand
compared with the fabric layers.   Nonetheless, mean overall
elimination was equivalent for faecal streptococci at 100 mm (88
per cent), 200 mm (> 96 per cent) and 500 mm (> 98 per cent).

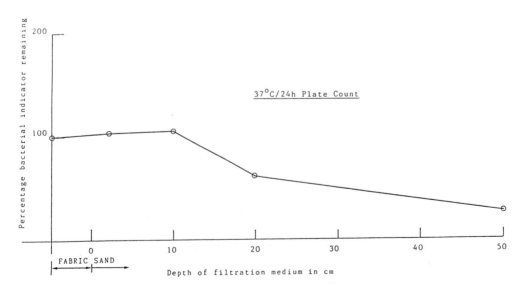

FIGURE 3. Percentage removal of bacterial indicator in a protected slow sand filter
           operated at a filtration velocity of $0.28mh^{-1}$ ○

Figure 3 depicts the slight increase and subsequent reduction in density of 37°C plate count organisms in interstitial water with respect to depth. In order to avoid sampling these organisms at the fabric-sand interface where populations way be subject to atypical influences, the first sample port was placed at a depth of 20 mm in the sand bed. It is noticeable that there was no reduction in numbers of these bacteria between the supernatant water and intersitial water at a depth of 100 mm in the sand bed. If anything, there was a slight, but relatively uniform increase. Below a sand depth of 100 mm, there was a marked attenuation with the result that at the filter outlet (500 mm), the overall reduction was 74 per cent.

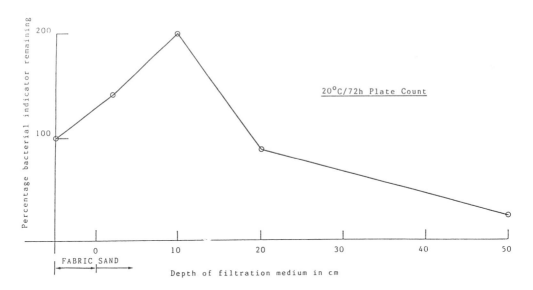

FIGURE 4. Percentage removal of bacterial indicator in a protected slow sand filter operated at a filtration velocity of 0.28mh^{-1} O

The results for 22°C plate count organisms (Figure 4) demonstrate an approximate doubling of bacterial populations between the supernatant water and interstial water at a depth of 100 mm. This increase in numbers appears to be approximately uniform across both fabric and sand horizons of the slow sand filter. However, below 100 mm, there was a marked decline in densities, resulting in similar overall reductions to those observed for 37°C plate count bacteria.

Attenuation of bacteriophage with respect to depth was investigated in an intensive experiment. Parallel slow sand filters protected by six layers of polypropylene fabric were operated at a flow rate of 0.28 mh^{-1} for a period of ten days to establish full biological maturity (reflected by a stable reduction of > 98 per cent in densities of thermotolerant coliforms). Preparations of five bacteriophage were simultaneously introduced into one filter in the following total numbers:

Serratia marcescens bacteriophage 1.54×10^9

Erwinia carotovora bacteriophage 1.52×10^{10}

Escherichia coli K12 bacteriophage 3.96×10^8

Enterobacter cloacae bacteriophage 1.51×10^9

Bacillus lincheniformis bacteriophage 1.24×10^8

Samples were taken from the supernatant (untreated) water and from interstitial water at the following depths in the sand bed: 0 mm (the fabric-sand interface), 100 mm, 200 mm and 500 mm (the filter outlet). Samples were taken at 10 minute intervals for 130 minutes, at 15 minute intervals for the following 165 minutes, 30 minute intervals for the following 240 minutes and on five occasions in the subsquent 24 hours.

The density of each bacteriophage passing each sampling point was related to flow, and the total number of each was then expressed as a percentage of the original inoculum. Results are described in Table 3 and depicted in Figures 5-9. Overall attenuations ranged from 99.23 per cent for Bacillus licheniformis bacteriophage to 99.84 for Erwinia carotovora bacteriophage.

Bacteriophage	Depth in Sand Bed (mm)			
	0*	100	200	500
Serratia marcescens	53.4	11.6	5.78	0.62
Erwinia carotovora	22.3	2.13	1.23	0.16
Escherichia coli K12	49.3	5.78	3.15	0.18
Enterobacter cloacae	63.7	13.5	7.06	0.65
Bacillus licheniformis	46.4	16.4	4.41	0.77

Table 3. Percentage of bacteriohphage remaining at successive sand depths in a slow sand filtration protected by six layers of fabric. *Depth 0 represents the fabric-sand interface.

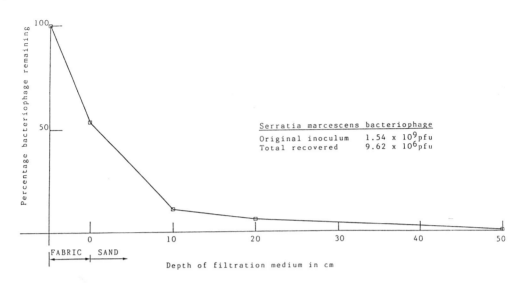

FIGURE 5. Percentage removal of bacteriophage in a protected slow sand filter operated at a filtration velocity of 0.28mh^{-1} □

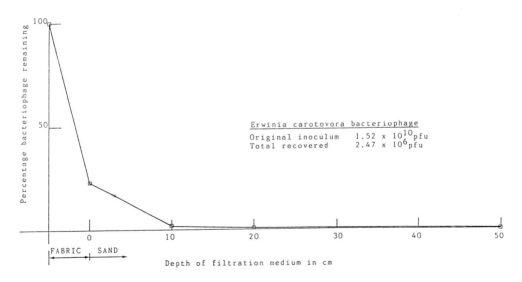

FIGURE 6. Percentage removal of bacteriophage in a protected slow sand filter operated at a filtration velocity of 0.28mh^{-1} □

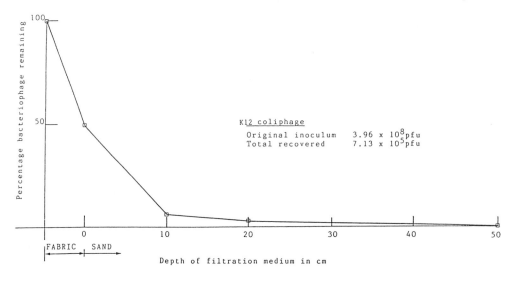

FIGURE 7. Percentage removal of bacteriophage in a protected slow sand filter operated at a filtration velocity of 0.28mh^{-1} □

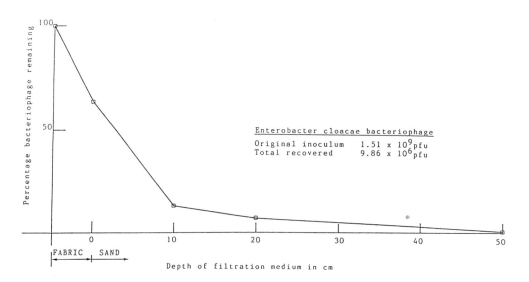

FIGURE 8. Percentage removal of bacteriophage in a protected slow sand filter operated at a filtration velocity of 0.28mh^{-1} □

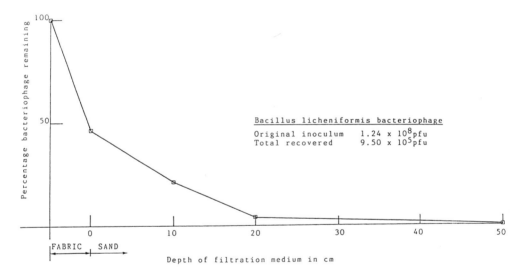

FIGURE 9. Percentage removal of bacteriophage in a protected slow sand filter operated at a filtration velocity of 0.28mh^{-1} □

Experiment II

The difficulty of securing a stable experimental system served by a raw water source with high levels of indigenous rotavirus led to the investigation of alternative systems which could provide comparable experimental data. The most appropriate system was considered to be a wastewater application site close to Lima, Peru : San Juan de Miraflores. The site occupies 20 hectares dedicated to 21 shallow, unlined lagoons and 375 hectares of flood-irrigated agricultural land and woodland. This former desert land receives 360 litres per second of raw sewage from three low-income residential areas in southern Metropolitan Lima. As a result, the permeable aeolian and alluvial sands (mostly 0.1 - 0.3 mm in diameter) allowed wastewater to penetrate at a rate of 0.07 - 0.10 m/day below lagoons and 0.03 - 0.05 m/day below irrigation channels (13).

Although the percolation rates were comparable, there was an important difference between microbial penetration through lagoon floors and penetration through irrigation channels. The former required passage through an anaerobic sludge of settled biomass approximately 50 - 100 mm in depth prior to transport through sand, whereas the latter represented direct application to the sand.

Samples were taken from sand horizons to a depth of 18 metres below the lagoons and irrigation channels. Results were calculated per 100 gram net weight of sand. Results are described in Figures 10 and 11. In the case of penetration below lagoons, there was a 4 $\log_{10}$ reduction in thermotolerant coliforms and faecal streptococci following infiltration through 5 m of sand. Rotavirus was not detected in levels greater than 60 fffu per 100 g more than 100 mm below the anaerobic sludge.

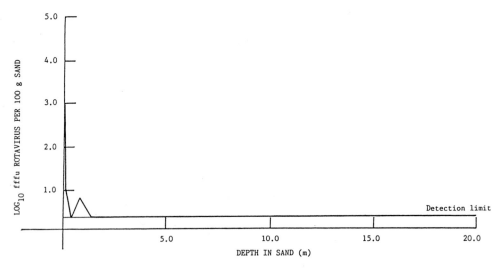

FIGURE 10. Attenuation of rotavirus with respect to depth of sand below wastewater lagoon.

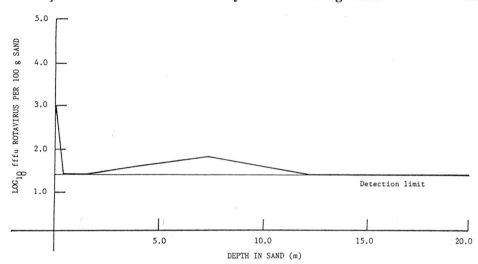

FIGURE 11. Attenuation of rotavirus with respect to depth of sand below irrigation channel.

Attenuation of faecal indicator bacteria through sand below irrigation channels was comparable to that below lagoons (3-4 log_{10} units). But in contrast with results obtained below lagoons, rotavirus was detected at levels of between 10^3 and 10^4 fffu per 100g to a depth of 7.5 metres. Rotavirus was undetectable (less than 20 fffu per 100 g) below 12.5 metres.

Experiment III

    Results of adsorption attenuation experiments are summarised in Table 4. They illustrate the fact that in matched conditions, non-sterile substrates incorporating biomass (schmutzdecke) achieved substantially more reduction in the density of simian rotavirus SA11 is raw water than any sterile substrate. Signficantly, sand and schmutzdecke taken from a conventional slow sand filter effected nearly a 1 log_{10} reduction in rotavirus densities in 60 minutes - an observation which confirms the potential value of this part of the medium in slow sand filters.

Time (mins)	Sterile Substrates			Non-Sterile Substrates		
				Schmutzdecke plus:		
	Acid Washed sand	Clean Sand	Polypropylene Fabric	Sand from Conventional SSF	Sand from PSSF	Fabric
10	NR	NR	NR	0.48	0.31	0.46
60	NR	NR	NR	0.94	0.36	0.47
1200	1.98	1.98	NR	2.58	2.58	2.58

Table 4 Log reduction of simian rotavirus (SA11) densities in water in intimate contact with various substrates. NR = No Reduction.

DISCUSSION

It has been noted in several reviews that the mechanisms of transport and removal of microorganisms (usually bacteria) by the slow sand filtration process are biological, physical and physico-chemical (19). However, because of the small size of viruses, some physical mechanisms of transport and removal are clearly of little importance, even if viruses are present as aggregates; these include straining and interception. In addition, gravitational sedimentation is not considered to be a major factor in the removal of particles of less than 1 um eg colloids, from aqueous solution. Thus, unless there is considerable association with suspended particulate matter, the three factors of greatest potential significance in the elimination of viruses in slow filters are microbial predation, adsorption to biomass or biofilms and absorption to non-biological surfaces.

Of these three factors, empirical observations such as those of Poynter and Slade (2) would suggest that two biologically-related phenomena : microbial predation and adsorption to biomass/biofilms have the greatest influence, particularly in those upper horizons of slow sand filters where mixed populations of bacteria, fungi, algae and scavenging protozoa and metazoa proliferate. However, it should be noted that the experiments of McConnell and co-workers (4) do not support this view.

Although their relative importance may vary, adsorption to both biological and non-biological surfaces may theoretically play a role at all depths of the filter. However, there is considereable evidence to suggest that virus adsorption to micro-bial biomass and biofilms may be a particularly important factor in removal in the upper horizons of slow sand filters.

The association between bacteria and protozoa has been understood for some time (20). And the presence of a zoogloeal "slime" around sand grains has long been postulated as an adsorbent for both solutes and colloids passing through slow sand filters (21). In nature, biofilms are noted for their porosity and adsorptive properties. Thus, it is entirely possible that extracellular polymers associated with microbial colonisation of sand play a very important role in providing local binding sites for viruses. Empirical evidence strongly supports this hypothesis. For example, experiments have shown that the efficiency of filtration is improved in the presence of polymers capable of enhancing bridging between colloidal particles and filter media (22)

It has been noted by Corpe (23) that virus adsorption to cells is mediated by cell fractions which can be experimentally inactivated by proteases. Thus mucoproteins, glycolipoproteins and lipoproteins are all candidates for virus receptors. The affinity of bacteriophage for peptidoglycans and other polymers such as lipopolysaccharides was also noted.

Adsorption of enteroviruses and bacteriophage to sewage floc has been investigated in detail. Balluz et al (24) showed that poliovirus 1 adsorbed strongly to bacterial floc in a benchscale activated sludge plant. Following sedimentation of this floc, the amount of virus remaining in the aqueous effluent of the plant was only 0.2 per cent of the original inoculum. In contrast, another member of the picornaviridae, f2 coliphage, was adsorbed and removed much less efficiently. More than 31 per cent of the original coliphage inoculum remained in the aqueous phase after sedimentation. Later work demonstrated that other enteroviruses, including coxsackie B5 and echovirus 1 adsorbed more or less as efficiently as poliovirus to sewage floc (25). Drury and Wheeler (26) demonstrated that high proportions of Serratia marcescens bacteriophage may be adsorbed by autoclaved activated sludge floc and subsequently removed from the aqueous phase by sedimentation. Gerba and co-workers (27) obtained 67 - 99.8 per cent adsorption of six enteroviruses to activated sludge floc.

Observations on full scale slow sand filters have confirmed that immediately after commissioning or cleaning, improvements in the bacteriological quality of filtrates may take several days to become established with reliable reductions in excess of 90 per cent (28). Thus the concept of a maturation period has been developed. Clearly, maturation is both microbiological and physico-chemical. The former requires the establishment of vigorous populations of protozoa, metazoa, fungi, bacteria and algae. The latter requires the establishment of appropriate physico-chemical properties throughout the filter medium in order to facilitate adsorption.

However, the removal of the top few centimetres of biologically active sand and biomass during conventional cleaning ("skimming") results in poor bacteriological removals for several days despite the presence of up to 1 metre depth of physico-chemically condtioned sand and associated biofilms. Thus it may be concluded that under normal circumstances, most of the activity of slow sand filters with respect to the removal of bacteria occurs within the uppermost horizons of the filter where microfauna and flora are most abundant. Thus, it is certain that for bacteria, it is in this area that both predation and adsoption are most effective .

CONCLUSIONS

In this study, the relative contribution of biological factors in the removal of indicator viruses has been indirectly assessed by establishing the pattern of removal with respect to depth in an operational filter. The pattern of removal was similar to that obtained for other micro-organisms of hygienic significance.

It has also been shown that an anaerobic microbial sludge overlying sand in a wastewater lagoon achieves much greater elimination of indigenous rotavirus than a similar environment where no biomass is deposited ie below irrigation channels. This was despite the fact that attenuation of bacterial indicators was comparable in the two environments. These two observations imply that the mechanisms of removal of bacteria and viruses (including bacteriophage and rotavirus) may not be identical, but that in real environments, the presence of microbial biomass is at least as important to the removal of viruses as it is for bacteria.

The preliminary results from laboratory adsorption experiments also confirm the importance of biomass in enhancing the removal of rotavirus from the aqueous phase. It is hoped that experiments currently being undertaken at the University of Surrey will further elucidate the nature of biological interactions responsible for the attenuation of viruses in sand filtration.

If it can be shown that the removal of agents with low infective doses eg enteric viruses by slow sand filtration is as reliable as the customary removal of indicator bacteria, then the process could be of considerable benefit in the reduction of mortabity and morbidity due to waterborne transmission of these agents. However, any unreliability or inefficiency in this respect may mean that the process will not achieve hoped for health benefits with respect to certain gastro-intestinal disorders.

ACKNOWLEDGMENTS

The authors wish to express their gratitude to the United Kingdom Overseas Development Administration for funding the research described in this paper, the Thames Water Authority for provision of facilities, the British Geological Survey for commisioning work in Peru and Ms Verity Larby for typing the Manuscript.

REFERENCES

1   Esrey, S.A., Feachem. R.G. and Hughes, J.M. (1985). Bulletin of the World Health Organisation, 63 (4), 757-772.

2   Poynter, S.F.B. and Slade, J.S. (1977). Prog. Water Tech., 9, 75-88.

3   Slade, J.S. (1978).J. Inst. Water Eng. Sci., 32, 530-536.

4   McConnell, L.K., Sims, R.C. and Barnett, B.B. (1984). Appl. Environ. Microbiol., 48 (4), 818-825.

5   Kool, H.J. (1979). In Biological Indicators of Water Quality
     eds. James, A. and Evison, L., 17.1-17.31.

6   Lloyd, B.J. and Morris, R. (1982) In Viruses and Disinfection
     of Water and Wastewater eds. Butler, M., Medlen, A.R. and
     Morris, R., University of Surrey, UK.

7   Logsden, G.S., Symons, J.M., Hoye, R. L. Jr. and Arozarenas,
     M.M. (1981). J. Am. Water Works Assoc., 111-118.

8   Kawata, K. (1982). Water Sci Tech., 14, 491-498.

9   Robeck, G.G., Clarke, N.A. and Dostal, K.A. (1962). J. Am.
     Water Works Assoc., 54 (10), 1275-1292.

10   Lefler, E. and Kott, Y. (1974). Israel J. Tech., 12, 298-304.

11   Pardon, M., Lloyd, B. and Wheeler, D. (1983). Waterlines,
     2(2), 24-28.

12   Lloyd, B.J., Wheeler, D. and Pardon, M. (1985). Water Sci.
     Tech., 17, 1367-1368

13   Geake, A.K., Foster, S.S.D. and Wheeler, D. (1987). Proc.
     Int. Conf. on Vulnerability of Soil and Groundwater to
     Pollutants, RIVM, The Netherlands.

14   Anon, (1982). The Bacteriologcial Examination of Water
     Supplies. Reports on Public Health and Medical Subjects N°71.
     HMSO, London

15   Tikhonenko, A.S. (1970). Ultrastructure of Bacterial
     Viruses. New York, Plenum Press.

16   Wheeler, D., Skilton, H.E. and Carroll, R.F. (forthcoming).J.
     Appl. Bact., in press

17   Adams, M.H. (1959). Bacteriophages. Interscience, New York.

18   Morris, W.R. (1986). Enteroviruses in Water and Wastewater.
     PhD Thesis, University of Surrey, Guildford, UK.

19   Van de Vloed, A. (1955). Proc. Int. Water Supply Congress,
     London, 7, 1-77.

20   Ball, B. H. (1969). In Research in Protozoology, ed. Chen,
     T.T., Pergamon Press, Oxford, 3, 565-718.

21   Lloyd, B.J. (1974). The Functional Microbial Ecology of Slow
     Sand Fiters. PhD Thesis, University of Surrey, Guildford, UK.

22   Sehn, P. and Gimbel, R. (1983). In Advances in Solid-Liquid
     Separation, ed Gregory, J., Academic Press, London.

23 Corpe, W.A. (1980) In Adsorption of Micro-organisms to Surfaces, eds. Bitton, G. and Marshall, K.C., Wiley and Sons, New York.

24 Balluz, S.A., Butler, M. and Jones, H. H. (1978).J. Hyg. Camb., 80, 237-242.

25 Bates, J. (1982). Viruses in Sewage Sludge. PhD Thesis, University of Surrey, Guildford, UK.

26 Drury, D.F. and Wheeler, D (1982). J. Appl. Bact., 53, 137-142.

27 Gerba, C.P., Goyal, S.M., Hurst, C.J. and Labelle, R.L. (1980). Water Res., 14, 1197-1198.

28 Burman, N.P. and Lewin, J. (1961). J. Inst. Wat. Eng., 15, (5), 355-367.

29 Bellinger, E.G. (1979). J. Inst. Water Eng. Sci., 33(1), 19-29.

4

Process performance

4.1 THE EFFECTS OF HIGH-CARBON AND HIGH-COLIFORM FEED WATERS ON THE PERFORMANCE OF SLOW SAND FILTERS UNDER TROPICAL CONDITIONS

J. M. Barrett and J. Silverstein — Department of Civil and Environmental Engineering, Campus Box 428, University of Colorado, Boulder, Colorado 80309, U.S.A.

ABSTRACT

Pilot-scale slow sand filters, operated at a temperature of 25C, were fed source waters characteristic of both unpolluted and polluted surface supplies. Coliform reductions, total organic carbon (TOC) removals, and hydraulic data are presented. Coliform reductions were high under non-polluted conditions. Filtration of polluted water supplies resulted in short run times due to rapid head loss. Growth of the biofilm responsible for the head loss was restricted to the surface of the filters.

INTRODUCTION

By the midpoint of the International Drinking Water Supply and Sanitation Decade (December 31, 1985) only 42% of the world's rural population and 77% of the world's urban population (excluding The Peoples Republic of China) had access to safe drinking water (1).
Slow sand filtration has successfully produced potable water in the United States and Britain for a century. Construction costs, usefulness of local materials, ease of operation and maintenance, and low energy requirements make slow sand filtration a water-treatment technology that is particularly well-suited for developing countries.
Extensive research has been performed on the efficiency of slow sand filters for treatment of water in temperate climates. When source waters are relatively low in temperature and turbidity, this process has been found to be effective in the removal of organics, certain ions, bacteria, pathogenic protozoa, and viruses (2,3,4). Slow sand filtration plants currently serve as sources of potable water in many tropical countries, including Columbia, India, Jamaica, Kenya, Sudan, and Thailand (5). This water-treatment process has also been recommended for schistosome cercariae control in north Cameroon village water, having been found an effective barrier to cercariae at very low hydraulic loading rates (6). Although slow sand filters are not uncommon in developing nations, many aspects of slow sand filter operation and maintenance under tropical conditions are not well understood. These aspects include ripening time (time from start-up to the production of potable water), length of runs (time to terminal head loss, hence frequency of cleaning), and the pathogen removal efficiency as a function of source-water chemistry under tropical conditions.
Tropical rivers and other surface waters which serve as receiving waters for waste discharges and runoff also serve as water supplies for downstream

communities. Warm temperatures and dramatic annual flow fluctuations present
particular challenges in rendering tropical surface waters potable. Because
slow sand filtration relies heavily on biological processes to purify water, it
is expected that temperature has a significant impact on filter behavior and
performance. Additionally, low flow rates of surface waters during dry seasons
will result in high levels of organic and microbial contamination from upstream
human waste, agricultural runoff, and discharges from agricultural and industrial processing facilities.

Research was performed at the University of Colorado to characterize the
behavior and efficiency of slow sand filters under tropical conditions. Initial
experiments involved the filtration of source water characteristic of unpolluted river water, that is, containing 1 ppm carbon from organic compounds
which are readily utilizable by microorganisms (7). In subsequent experiments
the concentration and nature of organic constituents of the filter feed water
were varied to simulate polluted supplies. Using glucose as a carbon source, a
carbon concentration range of 1 to 50 ppm, as carbon, was examined. In the
present paper, emphasis is placed on slow sand filter response to influents in
the lower portion of this carbon concentration range. To determine whether or
not filter performance was strongly impacted by humic materials, an experiment
was run in which humic compounds contributed 50% of the source-water organic
carbon.

As concentrations of fecal coliforms in untreated water supplies vary
widely in developing countries (8), polluted source waters were simulated by
maintaining a filter influent concentration of approximately 10^6 _Escherichia
coli_ cells per 100 ml.

A source-water temperature range of 25 to 27C was chosen, which corresponds to the annual range observed in the Orinoco River in Venezuela (9).

EXPERIMENTAL METHODS

Physical Configuration

A diagram of the pilot-scale slow sand filtration system is shown in
Figure 1. The experimental set-up consists of two identical filter columns,
though only one is diagramed. Each filter column is 3.4 m in height and 0.3 m
in diameter, and is constructed from two cylinders of polyvinyl chloride. The
cylinders can be disconnected just above the sand bed surface, allowing easy
cleaning and examination of the schmutzdecke. The sand bed is 105 cm in height,
and is supported by a 25-cm gravel underdrain. The sand had an effective size
(d_{10}) of 0.22 mm, and a uniformity coefficient (d_{60}/d_{10}) of 2.5.

Flow rate to the filters is controlled by a feed box which maintains a
constant head of liquid over fixed orifices. The inorganic, organic, and indicator organism feeds have been separated to minimize the growth of biofilm in
the feed tank and on the walls of the reservoir in the upper portion of tue
filter column. Carbon and coliform feed lines join the inorganic feed at the
bottom of the feed line from the constant-head box, and the combined flows are
fed into the filter reservoir at a height of 25 cm above the sand surface. The
combined flow to each of the filters is 243 ml per minute, which is equivalent
to a hydraulic loading rate of 0.2 m per hour.

Sight tubes have been installed from the reservoir above the sand bed, and
from ports drilled at depths of 1, 5, 10, 20, and 40 cm into the filter media.
These sight tubes allow measurements to be made of the static fluid heads at
any time during the filter run.

A thermocouple above the sand surface allows monitoring of the influent
temperature, which was maintained at 25 to 27C. A sample port was connected to
a tube that extended to a height of 2 cm above the sand surface. The proximity
of the sampling tube to the filter media allowed the exact characterization of
the feed water supplied to the sand bed. The effluent line ended at a spill
cup, which was fixed at 17 cm above the sand bed surface.

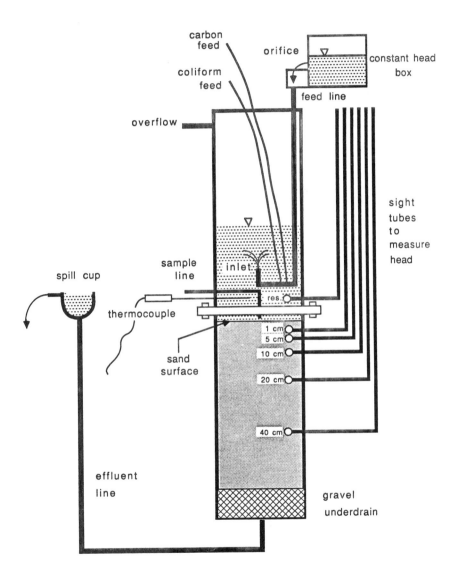

Figure 1. The pilot-scale slow sand filter.

Chemistry of the Feed Water
        The inorganic constituents of the simulated tropical source water were
chosen based on an average inorganic chemistry of South American rivers (10).
The exact formula for the inorganic feed water is presented in Table I. Sodium
thiosulfate (3 mg l^{-1}) was added to dechlorinate the inorganic feed water,
which was made from Boulder tap water.
        Two organic compounds were used in the feed-water formulation. Glucose was
used as an organic nutrient. Throughout this paper glucose is referred to as
the "labile" carbon source, that is, one which is readily utilizable by
microorganisms. A sodium salt of humic acid was used as the "refractory" carbon
source, that is, the organic compound which is much less biodegradable. The
sodium salt of humic acid is 40.27% carbon by weight, according to the
supplier*. All carbon concentrations given in this paper are as carbon.

Sequence of Events in a Filter Run
        At the end of each filter run the schmutzdecke was scraped. Sand was
removed to a depth of 1.5 to 2 cm. The scraped sand was replaced with new sand
to return the sand bed to its original height. The filter columns were then
backfilled slowly, until the liquid level was 1 to 2 cm above the surface of
the sand. Half a liter of chlorine bleach (5% sodium hypochlorite) was added to
each saturated sand bed, and the filters were disinfected for 24 hours. The
disinfection was followed by a 24-hour rinse with cool tap water.
        After the 24-hour rinse, the inorganic, carbon, and coliform feeds were
supplied to the filters. (Preparation of the coliform feed is detailed in a
subsequent section.) The run time for each experiment began as soon as the feed
waters were supplied to the filter columns. Immediately after supplying the
three feeds to the columns, the filters were inoculated, or seeded, to enhance
the growth of a diverse microbial community. Unless specified otherwise, the
inoculum used was 250 ml of water from the shallow zone of a eutrophic pond on
the University of Colorado campus.
        Influent and effluent samples were obtained for analysis of pH, turbidity,
coliforms, and TOC. These tests were performed daily, except for the TOC. Each
50-ml sample for TOC analysis was preserved with two drops of concentrated
sulfuric acid and chilled until the end of each run. All TOC analysis was
performed within two weeks of obtaining the samples. (Analytical methods, and
the equipment used, are described in the following section.) Fluid static heads
were measured every 20 to 24 hours.
        Each filtration experiment was terminated when the level of fluid in the
reservoir reached the overflow pipes (167 cm above the sand surface).

Analytical Methods
        Measurements of the static head of the water inside the columns were made
using the sight tubes described in the physical configuration section of this
paper. A scale with 1 mm divisions was used to estimate head to the nearest 0.5
mm. Hydraulic conductivities were calculated using Darcy's law (11):

----

    * Aldrich Chemical Co., P.O. Box 355, Milwaukee, Wisconsin 53201, USA

Table I.
Inorganic Chemistry of Feed Water
for Slow Sand Filters

Component	mg l^{-1}
NH_4HCO_3	10.3
K_2HPO_4	0.185
$NaHCO_3$	11.0
$KHCO_3$	5.13
$MgSO_4 \cdot 7H_2O$	12.0
$FeCl_3 \cdot 6H_2O$	0.482
$CaCl_2$	5.55

Table II.
Coliform Reductions by Slow Sand Filtration
of a 1-ppm Carbon Source Water at 25C

	Column 1	Column 2
Mean Reduction	99.9%	99.7%
Standard Deviation	0.099%	0.553%
Number of Samples	72	72

$$Q/A = -K(dh/dl)$$

or $$K = -Q/A(dl/dh)$$

where   K = hydraulic conductivity (cm min^{-1})
                Q = volumetric flow rate (cm^3 min^{-1})
                A = cross sectional area (cm^2)
                dh = head loss (cm)
                dl = depth of sand between points of head loss
                     measurement (cm)
          dh/dl = hydraulic gradient (cm head/cm sand)

All TOC samples were analyzed on a Beckman Model 915-B Total Organic
Carbon Analyzer**. Values of influent and effluent TOC, as well as percent
change in TOC, were plotted as a function of filter run times.
        Batch cultures of E. coli were grown for use as an indicator organism.
Sterile flasks containing 670 ml of lactose (4 g l^{-1}) and Nutrient Broth***
(1 g l^{-1}) were inoculated with a pure culture of E. coli bacteria. The flasks,
which contained magnetic stir bars, were incubated at a temperature of 35C for
24 hours. After incubation two densely-grown flasks of bacteria were poured
into a 20-1 carboy. The carboy was then filled to capacity with inorganic feed
water and placed on a magnetic stirrer to be agitated continuously. The resultant bacteria feed was pumped to the filters at a rate of 7 ml per minute per
column. This coliform feed system resulted in a continuous supply of approximately 10^4 bacteria per ml (10^6 per 100 ml) to each filter.
        On a daily basis, influent and effluent samples were collected in sterile
flasks, diluted 1:1000 with sterile buffered dilution water (12), and analyzed
for total coliforms according to the Membrane Filter method (12). Four diluted
aliquots - two of each of two sizes - were analyzed for each sample obtained.
        Turbidity was measured on a Hach 2100A Turbidimeter****, which gave values
in nephelometric turbidity units (NTU). An Orion 701A Digital Ionalyzer and
Orion***** combination pH probe were used to measure the pH.

RESULTS AND DISCUSSION

Head Loss and Hydraulic Conductivity
        Hydraulic conductivities through the near-surface and bulk zones of the
sand bed as a function of filter run time are given in Figure 2 for a labile
carbon feed water, and in Figure 3 for a humic feed water. These two figures
are representative of the hydraulic behavior observed in all of the slow sand
filtration experiments: the hydraulic conductivity decreases rapidly near the
sand bed surface, and does not change through the sand bed below a depth of a
few centimeters. That significant head loss is limited to the near surface is
depicted even more clearly in Figure 4, in which the hydraulic conductivities
of the individual zones are plotted as a function of filter run time. This
figure shows that there is little variation in the hydraulic conductivity
through each interior zone of the sand bed (5 to 10 cm, 10 to 20 cm, 20 to 40

** Beckman Industrial Division of Rosemont Analytical, 7500 West Mississippi,
   Suite D-1, Lakewood, Colorado 80226, USA

*** Difco Laboratories, Inc., Detroit, Michigan 48232, USA

****Hach Company, P.O. Box 389, Loveland, Colorado 80539, USA

*****Orion Research Inc., 529 Main Street, Boston, Massachusetts 02129, USA

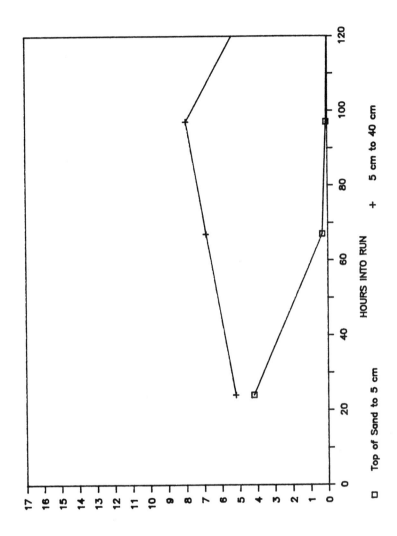

Figure 2. Hydraulic conductivity as a function of run time, through the
near-surface and interior zones (6-ppm glucose carbon feed).

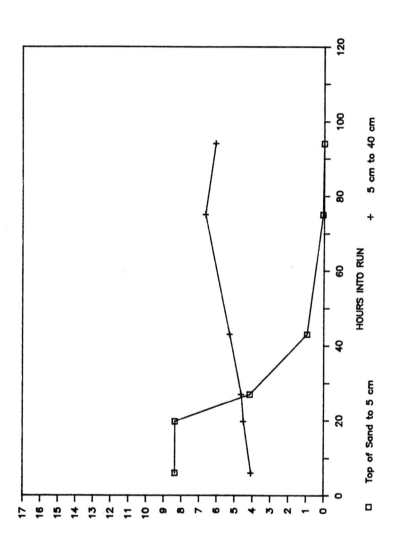

Figure 3. Hydraulic conductivity as a function of run time, through the near-surface and interior zones (6-ppm carbon - half humic, half glucose - feed).

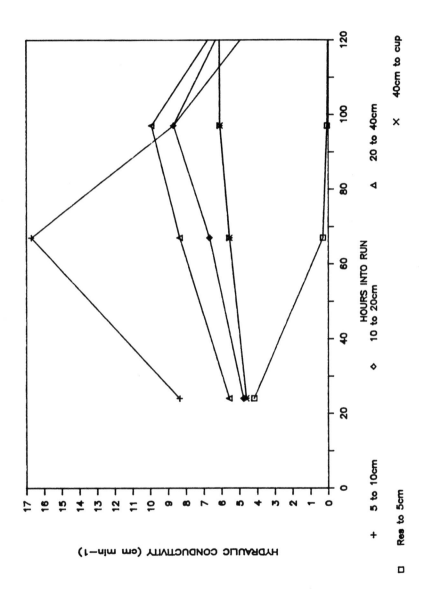

Figure 4. Hydraulic conductivity as a function of run time, through the
five zones of the sand bed (6-ppm glucose carbon feed).

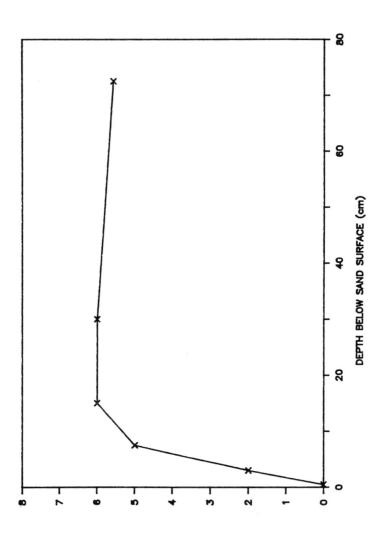

Figure 5a. Hydraulic conductivity as a function of filter media depth, at terminal head loss (6-ppm glucose carbon feed).

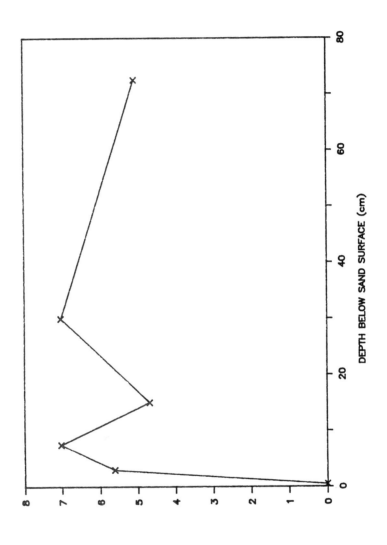

Figure 5b. Hydraulic conductivity as a function of filter media depth, at
terminal head loss (6-ppm carbon - half humic, half glucose - feed).

cm, and 40 cm to the effluent spill cup) except for an anomalous point at 67 hours for the 5-to-10 cm region. Calculated hydraulic conductivities in the interior zones of the filter media ranged from 4.5 to 10 cm min^{-1}.

The confinement of significant head loss to shallow regions of the filter media is important because head loss is a consequence of biofilm development. Hence, hydraulic data has been used to determine the depths to which biofilms have developed in the slow sand filters operated under simulated tropical conditions. From head loss data it was concluded that, regardless of the nature of dissolved organic component of the source water, biofilm development in the filters was limited to the sand bed surface. The observed localization of biological growth is evident from Figures 5a and 5b, in which hydraulic conductivities are plotted as a function of depth below the sand surface for a glucose and a humic feed, respectively.

The discovery in the present work of microbial growth being limited to the sand bed surface is contrary to the findings of Lloyd (13), who characterized populations in a mature slow sand filter to a depth of 20 cm, and Huisman and Wood (14), whose details of slow sand filter microbiology imply that microbial activity is high to depths of 40 cm.

There are several possible explanations for the unusual findings with respect to depth of biofilm development. They are: higher water temperature than used in previous studies, higher concentration of dissolved organic carbon, and low effective size and uniformity coefficient of the sand. Some combination of all three factors may be responsible for the lack of biological activity below approximately 10 cm into the filter medium. Perhaps the most crucial difference, however, between the slow sand filters in this study and those cited previously is the maturity of the filters. Throughout this study, the filter runs were short, and the filter beds were disinfected with a strong oxidant between experiments. Therefore, there was no accumulation of detritus, which reportedly creates a hospitable environment for a diverse collection of organisms (13).

Another consequence of frequent disinfection of the filters is the inability of predatory organisms to colonize the filter beds. Hence, removal of bacteria in the freshly disinfected filters of this study is believed to be due entirely to adhesion to the biofilm. (Bacterial removals obtained in this study are detailed later in this section.)

No hydraulic data was acquired for the 15- and 50-ppm carbon runs. Those experiments terminated in 4 days and 3 days, respectively. Experiments involving the filtration of 1-ppm carbon feed were terminated after three weeks, before they reached terminal head loss. No hydraulic data was obtained because the sight tubes were installed subsequently.

Total Organic Carbon

Removals of total organic carbon by the slow sand filters are shown in Figures 6a and 6b for glucose and humic feed waters, respectively. The changes in TOC, in percent, for the same two feed waters are presented in Figures 7a and 7b. The graphs indicate that there is a lag time before maximum TOC removal is reached. The lag time is approximately two days when the organic nutrient is labile, and two to three days when the feed water contains humic compounds. Maximum TOC removal is consistently around 80% when the influent TOC concentration is in the range of 4.5 to 7.5 ppm of readily utilizable carbon. By contrast, the maximum TOC removal has been found to be approximately 57% when the influent TOC concentration of 4.5 to 7.5 ppm contains a significant fraction of humic compounds.

Maximum percentage TOC removal was found to be sensitive not only to the nature of organic nutrient in the feed water (labile versus refractory), but also to the influent concentration of organic material. The filtration of source water containing 15 ppm glucose carbon resulted in a TOC reduction of 60 to 65%. However, when the influent contained 50 ppm glucose carbon, the TOC

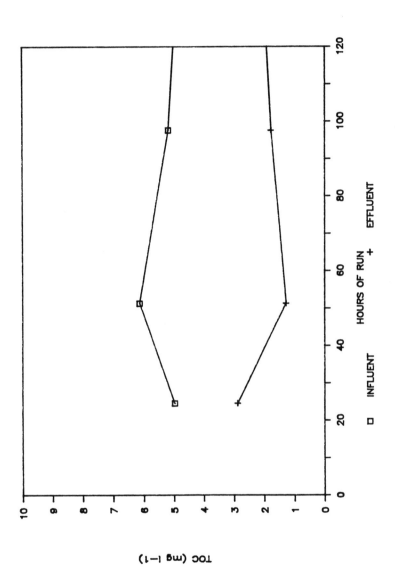

Figure 6a. TOC removal as a function of run time, for a 6-ppm glucose carbon
           feed.

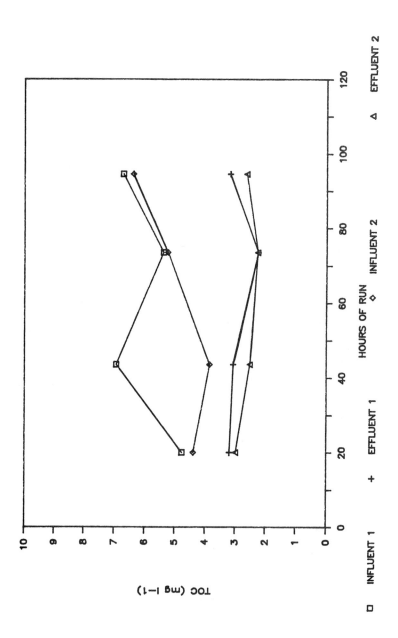

Figure 6b. TOC removal as a function of run time, for a 6-ppm carbon (half humic, half glucose) feed.

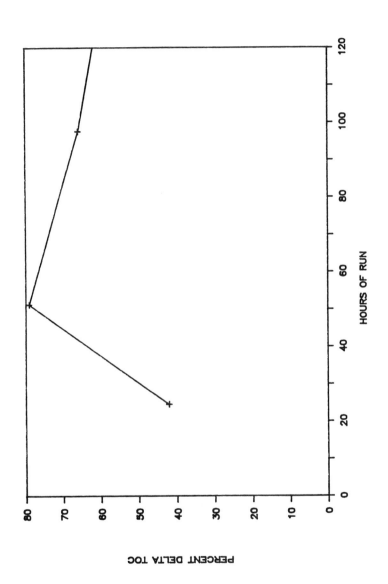

Figure 7a. Percent change in TOC as a function of run time, for a 6-ppm
          glucose carbon feed.

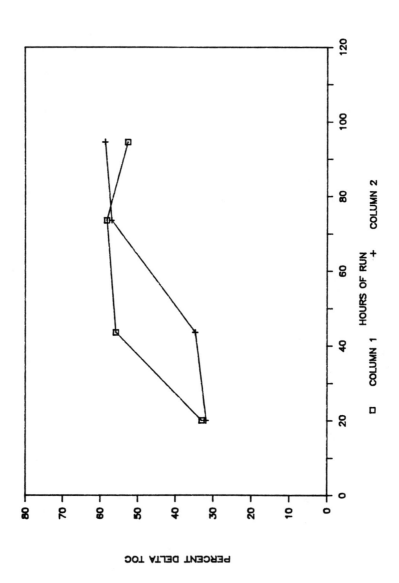

Figure 7b. Percent change in TOC as a function of run time, for a 6-ppm
carbon (half humic, half glucose) feed.

removals ranged from 12 to 40%. No TOC data was obtained for the experiments using 1 ppm-carbon influents because this concentration approaches the detection limits of the TOC analyzer.

Coliform Removals

The pilot-scale slow sand filters were more effective at removing coliforms from a 1-ppm carbon influent than from source waters containing higher concentrations of organic materials. Coliform reductions for the 1-ppm carbon feed, presented in Table II, were 99.9% for column 1 and 99.7% for column 2.

For experiments involving filtration of feed water with 15-ppm glucose carbon, coliform data indicate that the immature sand bed may act as a breeding site for the bacteria. In those experiments, filters were seeded with inocula from earlier schmutzdeckes. It is possible that the lack of an adequate predator population allowed the bacteria to reproduce according to the availability of nutrients.

The role of predation in the removal of indicator organisms was investigated by running experiments in which the schmutzdeckes were not seeded with an inoculum. The filters were fed a source water containing 8-ppm glucose carbon. By the fourth day of operation, coliform reductions of 88.9% and 96.8% were observed for the two filter columns. These results suggest that the absence of predator organisms does not affect coliform removals once the filters have ripened. Although seeding the schmutzdecke with pond water may not increase bacteria removal in the ripe filters, an inoculum may shorten the ripening time of the filter media. The observed reductions in bacteria concentrations appear to be due to adhesion, rather than predation. Additional conclusions that can be drawn from the experiments involving filtration of feed waters containing 8- and 15-ppm glucose carbon is that somewhere in that concentration range, slow sand filtration becomes inadequate for pathogen removal. That is, bacterial removal by adhesion cannot compensate for the re-growth of organisms in the presence of nearly 15 ppm labile carbon, which would result only from a severe pollution event.

When humic materials comprised approximately 50% of the 6-ppm carbon in the feed water, coliform reductions increased from 49.5% after 45 hours of filter run, to 80.7% at terminal head loss (96 hours). An analogous filter run involving a 6-ppm glucose carbon feed water produced an 87.0% coliform reduction after 51 hours of operation. Hence, high coliform removal efficiencies were achieved sooner for source waters containing only labile organic compounds.

To summarize, though coliform reductions were greater than 80% for all filter runs of source waters containing up to 8 ppm carbon, in no cases were effluent bacteria concentrations low enough to meet drinking water standards. However, an influent coliform concentration of 10^4 per ml (10^6 per 100 ml) is not expected to be common; such a high concentration of bacteria is indicative of a severe pollution event.

It is important to note that coliform removals in this study were obtained by the filtration of raw waters with low turbidities. Surface waters in tropical climates are likely to be more turbid, particularly during the rainy season. Since bacteria and other microorganisms tend to adhere to particles, it is expected that field operation of slow sand filters will demonstrate better coliform-removal efficiencies than in these experiments.

Turbidity, pH, and Color

In nearly all runs involving labile carbon, turbidities were decreased significantly by slow sand filtration. Typically, influents with a turbidity of approximately 1.2 NTU would be filtered to a finished-water turbidity of 0.20 to 0.30 NTU. The exception was a run of labile feed which followed a humic run. In that particular experiment, turbidities decreased, increased, then decreased again. In earlier experiments (15), silt and clay suspensions were added to the simulated tropical source waters which were fed to the slow sand filters. Raw-water turbidities of 11.0 NTU were clarified to a turbidity of 0.26 NTU.

Filters fed a source water containing humic compounds produced increasingly turbid effluents. While influent turbidities ranged from 1.80 to 2.10 NTU, effluents ranged from 2.30 NTU at the beginning of the run, to 9.20 NTU at terminal head loss. The observed increases in turbidities when slow sand filters were fed humic-laden water is not fully understood. The fact that increasing turbidities were accompanied by a steady reduction in TOC in the presence of humic compounds suggests that the turbidity may not have been caused by organic compounds. It is possible that inorganic components, such as salts of calcium and magnesium, are responsible for the increase in turbidity, although the high pH usually associated with calcium and magnesium precipitates did not occur. As the inorganic, organic, and bacterial feeds are combined just above the sand bed surfaces, contact time in the reservoir above the sand bed is minimal. However, the filter bed hydraulic residence time of approximately 5 hours is apparently sufficient for the flocculation, precipitation, or sloughing observed in the presence of humic materials. This would explain why the effluents were much more turbid than the influents.

The extent of anoxia in the depths of the filter bed has not been established. A reducing environment would lead to the formation of precipitates, such as iron sulfide, which would also increase the turbidity.

The efficiency of the pilot-scale slow sand filters at removing humic-derived color was found to decrease with length of run time. After one day, filters fed brown feed water (3-ppm humic carbon) produced a colorless effluent. However, effluents were visibly colored from the second day until terminal head loss was reached.

The filtration of humic compounds produced another peculiarity — the apparent adsorption of organic compounds from the filter media. Subsequent desorption of humics is evidenced by the negative values of percent change in TOC from the filtration run immediately following a humic run. Figure 8 indicates that, initially in the filtration run of a glucose solution, the effluent TOC concentrations were greater than the influent TOC concentrations. It is possible that desorption continued beyond the time when the percent change in TOC became positive (approximately 24 hours), but that the concentration of desorbed TOC was less than the TOC utilization by the filter's microorganisms (uptake in the filter). The desorption of humic compounds may be responsible for the fluctuations in effluent turbidities observed in this same run.

The apparent desorption of humic compounds during filter runs demonstrates that the method of disinfection and rinsing typically used between filtration experiments is inadequate for cleaning a filter bed which has been exposed to humic solutions. For laboratory-scale research, it is recommended that contact time with a disinfectant should be longer than 24 hours and/or the sodium-hypochlorite concentration should be greater.

In field operation, desorption of humic materials is not expected to be problematic. The minimal impact of humic compounds on filter parameters such as TOC removal, biofilm development (as evidenced by hydraulic conductivity), and coliform removal has been detailed earlier in this section.

The 24-hour flush prior to the start-up of a new experiment may also be inadequate for removing fine sand from the new replacement sand. In the pilot-scale operation, approximately 1.5 cm of sand is removed each time the schmutzdecke is scraped. To allow consistent hydraulic measurements, clean sand is placed in the filter columns to restore the sand beds to their original height. In field operations, there is no need to replace the scraped sand, as long as the depth of the filter media is greater than 0.5 to 0.7 m.

The breakthrough of color in the filters fed humic solutions indicates that the biofilm and filter media quickly became saturated with the humic compound.

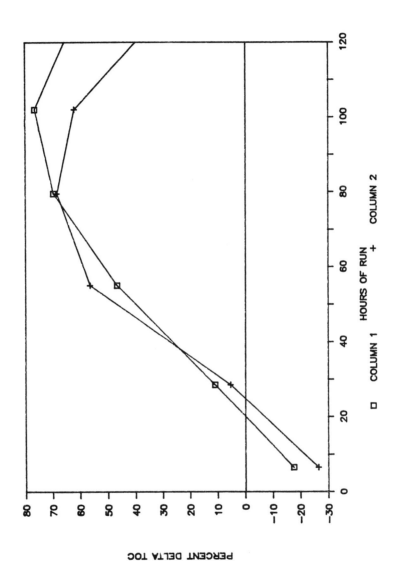

Figure 8. Percent change in TOC as a function of run time, for a 6-ppm
glucose carbon run which followed a humic carbon run.

No changes in pH were observed; the pH remained in the range of 7.8 to 7.9 for all filter runs.

In the pilot-scale slow sand filtration studies described herein, attempts were made to adhere to the specifications of Huisman and Wood (14) for the construction of slow sand filters. In particular, the sand used had a low uniformity coefficient and effective size. In field applications, it may be difficult to acquire filter media which meet these specifications. However, it is possible that the media specifications for optimum performance of slow sand filters in tropical climates differ from those prescribed for temperate-zone facilities. Visscher et al. (5) have suggested that locally available sand is often suitable, and have reported good performance in tropical developing countries of slow sand filters with sand uniformity coefficients of 2 to 5. Future pilot-scale investigations of slow sand filter performance under tropical conditions will examine the impact of changing filter media specifications.

CONCLUSIONS

Pilot-scale slow sand filters were operated at a temperature of 25C while being fed source waters characteristic of unpolluted and polluted water supplies. Under non-polluted conditions, that is, when source waters contained only 1 ppm carbon, coliform removals exceeded 99%, and run times were greater than three weeks. However, when influent carbon concentrations were greater than 1 ppm, filter run times were short, five days or less, due to a rapid increase in head loss. The head loss resulted from biofilm growth in the first few centimeters of the sand bed.

Given an influent bacterial concentration of approximately 10^4 cells per ml (10^6 per 100 ml), coliform removals were inadequate during the short run times. Removal of labile TOC reached a maximum of 80% in two days. When humic compounds comprised 50% of the influent TOC, maximum carbon removals averaged 57%.

The results of these investigations indicate that slow sand filtration facilities built to conventional specifications are not adequate when the source water to be treated is high in dissolved organic carbon and coliform bacteria. Future work will examine the impact of changing filter media specifications on the performance of slow sand filters which are fed warm source waters with high concentrations of TOC and bacteria.

ACKNOWLEDGEMENTS

The authors are grateful to the following persons for their contributions to this research: Dr. David Hendricks, Department of Civil Engineering, Colorado State University, for the loan of the filter columns; and William Hogrewe, Department of Civil and Environmental Engineering, University of Colorado, for technical support.

LITERATURE CITED

1. Rotival, A. H.. Status of the International Drinking Water Supply and
    Sanitation Decade. in Resource Mobilization for Drinking Water and Sanita-
    tion in Developing Nations. Montanari, F. W., et al. (ed.). American
    Society of Civil Engineers, New York, 1987.

2. Bellamy, W. D., Hendricks, D. W., and Logsdon, G. S.. Slow Sand Filtra-
    tion: Influences of Selected Process Variables. J.A.W.W.A. 77, 62-66,
    1985.

3. Poynter, S. F. B. and Slade, J. S.. The Removal of Viruses by Slow Sand
    Filtration. Prog. Wat. Tech. 9, 75-88, 1977.

4. Slezak, L. A. and Sims, R. C.. The Application and Effectiveness of Slow
    Sand Filtration in the United States. J.A.W.W.A. 76, 38-43, 1984.

5. Visscher, J. T., Paramasivam, R., and Santacruz, M.. IRC's Slow Sand
    Filtration Project. Waterlines 4, 24-27, 1986.

6. Kawata, K.. Slow Sand Filtration for Cercarial Control in North Cameroon
    Village Water Supply. Wat. Sci. Tech. 14, 491-498, 1982.

7. Seki, H.. Organic Materials in Aquatic Ecosystems. CRC Press, Inc., Boca
    Raton, Florida, 1982.

8. Cairncross, S. and Feachem, R. G.. Environmental Health Engineering in the
    Tropics. John Wiley & Sons, Chichester, 1983.

9. Nemeth, A., Paolini, J., and Herrera, R.. Carbon Transport in the Orinoco
    River: Preliminary Results. SCOPE/UNEP Sonderband 52, 357-364, 1982.

10. Livingstone, D. A.. Data of Geochemistry - Chapter G. Chemical Composition
    of Rivers and Lakes, United States Geological Survey Proffessional Paper
    440-G. United States Government Printing Office, Washington, 1963.

11. Freeze, R. A. and Cherry, J. A.. Groundwater. Prentice Hall, Englewood
    Cliffs, New Jersey, 1979.

12. Standard Methods for the Examination of Water and Wastewater. Sixteenth
    Edition, A.P.H.A., Washington, 1985.

13. Lloyd, B.. The Construction of a Sand Profile Sampler: Its Use in the
    Study of the Vorticella Populations and the General Interstitial
    Microfauna of Slow Sand Filters. Wat. Res. 7, 963-973, 1973.

14. Huisman, L. and Wood, W. E.. <u>Slow Sand Filtration</u>. World Health Organization, Geneva, 1974.

15. Barrett, J. M.. Slow Sand Filtration of Tropical Source Waters. in <u>Resource Mobilization for Drinking Water and Sanitation in Developing Nations</u>. Montanari, F. W., et al. (ed.). American Society of Civil Engineers, New York, 1987.

4.2 BENEFITS OF COVERED SLOW SAND FILTRATION

J. A. Schellart — Municipal Water Works of Amsterdam, Water Quality
Department, Leidsevaartweg 73,2106 NB Heemstede, The Netherlands

ABSTRACT

The Municipal Water Works of Amsterdam prepare drinking
water at two production plants with covered slow sand
filtration as the final purification step. In the past the
slow sand filters were not covered and in that time filter
runs were much shorter than today. In addition there were
much more operational problems and problems with the quality
of the finished drinking water than nowadays.

INTRODUCTION

The Amsterdam Water Works supply drinking water from two
production plants, the River (Rhine) -Dune Water Works
"Leiduin" at the west side of the city and the River-Lake
Water Works (Loenen-Weesperkarspel) at the south-east side.

The purification scheme of the River-Dune Water Works is
as follows: River Rhine-coagulation FeCl3/sedimentation -
rapid sand filtration - transport Nieuwegein/Heemstede
(60km) - dune infiltration (mean retention time 3 months) -
open winning -hardness reduction- aeration/activated carbon
powder/pH correction up to pH 8,2 - rapid sand filtration -
slow sand filtration - (safety chlorination until 1983) -
transport of drinking water (25 km) to Amsterdam-reservoirs-
distribution.

For the River-Lake Water Works the flow sheet of the
purification is: Bethunepolder/Amsterdam Rhine Canal -
coagulation FeCl3/sedimentation - Lake Loenderveen (mean
retention time 3 months) - rapid sand filtration - transport
Loenen/Weesp (18 km) - ozonation - hardness reduction -
coagulation FeCl3/sedimentation - pH correction up to pH 8.2
- rapid sand filtration - slow sand filtration - (safety
chlorination until 1983) - reservoirs - distribution.

As can be seen both plants apply slow sand filtration as
the final purification step. Originally the slow sand
filters were open (not covered). After bad experiences the
slow sand filters of the Leiduin plant have been covered in
the fifties.

In the south-east of Amsterdam ("Weesperkarspel") a new
plant was built in 1975 with roof covered slow sand filters.
Until 1983 a safety chlorination was operated after slow
sand filtration before distribution.
This consisted of a dose of 0.4-0.8 mg l^{-1} chlorine
(depending on temperature) which resulted in a level of 0.2
mg l^{-1} chlorine after 20 minutes contact time. The drinking
water was distributed without any free chlorine left in the
pipelines. Because of the good bacteriological quality of
the slow sand filtrate it was decided in 1985 - after a full
scale experiment - to stop safety chlorination under normal
conditions (1,2).

The present production (1988) of the Leiduin plant is about
$62.10^6 m^3$ $year^{-1}$; the Weesperkarspel plant produces $23.10^6 m^3$
$year^{-1}$.

TECHNOLOGICAL ASPECTS

Technical and operational data

The both plants "Leiduin" and "Weesperkarspel" have two
production streets each. The cross section of a slow sand
filter has been given in Fig. 1.

Fig. 1 Cross section of a slow sand filter (Leiduin)

water (~2 meter)		thickness of bed (m)	diameter particles (10^{-3}m)
dune sand[1]	}	1.35	< 0.4
river sand		0.1	0.4- 1.2
sand		0.08	1.2- 2.4
filter gravel		0.08	2.4- 4.8
filter gravel		0.08	4.8- 9.6
filter gravel		0.08	9.6-19.0
filter gravel		0.1-0.15	19.0-32.0
concrete supporting beams (0.06 m)			

1) Weesperkarspel: river sand (see Table 1)

The lay out of the supporting sand and gravel layers of the
slow sand filters at Weesperkarspel is about the same as
those at Leiduin.
Some technical and operational data of the slow sand filters
are summarized in Table 1.

Table 1   Some technical and operational data of the
           slow sand filters on the production plants
           "Leiduin" and "Weesperkarspel"

	River-Dune water (Leiduin)		River-Lake water (Weesperkarspel)	
	section 1	section 2	section 1	section 2
number of filters	12	10	6	6
surface per filter (m^2)	2000	1000	605	605
max. thickness of sand bed (m)	1.35	1.35	1.3	1.3
origin of sand	dune	dune	river	river
diameter of sand particles ($\emptyset$ m^{-3})	< 0.4	< 0.4	0.15-0.6	0.15-0.6
mean water level above sand bed (m)	2.0	2.0	2.0	2.0
operational filtration rate $(m.hour^{-1})$	0.1-0.2	0.2-0.3	0.3-0.4	0.3-0.4
cleaning frequency $(year^{-1})$	1-2	1-2	1-2	1-2

Cleaning procedures

The slow sand filters are cleaned mechanically by
scraping 0.02-0.03 meter of dirty sand from the surface. One
man can do the job within a day per filter. About once per
10 years new dune or river sand is brought on the filter
surface up to the original level of about 1.30 meter.

WATER QUALITY ASPECTS

According to the philosophy of the Municipal Water Works of
Amsterdam slow sand filtration is a hygienic and safety
barrier before distribution of drinking water.

Therefore extensive pretreatment of surface water is -
besides application of covered slow sand filtration -
absolutely required in order to produce drinking water of
high quality without final disinfection.

Organics

Table 2 gives an overview of turbidity and parameters for
organic mater at the two production plants "Leiduin" and
"Weesperkarspel" before and after slow sand filtration.

Table 2   Turbidity and concentration of organic matter[1])
           before and after slow sand filtration (SSF) at the
           production plants Leiduin and Weesperkarspel.

	River-Dune water ("Leiduin")		River-Lake water ("Weesperkarspel")	
	before SSF	after SSF	before SSF	after SSF
Turbidity (FTU)	0.27	0.15	0.19	0.12
Consumption of $KMnO_4$(mg litre[-1])	8	7	12	11
$DOC^{2)}$ (mg litre[-1])	2.1	1.9	4.7	4.3
$AOC^{3)}$ (ug acetate C eq.litre[-1])	∿ 8	∿ 6	∿16	∿ 12

1) Mean values
2) UV destruction analysis method; detection by auto analyser
3) Assimilable Organic Carbon; method according to v.d. Kooy (3)

As can be concluded from the table the turbidity decreases
50 per cent or more. Organic compounds ($KMnO_4$ and DOC),
however, decrease only to a small extent (about 10%).
Assimilable Organic Carbon (AOC) drops a third at optimal
conditions.
The AOC value gives a useful prediction of the
bacteriological regrowth potential. Low regrowth has been
observed in drinking waters with AOC values far below 10 ug
acetate C equivalents per litre (4). Since it is a problem
to maintain low AOC values in the River-Lake water plans
have been developed for additional granular activated carbon
(GAC) filtration before slow sand filtration at
Weesperkarspel (1,2) to remove organics (mainly humic
acids).

Because certain organic micropollutants (such as herbicides) can pass the treatment unchanged also plans are developed for the installation of GAC-filters at the Leiduin plant in the near future.

If final chlorination is applied the AOC increases again with 25-50 per cent because of oxidation reaction (1,2). The colour decreases in some degree (about 10 per cent) to values below 5 mg Pt.litre^{-1}).

Bacteria and viruses

In table 3 data are summarized on the bacteriological quality of slow sand filtrate at the two production plants.

Table 3   Bacteriological quality of the slow sand filtrate (SSF) compared to the Dutch law, and the E.E.C. criteria for drinking water and the reduction capacity for indicator organisms under optimal conditions (temperature > 4°C) of the filters.

Indicator organism	Numbers of bacteria in SSF	Reduction Capacity (log units)	Dutch law (1982)	E.E.C. Criteria (1978)
Coliforms/100 ml	< 0.1	2-3	neg[1]	neg
Thermotolerant Coliforms/100 ml	< 0.1	2-3	neg[1]	neg
Faecal streptococci	< 0.1	2-3	neg	neg
Spores of sulphite reducing Clostridia/ 100 ml	≤ 0.01	3	neg	< 5
Standard plate counts - at 37°C (CFU.ml^{-1}) - at 22°C (CFU.ml^{-1})	0-2 (5-10[2]) (5-30[3])	~ 1 ~ 1	< 10$^{4)}$ <100$^{4)}$	< 10 <100

1) At the waterworks: neg/300 ml
2) Median values Leiduin
3) Median values Weesperkarspel
4) Average for one year

At water temperatures below 4°C bacteriological reduction
capacity decreases dramatically. (reduction of 1 log unit
for coliforms or less). In very cold winters (1985, 1986 and
1987) higher faecal contamination is found in the raw water
of the River-Dune plant "Leiduin". It is caused by the many
birds in the open canals and collection pond in the dune
area after infiltration (comparatively warm water and no ice
during open winning). Under those conditions chlorination of
the slow sand filtrate is necessary during several weeks.
Contrary to the Weesperkarspel plant the Leiduin plant has
not yet ozonation as a disinfection step in its purification
system. (Ozonation and granular activated carbon filtration
are planned as additional purification steps before slow
sand filtration in the near future).
In recent years investigations have been done to the
presence of (opportunistic pathogenic) Aeromonas
bacteria.
The numbers of Aeromonas bakteria in CFU per 100 ml
range in the slow sand filtrate of River-Dune water from
below 1 in winter half year to 1-10 in summer half year. In
the slow sand filtrate of River-Lake water these numbers are
below 1 (winter) and 20-60 (summer).
Human enteric viruses not yet have been detected in volumes
of 500-1000 litre water after dune infiltration. However,
they are present in high numbers in river water and they
also can be detected in water samples of 1-10 litre before
dune infiltration (5).
From the Lake-River plant virus data are not yet available.
It may be expected, however, that the number of viruses in
the slow sand filtrate will be very low because of the
reduction capacity of the purification system (see flow
sheet in Introduction section).

Animal organisms

An impression of the median numbers of animal organisms in
the water before and after slow sand filtration has been
shown in Table 4.
It is self-evident that the numbers are generally higher in
summer and lower in winter time.

Table 4   Number of animal organisms per m^3 of water before
            and after slow sand filtration (SSF) at the
            production plants Leiduin and Weesperkarspel
            (filtration on 30 u netting; median values of 26
            observations in 1986).

	River-Dune water (Leiduin)		River-Lake water (Weesperkarspel)	
	before SSF	after SSF	before SSF	after SSF
Total animal organisms (numbers.m^{-3})	5800	140	730	170
Rotatoria	5300	30	510	55
Copepoda	5	20	65	20
Nauplii	270	20	90	20
Nematoda	5	25	<1	20
Ciliata	160	1	<1	<1
Testaceae	65	<1	15	<1
Cladocera	20	<1	<1	<1
Ostracoda	15	<1	<1	<1
Other organisms	<1	<1	<1	<1

It can be concluded from the table that Rotatoria take
generally between 70 and more than 90 per cent of total
organisms before slow sand filtration and that they are
considerably reduced - sometimes more than 99% - after slow
sand filtration. In the slow sand filtrate Rotatoria
take 25-35 per cent of total organisms. Their numbers rise
again in the distribution system, especially in summer time.
Only the numbers of Nematoda (and sometimes Copepoda)
are higher in the water after slow sand filtration.

Algae

Generally on both plants growth of algae does decrease rapid
sand filter runs in spring and summer time. The chlorophyl-A
concentration in the raw water before rapid sand filtration
ranged in 1986 from 2 to 13 ug.l^{-1} in the River-Lake water
and between 3 and 18 ug.l^{-1} in the River-Dune water. Growth
of algae in the upper water on the (roof covered) slow sand
filter beds never has been observed (no radiation of sun
light).
However, average once per 4 years in spring time slow sand
filter runs at Leiduin are significantly reduced (to one
month or less) after bloom of Diatomeae in the canals
and collection pond in the dune area after infiltration.

HISTORICAL EXPERIENCES

Before 1972 the Weesperkarspel plant had non-covered slow
sand filters. They were operated with low filtration rates
(about 0.2 m.hour^{-1}) and the cleaning frequency was high
(10-12 times a year per filter). Filter runs in summer time
were sometimes between 1 and 2 weeks only.
Coagulation (with $FeCl_3$) of water from the Bethune polder
after 1972 reduced the phosphate concentration in the
storage reservoir. (Lake Loenderveen) with about 80 per cent
which resulted in much lower biomass of algae. The cleaning
frequences of the slow sand filters decreased to 3-4 times a
year and filter runs in summer increased to at least 2
weeks. Chlorination of the slow sand filtrate always was
necessary because of the hygienically doubtful quality of
the finished drinking water caused by the presence of birds
on the filters. High fluctuating levels of chlorine (2 mg.l^{-1}
or more) were necessary in the past to maintain free
residual chlorine in the water after 20 minutes contact
time. In summer the oxygen concentration in the filtrate
varied between 0 mg.l^{-1} early in the morning and 12 mg.l^{-1}
late in the afternoon. Also the pH fluctuated and a temporal
increase of pH resulted in an increase of filter resistance
by biogenic softening. High colony counts and taste problems
of the distributed lake drinking water were normal
phenomenons during the sixties and early seventies. In 1976
a new plant was started up with roof covered slow sand
filters. These filters can be operated with higher
filtration rates (V= 0.3-0.4 m.hour^{-1}), whereas the cleaning
frequency decreased to 1-2 times a year. Filter runs in
summer range from 80 to 200 days. In 1983 it was decided to
stop routine chlorination of the prepared drinking water.
The history of slow sand filtration at the Lake-River water
plant "Weesperkarspel" is summarized in table 5.

Table 5   Some operational data in the history of slow sand filtration (SSF) at the plant Weesperkarspel

	Filtration rate (m.hour⁻¹)	Cleaning frequences per filter (year⁻¹)	Filter runs (weeks)	IN SUMMER:	
				Oxygen concentration in filtrate (mg.l⁻¹)	Final chlorination dose of chlorine (mg.l⁻¹)
Old plant (before 1976) non-covered SSF					
a) without coagulation of the raw water (before 1972)	0.2	10-12	1-2	0-12	> 2
b) with coagulation ($FeCl_3$) of the raw water (after 1972)	0.2	3-4	2	0-12	∼ 2
New plant (after 1976) with coagulation and roof covered SSF					
a) with final chlorination (before 1983)	0.3-0.4	1-2	> 12	∼ 10	0.4-0.8
b) without final chlorination (after 1983)	0.3-0.4	1-2	> 12	∼ 10	0

CONCLUSIONS

Summarized the following benefits of covered slow sand
filters can be pointed to:

a) No faecal contamination of birds and thus no introduction
of coliforms, pathogenic micro-organisms, fertilizing
nitrogen and phosphorus compounds.

b) No radiation by sunlight and thus no growth of algae
biomass, lower levels of organics (DOC and AOC) in the
filtrate, lower biological regrowth potential and less taste
problems of the finished drinking water.

c) Much lower cleaning frequency (less biomass on the filter
surface) and thus higher capacity all over the year. No
increase of pH by photosysnthesis on the filter surfaces and
thus no increase of resistance by biogenic softening.

d) No frozen filter surfaces during hard frost periods in
severe winters and thus higher capacity all over the year.

e) Higher filtration rates possible, lower filtration area
necessary and thus lower building expenses.

f) Rather constant and high oxygen concentration in the
filtrate and no differences in O_2 concentration between day
and night in summer half year.

g) Rather stable microbiological drinking water quality only
depending on water temperature.

h) Much lower or no necessity of final disinfection, lower
consumption of chlorine and thus lower levels of halogenated
organic compounds and lower expenses for disinfection.

REFERENCES

1) Schellart, J.A. (1987) The institution of water engineers
   and scientists. Summer Conference Torquay (G.B.), Paper
   no. 3, 19th-21st May.
2) Schellart, J.A. (1986) Wat. Supply, 4, (Mulhouse),
   217-225.
3) Kooy, D. van der, Visser, A. and Hijnen, W.A.M.. (1982)
   J. Am. Water Works Association, 74, 540-545.
4) Kooy, D. van der and hijnen, W.A.M. (1985) AWWA Water
   Quality & Technology Conference, Houston (USA), December.
5) Olphen, M. van, Kapsenberg, J.G., Baan, E. van de and
   Kroon, W.A. (1984) Appl. Environ. Microbiol., 47,
   927-932.

Fig. 2   Covered slow sand filter with
         cleaning equipment (Weesperkarspel)

Fig. 3 Cleaning equipment (Weesperkarspel)

4.3   COMPARISIONS BETWEEN ACTIVATED CARBON AND SLOW SAND FILTRATION IN THE TREATMENT OF SURFACE WATERS

J. Mallevialle and J. P. Duguet — Laboratoire Central de la Lyonnaise des Eaux, 38 rue du Président Wilson 78230 Le Pecq, France

ABSTRACT

The water purifying efficiency of a two–step system was investigated. The first process was a physicochemical process with coagulation, flocculation and sedimentation, followed by rapid sand filtration. The second process was slow sand filtration. Different types of carbon were evaluated and different filtration velocities in the second filtration stage were tested. The use of very slow filtration velocities and comparison with higher rates made it possible to study the mechanism of removal of organic matter. The relative importance of adsorption and biodegradation for the removal of organic matter using granular activated carbon also was assessed.

The plant at Mont Valérien produces potable water out of the Seine river water withdraw at Suresnes immediately downstream from the City of Paris (Figure 1). There the water is not appreciably more polluted than upstream from the city (Table I) because wastewaters from the area are collected and sent to a large secondary treatment plant at Acheres a few kilometers below Suresnes. Urban runoff, however, may contribute to the degradation of water quality when it flows through the French capital. Consumers' demand for a water of perfect quality and the scope of new

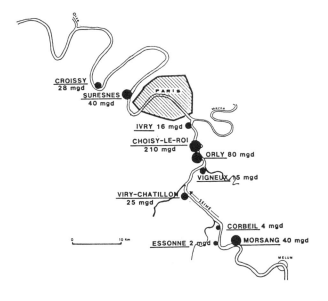

Figure 1. Potable water treatment plants in the Paris area.

Table I. Main Characteristics of the Seine River Water at the Suresnes
        Bridge over the Period of Experiment (Feb. 79 – Feb. 80)

Parameter	Average	Maximum	Minimum
Temperature (° C)	14.0	22.5	4.0
Dissolved oxygen (mg/L)	8.6	13.0	3.3
Turbidity (Ftu)	24.3	92	2.5
pH	7.75	8.20	7.21
NH_4 (mg/L)	0.52	1.5	0.2
N Total K (mg/L)	1.3	3.6	0.6
PO_4 (mg/L)	0.87	1.85	0.3
Cd (μg/L)	< 0.5	2	< 0.5
Pb (μg/L)	18.0	110	3
Hg (μg/L)	< 0.3	< 0.3	< 0.3
COD–$KMnO_4$ (mg/L O_2)	4.46	10.0	2.2
UV extinction (A^{1m}) 25.4 nm	10.42	25.5	1.94
Fluorescence (mV)	2.82	5.54	1.66
TOC (mg/L)	3.45	5.5	2.5
Total organic chloride (μg/L)	44.3	154	11

EEC water quality standards (1) prompted the need for the transformation
of the old Mont Valérien plant into an advanced one, capable of removing
trace pollutants effectively.

The plant at Mont Valérien consists of two different lines. The first
(Process A) consists of a slow sand filtration process with a capacity
of 25 MGD (100,000 m³/day). This line, built in 1904, includes coarse
gravel strainers and rougheners preceding slow (biological) sand filters
where the filtration velocity is about 5 m/day (0.2 m/h). The second
line (Process B) is a more recent physicochemical process with
coagulation, flocculation, and sedimentation in an upflow "pulsator"
clarifier, followed by rapid sand filtration at a velocity of 5 m/h.
This line, built in 1960, has a capacity of 15 MGD (50,000 m³/day).

These two processes have been operated in parallel but could, through
simple modification, be operated in series as well. The filtration
velocity could be raised appreciably so that the total flow rate would
be the same.

The objective of the study was to evaluate the efficiency of Process B
followed by Process A in a full-scale experiment. Slow sand filters
could be topped by a granular activated-carbon (GAC) layer and ozonation
could be introduced between the two processes, i.e., between rapid sand
filtration and slow GAC/sand filtration.

Different types of carbon were evaluated and different filtration velocities in the second filtration stage were tested. The GAC empty bed contact time (EBCT) was, however, kept the same in all filters.

The use of very slow filtration velocities and comparison with higher ones made it possible to study the mechanism of removal of organic matter. The second objective was to study the relative importance of adsorption and biodegradation for the removal of organic matter in GAC. This constitutes the core of the present paper.

EXPERIMENTAL

Flow Chart. Figure 2 shows the flow schematic for the study. The performances of the original two lines are measured at sampling points 0 and 6.

No prechlorination was applied during the experiment. The only chemical application consisted of coagulant and coagulant aide – average dosage of 40 ppm $Al(SO_4)_3$. 18 H_2O and 0.5 ppm of sodium alginate. The "pulsator" clarifier was operated at a low upflow velocity of 2 m/h. The first rapid sand filter 6, which fed all others in the experiment, was operated at about 5 m/h with air scouring and backwashing approximately every 30 h.

For this study, the following unit processes were tested :

1. Slow sand filtration at 0.625 m/h (15 m/day) in 50-m² filters.
2. Slow GAC filtration at 0.625 m/h. Norit PKST 1/4 – 1 GAC, 15 cm, was put on top of 65 cm of sand in 50-m² filters. Norit PKST is a
    nonreactivable peatbased broken carbon, of low cost with a total surface area of 800 m²/g (manufacturer's data).
3. Ozonation and slow sand filtration with same operating parameters as 1. Ozonation was carried out in a two-chamber contactor with a total contact time of 10 min. The average ozone dosage applied was 1.4 ppm with the residual after the second chamber being kept constant at 0.25 ppm.
4. Ozonation and slow PKST filtration. Operating parameters for the filtration and ozonation steps were the same as in 2 and 3.
5. Ozonation and rapid GAC filtration. Two different carbons were tested in 25-cm diameter pilot columns, Chemviron F-400 and Norit PKST. Chemviron is the European branch of Calgon corp. and their F-400 carbon is widely used in water treatment. It is a high quality coal-based broken carbon which can be reactivated. Its total surface area is higher than 1200 m²/g. The filtration velocity was 8.3 m/h, and the bed depth was 2 m, yielding a EBCT of 14,4 min. in both rapid and slow filters for each carbon.

Filter Operating Conditions and Media Characteristics. Table II summarizes the operating conditions for the filters. Control of the flow rates ensured constant operating conditions for all filters. The EBCT was equal to 14.4 min. in all GAC layers. Total EBCT 5gac = sand layers) was considerably longer in the slow filters than in the rapid ones. Previous studies incidated that the pilot sized rapid filters would correctly reproduce the performances of larger ones. Therefore, it is assumed a comparison between these small rapid filters and the large industrial (50 m²) slow ones is valid.

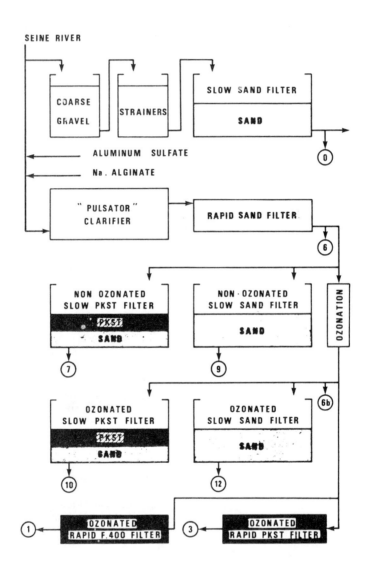

Figure 2. Flow schematic for the Mont Valérien experiment.

The small rapid filters were backwashed approximately every month for about 1 h (bed expansion of about 30 %) ; the large slow filters did not have to be backwashed during the year of experiment. Manipulation of the thin (15 cm) GAC layer could thus be avoided.

Table III represents the filter media characteristics. Rather similar particle sizes were sought for the different media, and the choice of the two carbons was primarily dictated by this consideration.

The particle size distribution curves were established and from them the average sizes were drawn.

The average external surfaces per volume of bed could thus be calculated. This characteristic was important for the interpretation of bioactivity measurements in the filter beds.

Water Quality Parameters. Most water quality parameters from the EEC potable water standards (1) were measured weekly at different stages of the processes. Table I which represents the main characteristics of Seine river water, includes only a part of those parameters. Lumped organic parameters and more particularly TOC are discussed in detail.

Table II. Filter Operating Conditions

Filter	Sampling Point Number	Filtration Velocity (m/h)	Filter Surface (m²)	Bed Height and Medium	EBCT (min.)
Rapid Sand	6	5	50	0.8 m sans	9.6
Slow Sand	9 (no ozone) 12 (ozone)	0.625	50	0.8 m sand	76.8
Slow PKST	7 (no ozone) 10 (ozone)	0.625	50	0.15 m PKST + 0.65 m sand	14.4 62.4
Rapid F-400	1	8.3	0.05	2 m F-400	14.4
Rapid PKST	3	8.3	0.05	2 m PKST	14.4

TOC measurements were carried out on a tocsin 2 apparatus. Organic carbon is converted to CO_2 and eventually to methane which is measured in a flame-ionization detector. The detection limit of the method is as low as 0.1 mg/L. No prefiltration of the samples was made.

Chemical oxygen demand (COD) was determined with $KMnO_4$ in hot acidic medium. The result is expressed in mg/L oxygen.

UV absorbance was measured by light extinction at 254 nm using a 10-cm cell and expressed for a 1-m cell. To be able to calculate and report cumulative removal profiles, the UV absorbance values were converted into µg/L of fluvic acid (Figure 3).

UV fluorescence also was measured, with an excitation wavelength of 320 nm and an emission wavelength of 405 nm. The fluorescence was converted into μg/L of salicylic acid (Figure 3). Expressed as such, these two parameters could be included in the material balance.

Biological examinations were conducted on samples from various depths of the filter beds including total viable cell (plate) counts, nitrifying bacteria counts, ATP measurements, and scanning electron microscopic observations. ATP measurements were made with the bioluminescence method ; 100 mg of GAC was treated with dimethyl sulfoxide, then MOPS was added to the solution, and the sample was quickly frozen. Readings were made with a Dupont 760 biometer.

The frequency of measurements for most parameters was once a week and the duration of the experiment was approximately 1 year from February 1979 to January 1980.

RESULTS AND DISCUSSION

Inorganic Parameters. Most inorganic pollutants and in particular heavy metals present at low concentrations were effectively removed during the first stage clarification process. The only noteworthy parameter is ammonia. It was effectively removed most of the time through biological oxidation in the sludge blanket of the clarifier and in the rapid sand filter, except for a short period in january 1980. Table IV indicates the results (eight measurements) for the different unit processes. When the level of ammonia in the raw water increases and the temperature decreases, GAC filters, particularly slow ones without ozonation, seem to be more efficient.

Organic Lumped Parameters. Breakthrough Curves. Figures 4 and 5 exhibit the typical patterns of breakthrough curves in terms of lumped parameters for such filters. Initially, the removal is rather high but, after a few months of operation, effluent concentrations increased rapidly and, thereafter, a relatively steady removal, between 10 and 20 % of the feed (filter 6), continued to occur throughout the experiment. The comparison of these two figures indicates the rapid filters are slightly more efficient than the slow ones. It is difficult, however, to make clear conclusions from such figures ; therefore, the results were averaged over the period of operation.

Average Removals. Table V summarizes the average percentages of removal for the 1-year experiment. These are averages of 45-50 data points and represent the reduction across each unit process, with the exception of the first two, which represent the reduction across the overall process up to the considered sampling point. The first two lines represent the reduction observed on the original processes of the plant. In the table, rapid sand filter 6 is the feed for nonozonated filters and rapid sand filter + ozone 6b is the feed for ozonated slow and rapid filters.

TOC. the initial large reduction is probably due to the removal of particulate organics. Table V indicates that the efficiency, for TOC reduction, fo the conventional slow sand filtration process is very similar to the physicochemical one.

Table III. Filter Media Characteristics

Filter	Effective Size (10%) (cm)	Uniformity Coeff. (60%/10%)	Average Diameter (50%)(cm)	Bulk Density (g/cm³)	Particle Density (g/cm³)	Average External Surface per Volume of Bed (cm²cm³)
Sand (slow and rapid)	0.048	2.60	0.109	1.14	2.54	58.84
Norit PKST 1/4-1	0.042	1.40	0.055	0.25	0.54	111.11
Chemviron F-400	0.066	1.52	0.093	0.41	0.75	63.83

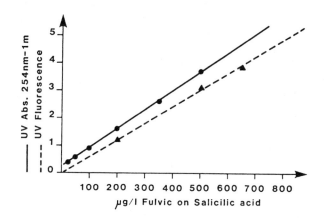

Figure 3. Plots of UV absorbance and fluorescence in terms of μg/L of fulvic and salicylic acid concentrations. Key : . fulvic acid and * salicylic acid.

Table IV. Average Ammonia Concentration in January 1980 (Eight
          Measurements).

Effluent from NH_4 (mg/L)

Rapid sand filter   0.31
Slow sand filter   0.28
Slow PKST filter   0.10
Ozonated slow PKST filter   0.11
Ozonated rapid PKST filter   0.30

Note : Mean water temperature was 5° C.

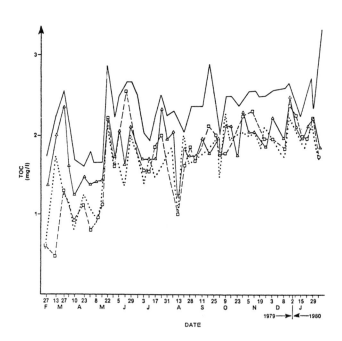

Figure 4. Breakthrough curves at Mont Valérien for slow filters.
          Key : -- rapid sand filter 6
                ▢ nonozonated slow PKST filter 7
                ...., ozonated slow PKST filter 10
                △ ozonated slow sand filter 12

TABLE V. Average Percent Organic Reduction across Unit Processes

Filter	Sampling Points	TOC	COD by KMnO$_4$	Im A 254nm	Fluoresc.
First stage					
Conventional slow sand	0	33	–	–	–
Rapid sand	6	32	70	63	35
Slow sand	9	10	17	10	9
Slow PKST	7	25	45	46	49
Ozone	6$_b$	5	25	41	54
Second stage					
Ozonated slow sand	12	12	12	0	12
Ozonated slow PKST	10	24	36	38	48
Ozonated rapid PKST	3	34	46	52	60
Ozonated rapid F-400	1	39	58	56	67

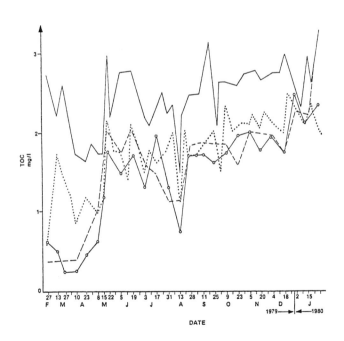

Figure 5. Breakthrough curves at Mont Valérien for ozonated GAC filters.
            Key : — rapid sand filter 6
                  .... ozonated slow PKST filter 10
                0 ozonated rapid F-400 filter 1
                --- ozonated rapid PKST filter 3

During the second stage of treatment, an additional 10 % of TOC removal is brought about by slow sand filtration 9. GAC filtration 7 removes 25 % which is 15 % more than the comparable sand filter 9.

Ozone itself, at the relatively low doses used, removes only 5 % of the TOC. The ozonated sand filter appears to remove a little more than the nonozonated one. The ozonated GAC slow filter seems to remove less organic material than the nonozonated one, but the difference is very small.

Rapid GAC filters remove appreciably more organics than the slow ones. For the same Norit PKST carbon, the additional TOC reduction is about 10 % and, with the best F-400 carbon, the average TOC reduction for the 1-year experiment is almost 40 % which is considerable given the 14.4-min EBCT. As an example of the improved water quality, the average TOC concentration in the effluent of rapid sand filter 6 was 2.33 mg/L (standard deviation 0.59) whereas the average TOC concentration in the water produced by the best filter 1 was 1.34 mg/L (standard deviation 0.42).

COD, Absorbance and Fluorescence. The other lumped parameters showed somewhat different percentages, particularly with ozonated waters. It has been shown (2) that ozonation decreases the size of the organic particles/molecules and therefore influences spectrophotometric measurements like UV absorbance or fluorescence. All data, however, are consistent with TOC values. The comparison made on the basis of TOC is confirmed using these additional parameters.

During the second stage, the slow sand filter 9 removed a limited amount of organic matter (through biodegradation ?). Ozonation does not appear to enhance this removal.

GAC is very efficient, particularly, if used in contactors with a rapid filtration velocity. Ozonation, on the other hand, seems to have a detrimental influence on overall GAC performance. However, the combination of the two processes, ozonation + GAC filtration, has a net positive effect over GAC filtration alone. Here again, these results are in agreement with previous observations (3).

The higher efficiency of the rapid GAC filters also is in agreement with accepted adsorption theories (4-6) : higher liquid film velocities outside the carbon leads to higher liquid film transfer coefficients. In the long term, these slow and rapid GAC filters with the same EBCT will remove the same amount by adsorption but the kinetics are apparently faster in the case of the rapid filters.

These differences in favor of rapid filters also suggest that adsorption plays a major role in the removal of organics by GAC even for an extended time period.

Cumulative Removals. The cumulative organic removals in a filter were plotted as a function of the cumulative organics applied. These curves are easier to interpret than the breakthrough curves and shed light on the respective influence of adsorption and biodegradation.

TOC. The first plot in Figure 6 represents the cumulative removal for the two slow sand filters (nonozonated and ozonated). The slope of these curves is almost constant throughout the operating period indicating that a small fraction of the TOC was removed by these slow sand filters right from the start of operation. Toward the end of the period of operation, a slightly greater removal of TOC occurred in the pre-ozonated slow sand filter 12, possibly indicating that the nature of the water changed and ozone had a more significant effect on organic removal.

The effluent from these slow sand filters was biologically stable, i.e., biodegradation was not observed on storage. The operation of these filters was not limited kinetically and therefore the same amount of TOC should be removed by bacteria in both sand and carbon beds. It was thus assumed that the fraction removed by the slow sand filters represents the total amount of biodegradable organics in each type of feed. All other cumulative plots were therefore prepared by subtracting from each data point the corresponding sand filter effluent concentration, e.g. the removal in the pre-ozonated slow sand filter 12 was subtracted from the TOC applied and removed in the case of the pre-ozonated slow GAC filter 10. The resulting curves were assumed to represent removal by adsorption only.

For the rapid GAC filter, the TOC removed was corrected for the difference in contact time (multiplied by 14.4/76.8) before subtraction from the TOC removed in the pre-ozonated rapid GAC filters 1 and 3. The very rough assumption was thus made that biodegradation was proportional to the overall EBCT.

Results are presented in Figures 7 and 8. On the whole, the curves appear to be clearly adsorption based. Their slope is constantly diminishing, particularly in the case of the rapid filters where more rapid exhaustion due to greater adsorption is expected. As shown in Figure 8, the filters were fully exhausted – slope near zero – before the end of the experiment. This finding also indicates that the kinetics of biodegradation are not significantly faster on carbon than on sand. GAC does not enhance biodegradation.

The fact that the slopes for the slow filters 7 and 10 (Figure 7) are not zero at the end may be explained by the fact that they had not reached their total adsorptive capacity (because of slower kinetics) at the end of the experiment. Eventually, they too would be expected to show a slope of zero.

COD, Fluorescence and UV Absorbance. Similar calculations were made for the other lumped parameters. These results are presented in Figures 9–11. The same general conclusions can be drawn and are best illustrated by Figure 9 for COD using $KMnO_4$. the case of UV absorbance is more difficult to interpret since the cumulative curves have an appreciable slope at the end of the experiment (Figure 11). One hypothesis is that GAC, because of its rough surface, retains more particulate organic matter than sand (as measured by UV).

Biological Examinations. Total Cells Counts. Cell counts are summarized in Table VI ; they constitute geometric means for eight measurements and indicate similar values in all filters. Counts seem to be slightly higher in ozonated GAC filters and more particularly in the rapid ones.

Bacterial counts were compared with scanning electron microscope observations ; numerous discrete bacteria were seen at the surface of the GAC particles, particularly in the vicinity of holes and crevices. Breakage of particles confirmed that bacteria were restricted to the outer surface of the particles.

ATP Measurements. ATP measurements of the samples collected at the surface of the filters were taken in femtograms (10^{15} g) per gram of filtration medium.

Table VI. Mean Geometric Total Viable Cell Counts (20° C) within the Filter Beds (cells per cm³ of media)

Filter	Sampling Point	Filter Bed Depths					
		Surface	15 cm	15 cm	20 cm	50 cm	75 cm
Slow sand	9	sand 2×10^6	sand –	sand –	sand 2.8×10^5	sand –	sand –
Slow PKST	7	PKST 3.3×10^6	PKST 1.4×10^6	sand $8. \times 10^5$	sand 2.8×10^5	sand 4.2×10^4	sand –
Ozonated slow sand	12	sand 2.5×10^6	sand –	sand –	sand 7.2×10^5	sand –	sand –
Ozonated slow PKST	10	PKST 9.3×10^6	PKST 5.2×10^6	sand 3.7×10^5	sand 3.6×10^5	sand 1.4×10^5	sand –
Ozonated rapid PKST	3	PKST 9.1×10^6	PKST –	PKST –	PKST –	PKST –	PKST 3.1×10^6

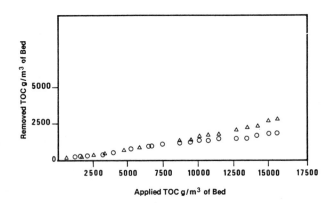

Figure 6. Cumulative plots for slow sand filters.
Key : O, slow sand filter 9
Δ, ozonated slow sand filter 12.

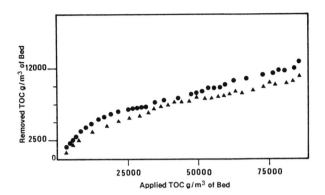

Figure 7. Cumulative plots for slow PKST filters.
Key : ● , slow PKST filter 7
       ▲ , ozonated slow PKST filter 10

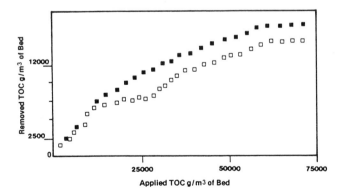

Figure 8. Cumulative plots for rapid GAC filters.
Key : ■ , ozonated rapid F-400 filter 1
       ▢ , ozonated rapid PKST filter 3

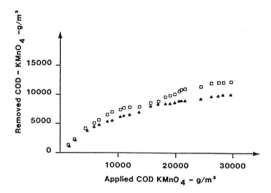

Figure 9. Cumulative plots for COD removal.
            Key : □ , ozonated rapid PKST filter 3
                  ▲ , ozonated slow PKST filter 10

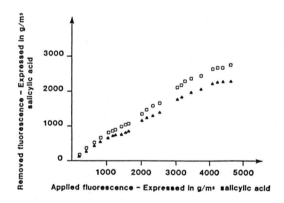

Figure 10. Cumulative plots for fluorescence reduction.
            Key : □ , ozonated rapid PKST filter 3
                  ▲ , ozonated slow PKST filter 10

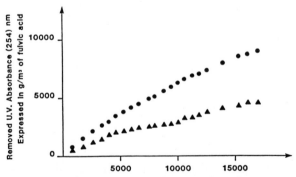

Figure 11. Cumulative plots for UV reduction.
            Key : ● , slow PKST filter 7
                  ▲ , ozonated slow PKST filter 10

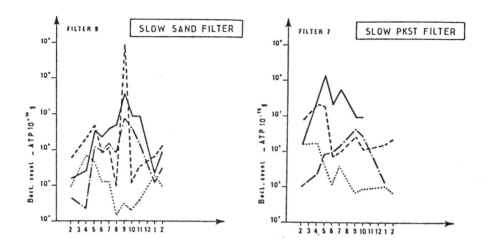

Figure 12. Variations in bacterial total plate counts and ATP measurements.
Key : —— —— , ATP on GAC surface
    - - - - , bacterial count on GAC surface
    _ . _ , ATP on sand surface
    · · · · , bacterial count on sand surface
Continued below.

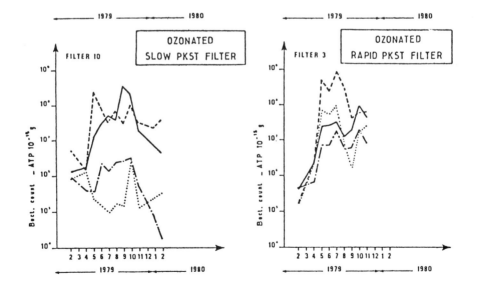

Figure 12. Continued.

ATP concentrations are higher in GAC than in sand (10-fold). They seem slightly higher after pre-ozonation and after rapid filtration.

Figure 12 also represents the total bacterial counts and clearly indicates that there was correlation with ATP concentrations.

Respirometric measurements also were performed but resulted in somewhat erratic values. Bacteria are present in the GAC beds in larger numbers than in the sand but no evidence could be drawn of their higher efficiency in the carbon than in the sand, for the removal of organics.

CONCLUSIONS

The Mont Valérien experiment also was used as a first industrial test for Benedek's (6) new adsorption predictive model. Predictions (2) were in good agreement with measurements, thus substantiating the slow adsorption theory. This finding brings further evidence that GAC does not significantly stimulate biodegradation.

After a year of experiment at Mont Valérien, the general following conclusions can be drawn. Slow sand filters treating clarified water in the second stage remove an average 10 % TOC primarily through biodegradation. Slow GAC filters remove an additional 15 % TOC primarily through adsorption. The total reduction is thus 25 %. Ozone removes 5 % but does not enhance GAC peformance.

Rapid GAC filters remove at least 10 % more TOC than the slow ones. The total reduction with the best carbon reached an average of 40 % for the whole year of experiment. Biodegradation does not seem to be significantly higher in GAC than in sand.

LITERATURE CITED

1. Fiessinger, F. (1980) Aqua, 9, 199.
2. Mallevialle, J. "Ozonation des Substances de Type Humique dans les Eaux". Second I.O.A. Symposium, Montréal (May 1975). (1976) Proceedings of the 2nd International Symposium on ozone Technology, IOA, 47.
3. McGuire, M., Suffet, I.H., Radziul, J.V. (1978) J. Am. Water Works Assoc., 70, 565.
4. Wilson, E.J., Geankoplis, C.J. (1966) In. Eng. Chem. Fund. 5 (1), 9.
5. Crittenden, J.C., Weber, W.J. (1978a) ASCE Jl Environ. Eng. Div., 104 (EE2), 185.
6. Benedek, A. Slow Adsorption Phenomenon. (March 1981) ACS. Environ. Chem. Div. Symposium on Activated Carbon, Atlanta.

4.4 MODIFICATIONS TO THE SLOW RATE FILTRATION PROCESS FOR IMPROVED TRIHALOMETHANE PRECURSOR REMOVAL

M. R. Collins and T. T. Eighmy — Environmental Research Group, Department of Civil Engineering, University of New Hampshire, Durham, New Hampshire (U.S.A.) 03824

ABSTRACT

The capabilities of the slow rate filtration process in removing trihalomethane (THM) precursor material were evaluated. The influence of filter cleaning, filter amendments, and loading rate was investigated. Physical, chemical, organic, and biological characterizations were made on water, filter media and schmutzdecke samples. Removal of specific fractions of aquatic organic matter was considered a function of both microbial activity and adsorptive capacity of the media. Granular activated carbon and anionic resin amendments can enhance removals and can be cost-effective if THMs must be controlled.

INTRODUCTION

At the turn of the century 25,000 deaths were attributed to typhoid fever – a waterborne disease. By 1935, a ten-fold drop in typhoid related deaths had occurred and improved water sanitation and chlorine disinfection was given much of the credit (1). The first realization that there may be some health hazards arising from the use of drinking water disinfectants came from the evidence that chloroform and other trihalomethanes (THMs) were produced as chlorination by-products of aquatic organic matter (AOM) (2,3).

Using such data, the US Environmental Protection Agency (US EPA) promulgated an amendment to the National Interim Primary Drinking Water Regulations in November 1979 establishing a maximum contaminant level (MCL) of 100 ug/L for total THMs in drinking water (4). Some individuals urge for a stricter MCL standard. Canada, Germany and the Netherlands have set lower standards for THMs in drinking water. In setting standards, US EPA has attempted to strike a balance between the risks associated with inadequate disinfection and the potential cancer risks associated with the resulting THM content in the treated water (5). The MCL has caused many smaller municipalities to consider filtration as a means of obtaining compliance. Yet these municipalities are faced with economic and management constraints that prevent them from utilizing newer filtration technologies.

One treatment method which has consistently been described as requiring little operating or maintenance skill yet is capable of producing an excellent finished water under certain conditions is slow sand filtration (6,7,8,9,10). Slow sand filters have consistently demonstrated their effectiveness in removing suspended particles with effluent turbidities below 1.0 NTU (11,12), in reducing bacteria by 98 to 99% (8,11) and in removing <u>Giradia</u> cysts when an established bacterial filter population is present (13). The use of oxidative pretreatment schemes can improve slow sand filtration removals of biodegradable AOM fractions

(14,15,16). However, little information is available in the literature that directly depicts the ability of the slow-rate filtration process to remove THM precursors as represented by AOM. Little is known about the indigenous microbial populations of slow sand filters and the role they play in the biodegradation and bioadsorption of AOM.

Dissolved AOM in natural waters contains aquatic humic substances (fulvic and humic acids). These constituents constitute 40 to 60 percent of the AOM in natural waters. The balance of the AOM is principally carbohydrate and protein (17) and is relatively biodegradable. The fulvic and humic acids, however, are metabolic end-products of lignin biodegradation (18) and have been considered to be recalcitrant (19). Recent evidence suggests that indigenous microbial populations are capable of cometabolizing or mineralizing this fraction (20,21,22,23,24,25,26,17). Thus, modifications to the slow sand filtration process which directly or indirectly enhance this potential may improve the removal of THM precursors.

The objective of research presented here was to evaluate selected modifications in the design and operation of slow sand filters for improved THM precursor removals. Characterizations of three municipal facilities were used to evaluate two different cleaning procedures (scraping versus harrowing). Pilot-scale slow sand filter studies were conducted to evaluate the influence of five filter amendments (aluminum oxide, anionic resin, granular activated carbon, anthracite and clinoptilolite) and two different flow rates (0.10 m/hour and 0.05 m/hour) on THM precursor removal.

Slow sand filters contained robust microbial populations that actively respire, mineralize benzoate, utilize AOM constituents, and locally control iron and manganese in the filter media. The type of filter cleaning, which affected population distributions, strongly influenced mass removal rates for nonpurgeable dissolved organic carbon (NPDOC), UV absorbance, and THM formation potential (THMFP). Filter amendments, notably anionic resins and granular activated carbon (GAC), promoted very high mass removal rates for NPDOC, UV absorbance, and THMFP. Transformation of AOM during filtration suggest utilization of specific fractions (notably hydrophobic and below 5,000 apparent molecular weight (AMW) fractions). The costs associated with amendments is not prohibitive in light of their superior treatment performance. Filtration rate had no influence on precursor removal for the rates that were investigated.

METHODS AND MATERIALS

Description of Municipal Slow Sand Filters

Covered slow sand filter facilities were visited at Springfield, Massachusetts and West Hartford and New Haven, Connecticut as shown in Figure 1. Two sets of untreated and treated water samples and filter media core samples from the three municipal slow sand filtration plants were acquired and completely analyzed during 1987. The first set of samples were collected during typical stabilized winter temperatures of late January and early February. The second set of samples was collected in early fall when the treatment plants typically experience higher raw water color.

Summaries of operating conditions for each slow sand filter at the time of winter and early fall sampling visits are shown in Table 1. All of the slow sand filters were operating at reduced loading rates in the winter. Both Springfield and New Haven operated at filtration rates consistently below a surveyed mean of 0.15 m/hr (12) whereas West Hartford operated considerably above the mean during the fall. Filtration loading rates were reduced during winter operations at Springfield, West Hartford and New Haven by roughly 20%, 50% and 57%, respectively, compared to fall operations. Correspondingly, filter

Figure 1 – Location of the municipal slow sand filtration facilities.

Table 1. Operating Conditions of Municipal Slow Sand Filters

	Date	Time From Last Cleaning (days)	Filter Depth (m)	Head- Loss (m)	Filtra- tion Rate (m/hr)	Filter Media Contact Time (hr)
WINTER VISIT:						
Springfield	1/23/87	200	0.79	--	0.040	20.0
West Hartford	2/11/87	--	0.67	1.0	0.131	5.1
New Haven	2/11/98	30	0.46	0.2	0.021	21.4
FALL VISIT:						
Springfield	9/24/87	90	0.70	--	0.049	14.5
West Hartford	9/24/87	25	0.64	1.7	0.250	2.6
New Haven	9/24/87	--	0.46	1.2	0.061	7.5

media contact times increased during winter operations at Springfield, West Hartford and New Haven by 140%, 195% and 285%, respectively.

All of the sampled slow sand filters, except for the winter visit at New Haven, had developed noticeable headlosses suggesting each filter had established mature schmutzdecke. Springfield could consistently operate the longest before reaching terminal headloss. By design, West Hartford "cleaned" their slow sand filters more frequently than either Springfield or New Haven slow sand filters.

Description of Pilot-Scale Slow Sand Rate Filters and Surface Amendments

Pilot-scale studies were conducted on a surface water supply for the City of Portsmouth, New Hampshire, and on a surface water supply for the town of Ashland, New Hampshire. The covered pilot scale filters were constructed from 30 cm (12 inch) diameter Schedule 40-PVC grey pipe. They were flanged and bolted 61 cm (24 inches) from the bottom to facilitate installation, cleaning and sampling of the filter media. A 0.64 cm (0.25 inch) PVC 0-ring was glued on the interior wall 7.6 cm (3 inches) below the media surface to deter sidewall channeling. A constant head of 91 cm (36 inches) over the top of the filter surface was maintained throughout the studies. The maximum headloss that could be measured by the piezometer tube for the 1.5 m (five foot) filter column was 142 cm (56 inches). Each of the filter columns were fed from a common constant head tank. Further details of the pilot filter are shown in Figure 2.

Initially, the filtration rates of the pilot scale filters were controlled manually by pinch clamps on the effluent line. Maintaining a constant flowrate by this method proved difficult and time-consuming. Later, multi-head peristaltic pumps were used to maintain constant effluent flowrates.

The pilot scale filters were comprised of several media layers as outlined in Figure 2. Typically, a 7.6 cm (3 inch) layer of surface amendment was supported by a 30 cm (12 inch) layer of sand which in turn was supported by a 15 cm (6 inch) layer of gravel. Specifications for the supported media and surface amendments are given in Table 2.

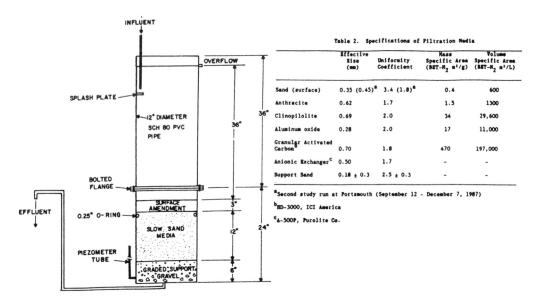

Table 2. Specifications of Filtration Media

	Effective Size (mm)	Uniformity Coefficient	Mass Specific Area (BET-N$_2$ m^2/g)	Volume Specific Area (BET-N$_2$ m^2/L)
Sand (surface)	0.35 (0.45)[a]	3.4 (1.8)[a]	0.4	600
Anthracite	0.62	1.7	1.5	1300
Clinopilolite	0.69	2.0	34	29,600
Aluminum oxide	0.28	2.0	17	11,000
Granular Activated Carbon[b]	0.70	1.8	470	197,000
Anionic Exchanger[c]	0.50	1.7	-	-
Support Sand	0.18 ± 0.3	2.5 ± 0.3	-	-

[a] Second study run at Portsmouth (September 12 - December 7, 1987)

[b] HD-3000, ICI America

[c] A-500P, Purolite Co.

Figure 2 – Schematic of pilot filters.
(note: 1 inch = 2.54 cm).

Physical-Chemical Analysis

Untreated and treated water samples were analyzed for pH, turbidity, non-purgeable dissolved organic carbon (NPDOC), UV absorbance and THM formation potential (THMFP). The pH measurements were made using a Fisher-Scientific C.A.T. system with a combination reference electrode that was standardized daily. Turbidity was determined by using Hach Model 2100A turbidimeters calibrated for each sample set with AEPA-1 turbidity standards (Advanced Polymer Systems, Redwood City, CA). Non-purgeable dissolved organic carbon (NPDOC) was analyzed by a UV-promoted sodium persulfate oxidation (Dorhmann DC-80 Total Organic Carbon Analyzer, Santa Clara, CA). Ultraviolet (UV) absorbance was measured at a wavelength of 254 nm, a cell pathlength of 1 cm, and a sample pH of 7 (Bausch & Lomb Spectronic 2000, Rochester, NY). THMFP was determined by quantifying the concentration of THM species present after a seven-day incubation at 20°C and pH 7 with a chlorine-to-carbon mass ratio of 5:1. THM species were quantified by gas chromatography (Perkin-Elmer Sigma 2000, Norwalk, CT) using a liquid-liquid extraction procedure (28).

Dissolved organic matter was operationally defined as all material passing a 0.45 um membrane filter. Dissolved organic matter was fractionated by apparent molecular weight (AMW) and by hydrophobicity. AMWs were determined using an ultrafiltration procedure (29) with nominal MW cutoffs of 500, 5000, and 10,000 (YC-05, YM-5, and YM-10, respectively, Amicon Corp., Danvers, MA). Hydrophobic separations were performed by non-ionic resin adsorption (30). Samples were adjusted to pH 2 and passed through a gel chromatography column packed with a synthetic polymer adsorbent (XAD-8, Rohm and Haas, Philadelphia, PA). Organics eluted from the column were considered hydrophilic. Hydrophobic organics were desorbed from the resin by eluting the column with a 0.1N NaOH solution of organic-free water.

Microbiological Analyses

     Grab samples of influent and effluent waters from municipal and pilot-scale
filters were collected to enumerate bacteria using epifluorescent microscopical
and spread plate culturing techniques which are described below.
     Municipal and pilot-scale filters were cored to enumerate bacteria with
depth in the filter media. A coring device was utilized as described
previously (31,32). Bacteria were extracted from the media using a 0.1% sodium
pyrophosphate solution buffered to pH 7.0 (33). Subsequent serial dilutions of
this first extract were made in sterile sodium pyrophophosphate for spread
plating.
     Acriflavine was used as the nucleic acid-specific flurochrome for
epifluorescent microscopical enumerations. The procedures developed by
Bergstrom et al. (34) were used to determine acriflavine direct cell counts
(AFDC) either per media dry weight or per ml of source or treated water.
Samples taken from the first sodium pyrophosphate extract were used for
enumerations of media-associated populations.
     A variety of selective media and spread plate methods were used to
enumerate nutritional specialists in the source or treated waters or on the
filter media. The R2A media developed by Reasoner and Geldreich (35) was used
to enumerate low carbon-requiring heterotrophs. The media developed by Clark et
al. (36) was used to enumerate heterotrophic iron precipitators. Ghiorse's
media (37) was used to quantify manganese oxidizers. The mineral media
developed by Hann (21) was used to enumerate AOM, benzoate, and catechol
utilizers. These substrates were individually added to Hann's media at 0.003
g/L as the sole carbon source.
     Filter media microbial biomass was determined using Folin-reactive material
(FRM) as a measure of biomass. The first sodium pyrophosphate extract was used
to quantify the protein content of filter media-associated bacteria according to
the methods developed by Lowry et al., (38).
     Extractable iron and manganese in the filter media was determined with
flame atomic absorption spectrophotometry after a nitric acid extraction. This
extraction recovers surface oxide and ligand-bound metals (39).
     Schumtzdecke and source water samples were evaluated for their respiratory
activity by examining the percentage of AFDC capable of reducing
2-(p-iodophenyl)-3-(p-nitrophenyl)-5-phenyltetrazolium chloride (INT). INT acts
as a terminal electron acceptor which is reduced during respiratory electron
transport. A reddish-purple deposit forms intracellularly when the tetrazolium
dye is reduced. Appropriate municipal filter schmutzdecke dilutions were
exposed to the tetrazolium dye INT for 20 minutes at a concentration of 0.2% INT
(20°C). The samples were fixed with formalin and then treated with acriflavin
according to the methods outlined by Bergstrom et al. (34) for AFDC
determinations. Bright field microscopy was used interchangeably with
epifluorescent microscopy to quantify the percentage of the AFDC reducing INT.
     Schmutzdecke associated populations from municipal filters were evaluated
for their ability to mineralize uniformly labeled ^{14}C benzoate to $^{14}CO_2$. The
methods of Subba-Rao et al. (40) and Simkins and Alexander (41) were adapted for
use in our mineralization studies. The mineralization assays were conducted
with sodium pyrophosphate extractions of schmutzdecke samples from appropriate
cores. Final cell concentrations in the incubation solutions were approximately
5×10^6 cells/ml. Uniformly labeled benzoate was used in the assay (Amersham,
120 mCi/mmole). Cold benzoate was used to augment the labeled benzoate at
higher benzoate concentrations. Initial DPMs in all incubations were
approximately 10,000/ml. Four benzoate concentrations were evaluated (1.0, 10,
100, and 1000 ng/ml). During the course of the incubations (20°C), 1 ml
aliquots were collected, acidified, and placed under vacuum to drive off $^{14}CO_2$.
An ethanolamine trap was initially used to verify the conversion of ^{14}C-benzoate

to $^{14}CO_2$. The counts remaining in solution represent free, bound, or incorporated benzoate and its metabolic by-products. Mineralization rates were determined by looking at DPM losses over time.

RESULTS AND DISCUSSION

Description of Municipal Filter Cleaning Methods

Cleaning of the slow sand filters at Springfield and New Haven is done manually by shoveling the top layer of media into a conveyer washing system. Normally, once a slow sand filter reaches terminal headloss the supernatant water is drained to just below the filter surface, and the exposed schmutzdecke and the top 1.3 cm (1/2 inch) of sand media are scraped off. A typical filter bed may be cleaned for several years by the scraping method before the depth of the filter bed is so depleted that the bed needs recharging with clean stored sand. Springfield filter scrapings (2.5–5.0 cm layers) are generally done twice a year. Complete filter resanding is done when there is 30.5 cm (12 inches) of sand remaining after repeated scrapings. This occurs roughly every eight years. All sand is removed from the filter being resanded; stored sand is then used to replenish the original 107 cm (42 inches) filter bed depth. The operators at New Haven discontinued replacing the scraped sand (0.6–1.2 cm layer) since 1972 because of complaints about the washwater. Consequently the filter media depth continues to decrease with each cleaning at this facility.

Operators at West Hartford have a unique method of cleaning slow sand filters. The supernatant water is drained to a height approximately 45 cm (18 inches) above the sand media at terminal headloss. A rubber-tired tractor equipped with a comb harrow is placed on top of the filter to rake the sand media to a depth of 30 cm (12 inches). The filter sump well discharges are kept open, causing a steady flow of water toward the surface sump. As the harrow is dragged over the sand, colloidal debris in the top 30 cm (12 inches) sand layer is loosened and is caught by the moving stream water toward the sump and is eventually discharged. When the depth of water on the filter sand drops below 7.6 cm (3 inches), operations are suspended until the filter has been refilled by reverse flow to a depth of 45 cm (18 inches) when harrowing is again resumed. The process is repeated until the entire filter surface has been harrowed.

Only fine clay colloids and other small particulate debris are removed by this method but there appears to be some major treatment advantages. The harrowing method typically requires significantly less time to complete than the usual scraping method. Moreover, harrowed filters are put back on line within hours instead of days or weeks. The method apparently causes a majority of the debris of the schmutzdecke to be washed away while the bacteria of the schmutzdecke that are attached to sand grains are raked into the depths of the filter.

The distribution of bacterial populations in the municipal slow sand filters is shown in Figure 3. The results show the effects of cleaning procedure on the bacterial population distributions in the media. The results presented here are limited to samples collected in the winter visit, similar results were obtained in the fall visit.

The analysis of cores taken from the Springfield facility (Figures 3a–3b) show population, biomass, and iron and manganese distributions from a filter close to terminal head loss where scraping is employed and cleaned sand is returned to the filter. High levels of AFDC and spread plate counts are present in the schmutzdecke and then drop by one to three orders of magnitude directly below the schmutzdecke. Distributions below the schmutzdecke are fairly uniform. Biomass (as FRM) and iron and manganese were significantly correlated to AFDC. Visual observations of cores from this facility showed a dark, dense schmutzdecke (1 cm thick) on top of relatively clean sand.

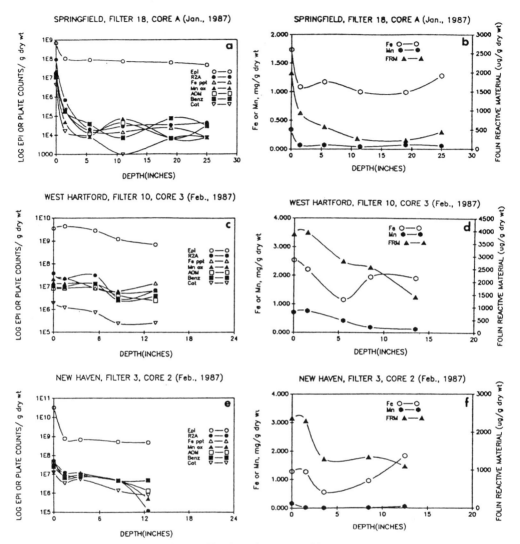

Figure 3 - Distribution of acriflavine direct cell counts, nutritionally specific spread plate counts, biomass (as FRM), and iron and manganese in cores from the three municipal facilities (note: 1 inch = 2.54 cm).

The analysis of cores taken from the West Hartford facility (Figures 3c-d) show population, biomass, and iron and manganese distributions from a filter close to terminal headloss where harrowing is practiced. Schmutzdecke-like population densitites (10^9 AFDC/g dry wt) and FRM distributions (2,500 to 4,500 mg/g dry wt) are present with no marked change in density or distribution between the schmutzdecke and underlying filter media. Metals and biomass, however, were still significantly correlated to AFDC. Visual observation of the core material from this facility supports the observation that a loose schmutzdecke-like appearance was present to a depth of 30 cm (12 inches).

Table 3. Acriflavine Direct Count (AFDC) and Folin Reactive Material (FRM)
By Filter Depth at Municipal Slow Sand Filters

	AFDC Concentration Factor[a]	FRM Concentration Factor[b]
Winter		
Springfield	9.44	4.22
West Hartford	10.41	4.72
New Haven	-	-
Fall		
Springfield	9.79	4.34
West Hartford	10.05	4.34
New Haven	10.01	4.17

[a] $\log \sum_{i=1}^{n} (AFDC_i \times depth_i)$

[b] $\log \sum_{i=1}^{n} (FRM_i \times depth_i)$

Table 4. Evaluation of the Ratio of the Numbers of Respiring/AFDC
in Water and Schmutzdecke Samples from Springfield, West
Hartford, and New Haven (Jan-Feb., 1987)

		Location	
Treatment	Springfield[a]	West Hartford[b]	New Haven[c]
	Percent INT/AFDC (± SD)		
Raw Water Samples[d]	1.8 (± 1.4)	2.1 (± 1.3)	1.7 (± 1.1)
Schmutzdecke Extract Samples[d]	4.4 (± 1.9)	14.3 (± 2.9)	5.8 (± 2.3)

[a] Filter 17 and Core 8

[b] Filter 10 and Core 1

[c] Filter 3 and Core 1

[d] Based on means of counts from 8 fields under oil immersion for epifluorescent and light microscopical enumeration

The analysis of cores taken from the New Haven facility (Figures 3e-f) show population, biomass, and iron and manganese distributions from a filter close to terminal headloss where scraping without replacement is practiced. Very high levels of AFDC and spread plate counts were observed in the schmutzdecke, and then declined with depth directly below the schmutzdecke. Metals and biomass were significantly correlated to AFDC.

As shown in Table 3, integrating the AFDC and FRM distributions with depth for each filter and taking the log of that summation, it is apparent that the harrowing practiced in the West Hartford facility resulted in more AFDC and biomass being present with depth in that facility than in either the Springfield or New Haven facilities.

The relative bacterial population distributions in each filter were similar with regard to the relationships between AFDC and all other spread plate enumerations. The selective enumeration methods used in this study appears to only recover a small percentage of the AFDC.

Similar bacterial distribution profiles with depth, and low recoveries of plate counts compared to epifluorescent direct counts have been observed in GAC filters which treat surface waters (14) and in subsurface and saturated soil systems (42). Both systems are characterized by high populations in association with organic rich particulate material.

The data on the percent respiring population of the schmutzdecke AFDC is shown in Table 4. West Hartford schmutzdecke bacteria exhibited a significantly higher respiratory population compared to the other two facilities. The levels that were observed (14% of the AFDC reduced INT) are typical for marine and soil bacteria (42,43,44). The balance of the AFDC therefore appears to be nonculturable and nonrespiratory, but possesses intact cyctoplasm with nucleic acid material which stains with acriflavine.

The schmutzdecke bacteria from West Hartford also exhibited significantly more robust ^{14}C-benzoate mineralization ability than bacteria from the other facilities (Figure 4). Mineralization was biphasic and up to 70% of the substrate was mineralized, regardless of the concentration of benzoate. Similar results have been shown by Subba-Rao et al. (40) and Rubin et al. (45) for wastewater and lake isolates. Previous work by Hann (22,23,24) has shown that the ability of indigenous lake isolates to utilize benzoate is correlated with fulvic acid utilization. Thus, schmutzdecke bacteria, particularly the West Hartford population, exhibit a propensity to degrade substituted monoaromatic compounds.

Effects of Filter Cleaning Methods on Treatment Performance

Overall treatment performances achieved at the various municipal plants during the winter and fall visits are presented in Table 5. All of the untreated waters were derived from low turbidity sources containing low to moderate levels of dissolved AOM. The general nature of the dissolved organic

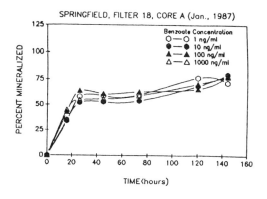

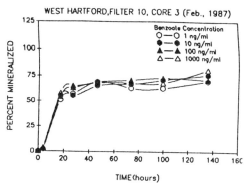

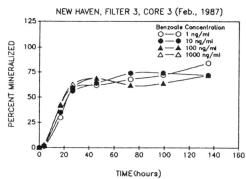

Figure 4 – Mineralization of ^{14}C-benzoate by schmutzdecke bacteria. The figures show the percent of the initial substrate concentration that was mineralized over time.

matter was found to vary slightly from source to source as evidenced by a relative comparison of UV absorbance and NPDOC levels. All of the water sources experienced higher organic loadings during the warmer temperatures of the early fall versus the cold temperatures of mid-winter.

Each of the slow sand filtration plants were successful in turbidity removals with percent reductions ranging from 50 to 86 in the winter and 57 to 94 in the early fall as typically reported in the literature (8,12). Springfield's slow sand filter consistently achieved the highest turbidity reductions and the lowest treated water turbidity with values frequently below 0.1 NTU. Each of the municipal slow sand filters emphasized particulate removals.

Conventional slow sand filters are not normally effective in removing dissolved organic matter as compared to particulates (10,46). Each of the sampled slow sand filters was only moderately successful in THM precursor removals with NPDOC, UV absorbance and THMFP reductions ranging from 13-33, 17-33 and 9-27 percent, respectively, during winter conditions and 12-31, 17-43 and 14-27 percent, respectively, during fall operations. With a few exceptions, West Hartford's slow sand filter consistently achieved higher precursor removals despite having the highest filtration loading rates and shortest filter media contact times. Apparently, West Hartford's unique cleaning procedure enhanced treatment performance since differences in filter media or raw water chemical quality between the slow sand filters were not considered significant.

A new method of expressing organic removal rates was necessary in order to take into account the vastly different operating conditions of each of the slow sand filters at the time of sampling. A mass removal flux term takes into account the total quantity of material removed at a given filtration loading rate. Algebraically, the flux is equal to the influent concentration minus the effluent concentration times the filtration loading rate ($\Delta m/L^3 \cdot L/T = m/L^2T$).

Table 5. Treatment Performance Achieved at the Municipal Slow Sand Filtration Plants –
(a) Winter and (b) Fall

Site: Water Sample	pH	Turbidity (ntu)	Non-Purgeable Dissolved Organic Carbon (mg/L)	UV absorbance (254nm, pH 7)	Trihalo-methane Formation Potential (ug/L)	Acriflavine Direct Cell Count (#/ml)x10⁶	Heterotrophs (CFU/ml)x10⁷
a. WINTER VISIT							
Springfield:							
Untreated	6.6	0.64	2.34	0.088	152	—	5.0
Treated	6.5	0.09	2.00	0.063	138	—	1.0
West Hartford:							
Untreated	6.9	0.50	1.82	0.042	180	—	37.5
Treated	6.7	0.25	1.22	0.032	132	—	1.3
New Haven:							
Untreated	6.8	1.50	1.16	0.042	108	—	94.0
Treated	6.6	0.30	1.01	0.035	98	—	36.5
b. FALL VISIT							
Springfield:							
Untreated	6.5	0.36	2.96	0.093	88	7.37	19.9
Treated	6.5	0.26	2.61	0.073	76	0.28	57.4
West Hartford:							
Untreated	6.6	0.50	2.26	0.056	62	4.67	15.3
Treated	6.4	0.15	1.56	0.032	45	0.34	88.4
New Haven:							
Untreated	7.8	1.30	3.18	0.079	101	7.13	19.1
Treated	7.6	0.17	2.48	0.067	85	0.29	40.0

The West Hartford slow sand filters dramatically achieved higher mass removals of NPDOC and THMFP than the New Haven or Springfield slow sand filters as shown in Figure 5. The removal trends between NPDOC and THMFP were similar but more pronounced during the higher organic loading conditions of the Fall.

Filter loading rates and subsequent media contact times did not appear to have a significant impact on organic precursor mass removal rates. The highest removals were achieved at West Hartford despite having the highest loading rates and shortest media contact times. Moreover, the best removals at New Haven were achieved in the fall when loading rates were three times higher and contact

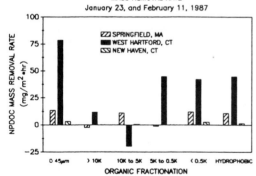

Figure 5 – Mass removal rates for NPDOC and THMFP in municipal filters.

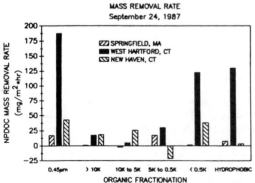

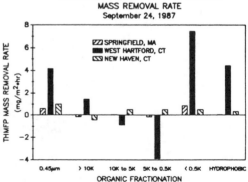

times one-third less their winter values. Higher filtration rates may enhance transfer mechanisms and sticking opportunities for colloidal and dissolved organic constitutents.

The municipal slow sand filters demonstrated a preference for removing organic precursor material in the lower AMW ranges. These lower AMW species contain substituted monoaromatics which schmutzdecke isolates are adept at biodegrading. The West Hartford filter achieved greatest NPDOC mass removals below an AMW of 5,000 with significant AMW reductions below 500 in the fall. Springfield experienced slight removals in the 10,000 to 5,000 and less than 500 AMW range during the fall. The New Haven slow sand filter did not experience NPDOC removals in the winter but developed significant mass removals in all of the AMW ranges in the fall except in the 5000 to 500 range which showed a production.

The production of organic material in a given AMW range may occur by either (i) microbially mediated enzymatic cleavage of large organic molecules into smaller molecules, thereby constituting production in the lower AMW range; or (ii) collection of smaller molecules in sufficient concentrations to polymerize into larger organic molecules, thereby constituting production in a higher AMW range. Recent evidence of microbially-mediated transformations in lake water (47,48) suggests that these microbially mediated depolymerizations and repolymerizations may also be occurring in slow sand filters.

Slow sand filtration appears to preferentially remove hydrophobic, or ostensibly, humic material over other constituents of AOM; a finding consistent with more conventional water treatment practices (29,49). Mass removals for the hydrophobic fraction of AOM followed the same trends noted previously for total dissolved organic matter (0.45 µm) although at reduced rates.

Relationship Between Filter Microbiology and Treatment Performance

The superior THM precursor removal performance by the West Hartford filters was attributed to higher bacterial population densities throughout the entire filter bed. It is also thought the microbiological maturity of the sand bed rather than the presence or absence of a schmutzdecke is also an important variable in the removal of Giardia cysts and coliform bacteria (13). The bacteria in the West Hartford facility were also more active with regard to respiratory activity and ability to mineralize substituted monoaromatics.

Quantifications of filter media biomass and organic mass removals for the municipal slow sand filters were analyzed for correlations between filter bacterial AFDC populations or biomass (as FRM) and precursor removal rates. Linear regressions between NPDOC and THMFP mass removal rates versus filter media AFDC and FRM are shown in Figures 6.

Development of a definitive relationship between filter biomass and precursor removals was limited to the number of data points used and the diversity of the sampled treatment operations. Realizing these limitations, there were positive correlations developed between the municipal filter AFDC and NPDOC mass removal rates as shown in Figure 6a-b. Although the correlation was not as high for the hydrophobic material, the regression trends for the AMW fractions supports the preference of slow sand filters for removing the smaller organic molecules with AMW below 5000 over the larger organic molecules. It is not uncommon for production of NPDOC to occur in a given AMW fractions as noted for the 10,000-5,000 range.

There was a statistically significant relationship between THMFP mass removals and filter media biomass as quantified by both AFDC ($r = 0.85$) and FRM ($r = 0.84$) as shown in Figure 6c-d. Apparently, the microorganisms inherent in slow sand filters preferentially removed the more reactive fraction of AOM. Higher THM precursor material should be expected to be removed by slow sand filters having higher bacterial populations.

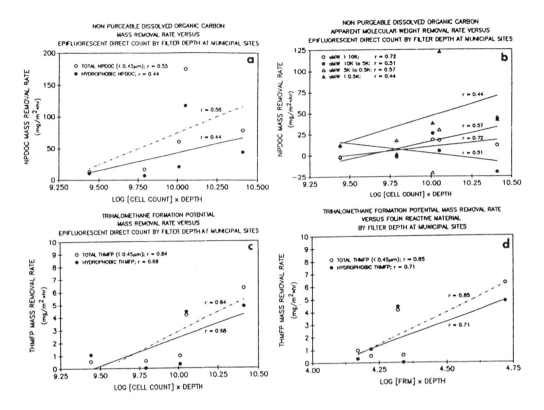

Figure 6 - Correlations between NPDOC (a-b) and THMFP (c-d) mass removal rates. The values used on the x-axis were obtained from Table 3.

Effects of Filter Media Amendments

The pilot slow rate filters in the first study run at Portsmouth were constructed with the following surface media configurations:

Surface Amendment	Filter Replication
Sand (baseline)	5
Anthracite	1
Clinoptilolite	1
Aluminum Oxide (Al$_2$O$_3$)	1
Granular Activated Carbon (GAC)	2
Anionic Resin	1

Media replicates were used to provide additional data for comparing various filters by taking into account the variability inherent with column studies. This first study run lasted for a total of 89 days beginning on May 23, 1987.

All filters were shut down for extensive cleaning on Day 45 with the second
filter run beginning on Day 47 (July 9, 1987).

Raw water quality varied throughout the first study run at Portsmouth. The
raw water organic parameters NPDOC, UV absorbance and THMFP averaged 11 $\pm$ 3
mg/L, 0.450 $\pm$ 0.100 (254 nm & pH 7) and 750 $\pm$ 250 ug/L, respectively, with the
higher levels generally occurring near the end of the filter run. Raw water
turbidities average 2 $\pm$ 1 NTU with the higher values again near the end of the
run.

The pilot slow sand filters used during the filter amendment studies were
evaluated for (i) their efficacy in removing bacteria from the source water
(quantified with AFDC and R2A spread plate procedures), (ii) the development of
the schmutzdecke population using AFDC as a development parameter, and upon
completion of the final filter run, (iii) the population distributions with
depth in the filter media. Results from the microbiological analysis of the GAC
surface amendment, purolite surface amendment, and baseline sand filters are
presented because the GAC and purolite were the most effective treatments, the
others being non-distinctive with the replicated baseline sand filters.

As shown in Figure 7a, the baseline sand filters sometimes effectively
removed bacteria from the source water and at other times acted as a source for
bacteria. Production tended to occur after start-up (day 0) and filter cleaning
(day 45). The schmutzdecke population responded rather dynamically to changes
in influent bacterial loading and to cleaning. Core bacterial profiles (Figure
7b) were similar to profiles seen with undisturbed municipal filters, except
that sub-schmutzdecke media tended to be less mature. Similar cell densities, FRM
content, and extractable iron and manganese content were observed (Figure 7c).

Data from the GAC surface amendment pilot-filter are shown in Figure 8.
The GAC tended to support a higher schmutzdecke population than the baseline

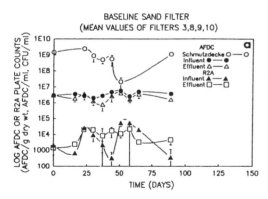

Figure 7 - Microbiological data
from the baseline sand pilot-scale
filters. (a) schmutzdecke development
and bacterial removal or production
over time, (b) population distributions
with depth, and (c) biomass, iron, and
manganese distributions with depth
(1 inch = 2.54 cm).

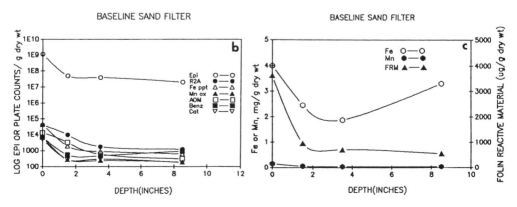

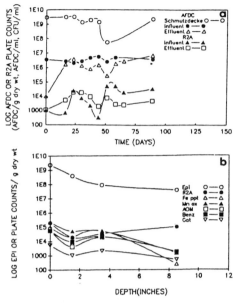

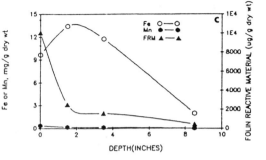

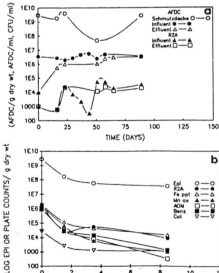

Figure 8 - Microbiological data from the GAC surface amendment pilot-scale filter   (a) schmutzdecke development and bacterial removal or production over time, (b) population distributions with depth, and (c) biomass, iron and manganese distributions with depth (1 inch = 2.54 cm).

sand filters (Figure 8a), which also behaved dynamically in response to bacterial loading and cleaning. The GAC amended filter also sometimes acted as a net producer of bacteria. Population profiles were similar to municipal and baseline sand pilot fillers. The GAC amendment distinguished itself by posessing significantly higher levels of FRM.

The purolite surface amendment was similar to the GAC in schmutzdecke development, bacterial removal and production, and in the population, biomass, and iron and manganese distributions with depth (Figure 9a–c). Like the GAC amendment, the purolite adsorbed higher levels of FRM. Both surface amendments are adsorptive surfaces that appeared to remove and support high levels of cell biomass.

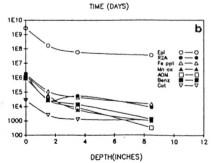

Figure 9 - Microbiological data from the purolite surface amendment pilot-scale filter   (a) schmutzdecke development and bacterial removal or production over time, (b) population distributions with depth, and (c) biomass, iron, and manganese distributions with depth (1 inch = 2.54 cm).

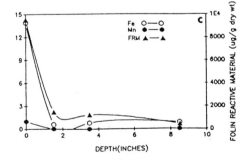

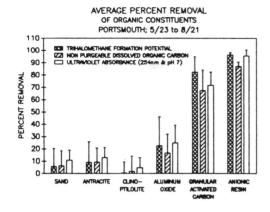

Figure 10 – Average percent removals of THMFP, NPDOC, and UV absorbance by various pilot-scale filters.

THM precursor removals by slow rate filters are significantly influenced by the characteristics of the filter media. Average NPDOC, UV absorbance and THMFP percent removals and their corresponding standard deviations for the various amended pilot filters compiled during the first study run at Portsmouth are shown in Figure 10.

Organic precursor removals for the anionic resin and GAC amended filters frequently exceeded 90 and 75 percent, respectively. These values are much higher than those reported for municipal slow sand facilties. Slow GAC filters have been known to achieve excellent organic precursor removals, frequently exceeding 90 percent (46). GAC removal in this study was limited by the relative small depth [7.6 cm (3 inches)] of amendment used. Higher precursor removals would be expected from deeper beds of GAC. Conversely, treatment by the anionic resin did not appear to be limited by bed depth. The kinetics of precursor removal by ion-exchange is apparently faster than the transport and attachment mechanisms inherent to GAC.

Treatment efficiencies of the aluminum oxide, anthracite and clinoptilolite amended filters could not be clearly distinguished from the more conventional slow sand filters. Typical organic precursor removals by these filters frequently averaged 5-25 percent which agrees with the approximately 20 percent removals reported by other researchers (46). The slightly higher percent removals (~5%) noted for the aluminum oxide amended filter might be attributable to the smaller effective size of the amendment. The poor removals noted for the clinoptilolite amended filter was attributed to the lack of sunlight-induced algal growths typically observed on the schmutzdecke of uncovered slow clinoptilolite/sand filters (50).

Important organic mass removal rate comparisons for the selected organic parameters between the various slow rate filters are summarized in Figure 11. Error-bars are shown for the replicated sand filters in order to show inherent variability between filters. No significant removal rate differences between the GAC surface amended and sublayering filters were consistently detected so their results were averaged.

The anionic resin and GAC amended filters consistently achieved higher precursor mass removal rates than other filter media combinations studied. Although fluctuations were apparent throughout the study run, NPDOC $(mg/m^2 \cdot hour)$, UV absorbance $(abs \cdot 10^4 \cdot m/hour)$ and THMFP $(mg/m^2 \cdot hour)$ mass removal rates for the anionic resins averaged over 1000, 400 and 75 respectively, whereas GAC averaged roughly 750, 300 and 70, respectively. These removal rates were significantly higher than those observed in the municipal filters.

The superiority of the anionic resin over the GAC was especially noteworthy after all filters were cleaned at the halfway point of the study run (after Day 45). The only difference in cleaning procedures between media was that the

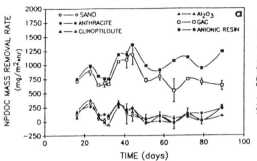

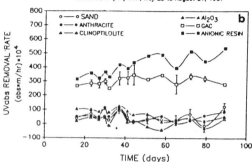

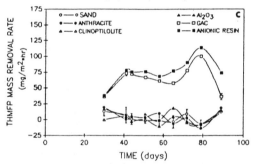

Figure 11 - (a) NPDOC, (b) UV
absorbance, and (c) THMFP mass
removal rates for various
pilot-scale filters.

anionic resin was soaked overnight in a dilute salt solution which regenerated
its chloride-based exchange functional group. There is a good possibility that
had the GAC media been regenerated or replaced with virgin carbon the removal
differences between the filters would not have been as significant.

Particulate removals by the slow rate filters, as shown in Figure 12,
correlated well with the removal trends observed for THM precursor material.
Both the anionic resin and GAC amended filters generally achieved higher
turbidity reductions compared to the other filters. However, this difference
was not as significant as compared with the precursor mass removal rates between
the anionic resin and GAC amended filters and the other filters. Raw water
turbidities increased over the study run varying from 1.0 NTU at the beginning
to over 3.0 NTU near the end. Effluent turbidities were consistently below 1.0
NTU during the first half of the run but fluctuated slightly over 1.0 NTU as the
raw water turbidity increased. The generation of turbidity by the
clinoptilolite amended filter during the early stages of the study run was
possibly due to the fractionation of the amendment during column installation.
The turbidity removal rate for each filter also increased as the raw water
turbidity increased.

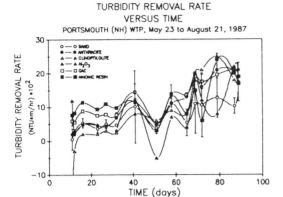

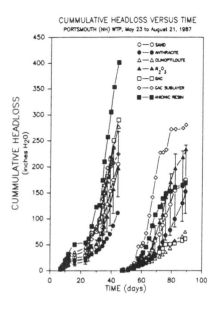

Figure 12 (above) - Turbidity
removal over time by various
pilot-scale filters.

Figure 13 (right) - Headloss
development by various pilot-
scale filters (1 inch = 2.54 cm).

The excellent THM precursor removals demonstrated by the anionic resin and
GAC amended filters was tempered by their excessive head loss development as
shown in Figure 13. Cumulative headloss was used to minimize the scatter and
organize the data into comparable trends between the filters. The shape of the
headloss curves followed exponential patterns of cake filtration commonly
reported in slow sand filtration (11). The anthracite and clinoptilolite
amended filters developed slower headloss curves during the filtration run which
was attributable to the larger effective size of the amendment and their poor
organic and turbidity removals.

Effects of Filter Loading Rate

The influence of two different filtration loading rates, 0.10 m/hr and 0.05
m/hr, on slow sand filtration performance was evaluated on the two different raw
water sources that we studied (Portsmouth and Ashland, New Hampshire). One
study run was conducted in Ashland and lasted 132 days beginning on 7/29/87 and
ending on 12/8/87. Another study run was conducted in Portsmouth and lasted for
87 days beginning on 9/12/87 and ending on 12/7/87.

Treatment performances between the "fast" and "slow" filters at Ashland and
Portsmouth were indistinguishable from each other as shown in Figure 14.
Moreover, the pilot sand filters were not able to achieve precursor removals
beyond a certain efficiency which in this study averaged between 5-20 percent.
Although precursor loading rates were typically three times greater at
Portsmouth than at Ashland, mass removal rates were adjusted by the pilot
filters in order to produce comparable percent reductions. In another study,
slow sand filters treating different water sources achieved similar precursor
removals of 15-19 percent (46). The higher percent removals for an anionic
resin amended slow rate filter and the municipal conventional plant at
Portsmouth are included in Figure 14 for comparison.

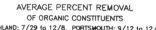
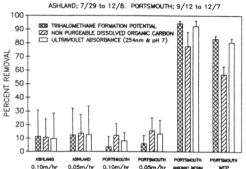

Figure 14 - Average percent
removals of THMFP, NPDOC, and
UV absorbance by various
pilot-scale filters and conventional
treatment facilities.

Organic precursor removals did not appear to be significantly dependent on filter loading rates from the same source water, although fluctuations were apparent throughout the study runs. Moreover, changes in mass removal rates did not appear to change as the filters matured during the study run. Since the "fast" filters were assumed to reach maturity before the "slow" filters, organic precursor removals were not considered to be significantly related to the age of the schmutzdecke for the low alkalinity-low hardness water sources utilized in this study. As noted previously, the microbiological maturity of the sand bed rather than just the schmutzdecke is the most important determinant in slow sand filtration removals (13). Apparently, the aging process for the "fast" filters was limited to the schmutzdecke and was not operating long enough to affect the filter bed.

The anionic resin amended slow rate filter consistently achieved higher precursor removals of THM precursor material than the baseline slow sand filters operating at different loading rates or the alum based conventional treatment process throughout the entire study run as shown in Figure 15. The anionic

Figure 15 - Treatment performance
of various pilot-scale filters and
conventional treatment facilities.
(a) NPDOC levels, (b) UV absorbance
levels, (c) THMFP levels.

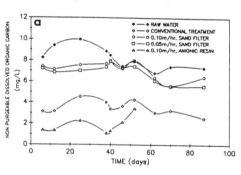
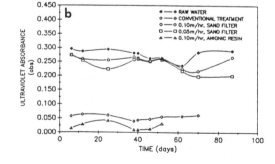
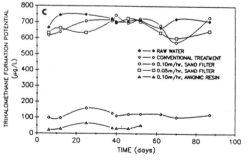

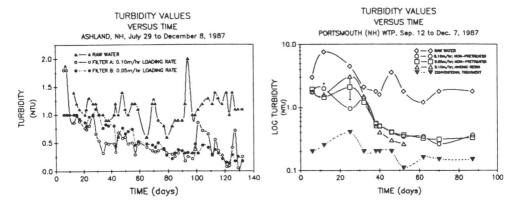

Figure 16 - Turbidity removal by various pilot-scale filters and conventional facilities.

resin was able to consistently satisfy current THM drinking water MCL of 100 ug/L and, with few exceptions, the proposed maximum containment level of 50 ug/L. The conventional treatment process rarely exceeded current standards. As reported previously, the slow sand filters did not achieve significant removal of organic precursor material.

The slow sand filter, regardless of loading rate or surface amendment, achieved similar removals with average effluent turbidity values around 0.3 NTU as shown in Figure 16a-b. Typical raw water turbidity values for both Ashland and Portsmouth study sites averaged 1.5 NTU.

The slow sand filters had to have a significant ripening period before optimum removals were reached. The filters at Portsmouth seemed to have achieved maturity more quickly than the filters at Ashland by reaching the lower effluent turbidity values after only 40 days of operation compared to over 80 days required for the Ashland filters. This maturation process may be a function of the organic content of the raw water.

Economic/Construction Costs of Selected Modifications

Since an economic analysis is strongly dependent on local market conditions and availability of building material, the following discussion will be limited to construction and operation and maintenance cost estimates for a proposed slow rate filtration plant in the Town of Ashland (design population of 3000). The design criteria utilized in the costing of the slow sand filters were 0.125 m/hour (0.052 gpm/sq.ft.) filter loading rate, 0.013 m^3/s (0.3 MGD) flow capacity, two connecting filter beds (each - 12.2 m (40 feet) wide x 15.2 m (50 feet) long x 4.9 m (16 feet) high), filter sand depth of 1.2 m (4 feet), filter support gravel of 0.6 cm (2 feet) and complete concrete enclosure.

Unit cost analysis based on square foot of constructed filter, 1000 gallons of treated water, and percent precursor removal were developed to comparably evaluate the costs associated with modifications to improve slow rate filtration performance. The slow sand filtration comparisons, as outlined in Table 6, include a 5.1 cm (2 inch) surface layer of GAC, a 2.5 cm (1 inch) layer of anionic resin, and a packaged water clarification treatment plant with coagulation, flocculation, sedimentation, filtration.

Slow sand filters are cheaper to build than a comparably sized packaged clarification treatment plant. The cost of the packaged plant was based on updated data from the EPA Cost Manual (51) using a 4.9 m/hour (2 gpm/sq.ft.) filtration rate, a 1988 means index of 201.5 and a city index of 92. The GAC , and anionic resin amended filters increased the construction costs of slow sand filters by approximately 10% and 50%, respectively. A thinner layer was specified for the anionic resin surface amendment because of the faster removal kinetics of ion-exchange over surface adsorption, as determined from unpublished studies (52).

Higher operation and maintenance costs were associated with the packaged treatment plant and the GAC amended filters. Most of the packaged plant unit cost was associated with labor, electrical and chemical costs. It should be recognized that there is a growing trend to automate packaged treatment plants to the extent that labor costs can be significantly reduced. The GAC amended filter unit costs were associated with an increased cleaning frequency and filter media replacement costs. No attempt was made to regenerate the GAC. Recycled GAC could cost less than half of virgin GAC; however, the savings of using recycled GAC may be offset by increases in filter cleanings necessitated by using smaller sized GAC resulting from the regeneration process.

The lower operation and maintenance costs of the conventional and anionic resin amended slow sand filters reflect the simplicity of these two treatment processes. The total labor and unit operation and maintenance costs for the slow sand filters are in agreement with values reported in the literature (53). The anionic resin amended filter unit operation and maintenance costs reflect a significant increase in filter cleanings and the regeneration process required for the resin.

Handling of the anionic resin layer may be greatly simplified by containing the resin in a synthetic quilted filter blanket. The permeable blanket may be sectionally placed and removed from the filter media surface. The regeneration process would consist of water rinses followed by short soaking periods in a chloride solution.

Comparisons of construction and yearly operation and maintenance costs as a function of precursor removal performances are also summarized in Table 6. The average percent precursor reductions for the different treatment processes are based on the general trends noted during the pilot filter studies conducted at Portsmouth.

Table 6. Unit Construction, Operation & Maintenance and Performance Costs
of Amended Slow Sand Filters (SSF) at Ashland, NH (0.3 MGD capacity)

Description	Filter Construction Costs	Total Unit Construction Costs ($/sq.ft.)	No. of Filter Cleanings per Year	Total Labor Costs ($/year)	Filter Media Replacement Costs ($/year)	Total Operation and Maintenance Unit Cost ($/1000 gallon)	Average Percent Reductions in THM Precursors (%)	Construction Costs per Percent Precursor Removal ($/% Removal)	Yearly Operation and Maintenance Costs per Percent Precursor Removal ($/% Removal)
Conventional Slow Sand Filter (SSF)	$200,000	50	2	5,650	185	0.053	15	13,333	387
GAC Amended SSF[a] (2-inch layer)	$224,000	56	4	11,300	46,640	0.529	70	3,200	828
Anionic Resin Amended SSF[b] (1-inch layer)	$300,000	75	4-6	18,000	185	0.166	85	3,571	179
Packaged Clarification Treatment Plant[c]	$500,000	-	-	32,000	36,000[e]	0.621	80	4,163	850

[a] Assumes GAC cost of $35/cubic ft.
[b] Assumes anionic resin cost of $300/cubic ft.
[c] Based on updated costs from EPA manual (1979). Includes coagulation, flocculation, sedimentation, and filtration.
[d] Assumes $10/hour rate
[e] Includes electrical and chemical costs

The most cost effective precursor treatment scheme to construct appears to be the GAC amended slow sand filter, followed closely by the anionic resin amended filter and the packaged treatment plant. The least favorable treatment scheme for precursor reductions is the conventional slow sand filter; such large treatment costs reflect the difficulty of conventional slow sand filters in removing precursor material. The most efficient precursor treatment process to operate appears to be the anionic resin amended slow sand filter followed by the conventional slow sand filter. Both the GAC amended filter and the packaged treatment plant are expensive to operate and maintain at optimum performance.

CONCLUSIONS

Filter Cleaning Methods (Municipal Filters)

o   The municipal filters had robust microbial populations; relative bacterial distributions were a function of filter management and cleaning method.

o   All municipal slow sand filtration plants were successful in turbidity removals ranging from 50-86% in the winter to 57-94% in the early fall. Treated water turbidity consistently achieved values below 0.1 NTU. The sampled slow sand filters were only moderately successful in precursor removals with NPDOC, UV absorbance and THMFP reductions averaging 12-32%, 17-40% and 9-27%, respectively.

o   West Hartford dramatically achieved higher mass removal rates of NPDOC, UV absorbance and THMFP than the New Haven or Springfield slow sand filters. The trend was more pronounced during the higher organic loading conditions in the fall.

o   Municipal slow sand filters demonstrated a preference for removing specific fractions of organic precursor material. Most of the AOM removals from slow sand filters were concentrated in the 5000-500 and less than 500 AMW ranges. Hydrophobic, or ostensibly, humic material was preferentially removed over other constituents of AOM.

o   West Hartford filters consistently achieved higher precursor removals despite having the highest loading rates and shortest media contact times. By harrowing the schmutzdecke into the filter media, much higher cell concentrations over the entire filter depth were observed. This filter cleaning method enhanced NPDOC, UV absorbance, and THMFP mass removal rates at this facility. The bacteria in West Hartford schmutzdecke samples also exhibited the highest INT reducing/AFDC ratios and the most robust mineralization activity. Thus, filter management and source water characteristics both govern the population distributions, activity, and treatment performance provided by the filters.

Filter Media Amendments (Pilot Filters)

o   Pilot-scale microbial population distributions were similar to municipal filters.

o   The anionic resin and GAC amended filters consistently achieved higher organic precursor removals than any other filter media combination studied. Organic precursor removals for the anionic resin and GAC filters frequently exceeded 90 and 75 percent, respectively. However, these excellent organic material removals were tempered by their excessive headloss development. Mass removal rates were significantly higher for these surface amendments.

o   Treatment efficiencies of the aluminum oxide, anthracite and clinoptilolite
         amended filters could not be clearly distinguished from the more
         conventional slow sand filters. Precursor removals only averaged 5-25
         percent. Organic mass removal rates for the pilot slow rate filters (other
         than anionic resin and GAC amended filters) were remarkably similar to the
         municipal slow sand filter.

o   Organic precursor removals in slow rate filtration is a function of both
         the microbiological maturity and adsorptive capacity of the schmutzdecke
         and filter bed.

Filter Loading Rates (Pilot Filters)

o   Organic precursor removals did not appear to be significantly dependent on
         the filter loading rates used in this study. Moreover, changes in organic
         removal rates did not appear to improve as the filters matured during the
         filtration runs.

o   Particulate removals by slow sand filters appeared to be independent of
         typical loading rates but continued to improve as the filters matured
         during the filtration run.

o   Slow sand filters treating high organic waters achieved maturity and
         reached optimum particulate removals significantly more quickly than
         filters treating raw waters of lower AOM.

o   The anionic resin amended slow rate filter consistently achieved higher
         removals of THM precursor material than the slow sand filters operating at
         different loading rates or the alum-based conventional treatment process
         throughout the entire study run. The anionic resin was able to
         consistently satisfy current THM drinking water standards of 100 ug/L and,
         with few exceptions, a proposed MCL of 50 ug/L. The conventional treatment
         process rarely exceeded current standards.

o   The conventional treatment process achieved the best particulate removal
         with turbidity values consistently below 0.2 NTU. The slow sand filter,
         regardless of loading rate or surface amendment, achieved effluent
         turbidity values around 0.3 NTU. Slow sand filters required a significant
         ripening period before optimum removals were reached.

Economic/Construction Cost

o   Where land area requirements are not a concern, the most cost effective
         precursor treatment scheme to construct for small communities appears to be
         the GAC amended slow sand filter followed closely by the anionic resin
         amended filter and a packaged treatment plant.

o   The most efficient precursor treatment process to operate appears to be the
         anionic resin amended slow sand filter followed by the conventional slow
         sand filter. Both the GAC amended filter and the packaged treatment plant
         are expensive to operate and maintain at optimum performance.

ACKNOWLEDGMENTS

Research funding was provided by the American Water Works Association Research Foundation (AWWARF #208-86). We thank Debbie Brink of AWWARF for guidance as project officer. We thank the following operators and administrators of the water treatment facilities for their assistance: Robert Hoyt, Springfield, MA; Dick Allen, West Hartford, CT; Howard Dunn and Ray Creaser, New Haven, CT; Wayne Hughes and Ray Marchand, Ashland, NH; and Al Leathers and Jennifer Royce, Portsmouth, NH. Lastly, we thank our graduate students Jim Fenstermacher and Steve Spanos for their contributions and efforts which made this work possible.

REFERENCES

1. E.W. Akin and J.C. Huff in Water Chlorination: Chemistry Environmental Impact and Health Effects, Vol. 5, R.L. Jolley, R.J. Bull, W.P. Davis, S. Katz, M.H. Roberts, Jr. and V.A. Jacobs, ed., Lewis Publishers, Chelsea, MI, p. 99 (1984).
2. J.J. Rook, Water Treat. Exam., 23, 234-243 (1974)
3. R.J. Bull and L.J. McCabe in Water Chlorination: Chemistry Environmental Impact and Health Effects, Vol. 5, R.L. Jolloy, R.J. Bull, W.P. Davis, S. Katz, M.H. Roberts, Jr. and V.A. Jacobs, ed., Lewis Publishers, Chelsea, MI, p. 110 (1984).
4. Federal Register, 44, No. 231, 68624-68707 (1979).
5. N.B. Munro and C.C. Travis, Environ. Sci. & Technol., 20 768-769 (1986)
6. G.M. Fair, J.C. Geyer and D.A. Okun in Water and Wastewater Engineering, Vol.2, John Wiley & Sons, New York, p. 27-5 (1968).
7. L. Huisman, and W.E. Wood, (1974) in Slow Sand Filtration, WHO, Geneva, 122 pp. (1974).
8. J.L. Cleasby in Physicochemical Processes, W.J. Weber, Jr., ed., John Wiley & Sons, New York, p. 139 (1972).
9. E.W. Steel and T.J. McGhee in Water Supply and Sewerage, 5th Ed, McGraw-Hill, New York, p. 240 (1979).
10. J.M. Montgomery in Water Treatment Principles and Design, John Wiley and Sons, New York, p 545 (1985).
11. Cleasby, D.J. Hilmore and C.J. Dimitracopoulos, J. Amer. Water Works Assoc., 76, 44-55 (1984)
12. L.A. Slezak and R.C. Sims, J. Amer. Water Works Assoc., 76, 38-43 (1984)
13. W.D. Bellamy, G.P. Silverman, D.W. Hendricks and G.S. Logdon, J. Amer. Water Works Assoc., 77, 52-60 (1985)
14. H. Sontheimer and C. Hubele, (1986) In Proceedings of the Second National Conference on Drinking Water, Edmonton, Canada, p. 45 (1986).
15. D. Van der Kooij, A. Visser and W.A.M. Hijnen, J. Amer. Water Works Assoc., 84, 540-545 (1982)
16. J. Mallevialle, and J.C. Counarie, Ozone Sci. Technol., 4, 33-44, (1982)
17. G.L. Amy, M.R. Collins, C.J. Kuo and P.H. King, J. Amer. Water Works Assoc., 79, 43-49 (1987)
18. R.F. Christman and R.T. Oglesby, in Lignin: Occurenece, Formation Structure, and Reactions, K.V. Sarkanen and C.H. Ludwig, eds., Wiley-Interscience, N.Y., P. 769 (1971).
19. M. Alexander, Adv. Appl. Microbiol., 7, 35-80 (1965)
20. T.L. Bott, L.A. Kaplan and F.T. Kuserk, Microbial Ecol, 10, 335-344 (1984)
21. H. de Hann, Freshwater Biol., 4, 301-310 (1974)
22. H. de Hann, Plant Soil Sci., 45, 129-136 (1976)
23. H. de Hann, Limnol. Oceanogram, 22, 129-136 (1977)
24. H. de Hann in Aquatic and Terrestrial Humic Materials, R.F. Christman and E.T. Gessing, eds., Ann Arbor Science, Michigan, p. 165 (1983).

25. A. Geller, Arch. Hydrobiol 99, 60-79 (1983)
26. N. Rifai and G. Bertru, Hydrobiologia, 75, 181-184 (1980)
27. L.J. Tranvik and M.G. Hofle, Appl. Environ. Microbiol., 53, 482-488 (1987)
28. A.R. Trussell, M.D. Umphres, L.Y.C. Leong and R.R. Trussell, J. Amer. Water Works Assoc., 72, 385-388 (1979)
29. M.R. Collins, G.L. Amy and C. Steelink, Envir. Sci & Technol, 20, 1028-1033 (1986)
30. G.R. Aiken in Humic Substances in Soil. Sediment and Water, G.R. Aiken, D.M. McKnight, R.L. Wershaw and P. MacCarthy, eds. Wiley-Interscience, New York, p. 363 (1985).
31. T.T. Eighmy. M.R. Collins, S.K. Spanos and J. Fenstermacher, in Proceeding of the International Converence on Water and Wastewater Microbilogy, D. Jenkins, ed., p. 47-1 (1987).
32. J.M. Fenstermacher, S.K. Spanos, M.R. Collins and T.T. Eighmy, in Proceedings 1988 American Water Works Association Annual Conference (1988).
33. D.L. Balkwill and W.C. Ghiorse, Appl. Environ. Microbiol., 50, 580-588 (1985)
34. I. Bergstrom, A. Heinanen and K. Salonen, Appl. Environ. Microbiol., 51, 664-667 (1986)
35. D.J. Reasoner and E.E. Geldreich, Appl. Environ. Microbiol., 49, 1-7 (1985)
36. F.M. Clark, R.M. Scott and E. Bone, J. Amer. Water Works Assoc., 59, 1036-1042 (1967)
37. W.C. Ghiorse, Ann. Rev. Microbiol., 38, 515-550 (1984)   .
38. O.H. Lowry, O.H., N.J. Rosebrough, A.L. Farr and R.J. Randall, J. Biol Chem., 193, 265-275 (1951)
39. W.B. Lyons, P.B. Armstrong and H.E. Gaudette, Mar. Pollut. Bull. 14, 65-68 (1983)
40. R.V. Subba-Rao, H.E. Rubin and M. Alexander, Appl. Environ. Microbiol., 43, 1139-1150 (1982)
41. S. Simkins and M. Alexander, Appl. Environ. Microbiol., 47, 1299-1306 (1984)
42. R.M. Beloin, J.L. Sinclair and W.C. Ghiorse, Microbial Ecol, 16, 85-97 (1988)
43. P.S. Tabor and R.A. Neihof, Appl. Environ. Microbiol., 48, 1012-1019 (1984)
44. R. Zimmermann, R. Iturriaga, and J. Becker-Birck, Appl. Environ. Microbiol., 36, 926-935 (1978)
45. H.E. Rubin, R.V. Subba-Rao and M. Alexander, Appl. Environ. Microbiol., 43, 1139-1150 (1982)
46. K.R. Fox, R.J. Miltner, G.S. Logsdon, D.L. Dicks and L.F. Frolet, J. Amer. Water Works Assoc., 76, 62-68 (1984)
47. L.A Kaplan and T.L. Bott, Freshwater Biol., 13, 363-377 (1983)
48. A. Geller, Schweiz. Z. Hydrol, 47, 27-44 (1985)
49. M.J. Semmens and A.B. Stables, J. Amer. Water Works Assoc., 78, 76-80 (1986)
50. D.R. McNair, R.C. Sims, D.L. Sorensen and M. Hulbert, J. Amer. Water Works Assoc., 79, 74-81 (1987)
51. USEPA, in Estimating Water Treatment Cost-Vol 3, EPA-60012-79-162c, MERL, Cincinnatic, OH (1979).
52. M.R. Collins, Unpublished Data, University of New Hampshire, Durham, NH (1988)
53. T.R. Cullen and R.D. Letterman, J. Amer. Water Works Assoc., 77, 48-55 (1985)

5

Process developments

5.1 PILOT PLANT EVALUATION OF FABRIC-PROTECTED SLOW SAND FILTERS

T. S. A. Mbwette and N. J. D. Graham — Imperial College of Science & Technology, Department of Civil Engineering, London SW7 2BU

ABSTRACT

A brief review of the potential benefits of using Non-Woven Synthetic Fabrics (NWF) with slow sand filtration is given, together with a summary of previous studies concerned with this innovation. The results of a recent pilot-plant investigation of fabric protected slow sand filters treating stored, surface water is presented. By comparing directly the performance of fabric-protected units with a conventional unit, it has been found that filter run times can be increased by a factor of four or more. By selecting an optimal type and thickness of NWF it is possible to contain, to a large degree, the treatment zone, and thus the loss of hydraulic head, within the fabric layers.

INTRODUCTION

Besides all recent technological advances in the field of filtration of drinking water, at present Slow Sand Filtration (SSF) remains the most efficient individual filtration step in view of its ability to improve the physical, biological and chemical quality of water. Literature (1) indicates that the first Slow Sand Filter (SSF) to supply water (by carts) to a whole town was commissioned at Paisley, Scotland in 1804. In 1827, SSF designed by Robert Thom were used with little success at Greenock, Scotland. However, in 1829, James Simpson commissioned the first successful SSF at Chelsea water works, London. In 1838, James wrote a short article describing the filter in a technical journal (2) . Since then, improvements in basic SSF design have been minimal and mostly concerned with improved methods for handling filter media or changes in filter wall construction materials.

Although following the introduction of rapid sand filters in the 1880's interest in the application of SSF in urban communities dwindled, at present, the cities of Zurich, Amsterdam and London metropolitan still use SSF as a secondary filtration step. Even in the U.S.A. where the use of rapid filtration is universal, SSF have continued to be successfuly used in small scale water supply schemes serving rural communities (3). The resurgence of interest to carry out research work on SSF in the U.S.A. (3-10), almost a century after the introduction of rapid sand filters there, is an encouraging development. In the U.K., it is estimated that at present between 20-25% of potable water is treated by SSF (11,12) either as a secondary or as the main filtration process in small waterworks.

In the Thames Water Authority Region more than 70% of water treatment capacity utilizes SSF as a secondary filtration stage (11). Literature (13-18) describes experience in the use of SSF within the Thames Water Authority(formerly The Metropolitan Water Board) especially in relation to bacterial and virus removal over a long period. The use of SSF in developing countries is also well documented (19-33). In developing countries, the use of SSF is considered to be particularly suitable for rural water supply schemes in view of the ease and simplicity of operation and maintenance. The most common limitation in the applicability of SSF is the direct filtration of very turbid water (34,35). In such cases, improvement of existing pretreatment units or introduction/addition of new pretreatment units is usually the only answer. However, often problems related to lack of proper training of operators or community participation are not unusual (19).

It has been reported that in the Thames Water Authority region up to 70% of direct operational costs for SSF are associated with filter cleaning and resanding (11). During spring and summer, massive die off of algal blooms or proliferation of the blocking type of algae(Anabaena,Asterionella,Stephanodiscus) can at times reduce the filter run lengths down to 12 days (36). This accompanied with the existence of blanket weeds such as Melosira varians,Cladophora and Spirogyra require considerable labour to remove them prior to skimming. In cases where very turbid water is fed to SSF, the operational result is again very short filter run times which also places a big demand in terms of labour and effort in filter cleaning. In some instances SSF units are reported to have been by-passed by operators in order to avoid the cleaning workload (19,22).

This paper summarises the results of recent research work concerned with the application of Non-Woven Synthetics Fabrics (NWF) to the top surface of SSF units. The optimal use of NWF is expected to substantially increase the filter run time and also concentrate most of the treatment process in the fabrics layer, hence reducing the frequency of filter cleaning and resanding. The net effect is therefore a significant reduction in the operating costs of SSF.

NON-WOVEN SYNTHETIC FABRICS (NWF)

Following the discovery of Polymer chemistry in the 1930's, commercial production of synthetic fabrics started in 1942 when the first few thousand metres were produced in the U.S.A. (37). Synthetic fabrics are man-made textiles which consist of organic compounds made up of long, chain-like molecules with repeating molecular units linked by covalent bonds (38). A distinct definition of a non-woven fabric is very hard to find in the literature (39-41), however, all fabrics which are manufactured directly from fibres or webs without the need to produce yarns are regarded as non-woven fabrics. A synthetic fabric which is not produced by either weaving or knitting or both can be regarded as a NWF although composite fabrics made of a woven and non-woven part are not regarded to be NWF .

Three synthetic non-woven product categories can be identified (37,40):

<u>Wet laid NWF</u>-Production involving water extraction,drying and rolling of a dispersion of synthetic fibre staples in water. Binders are or can be added at a convenient stage of the production process.

<u>Dry laid NWF</u>-By far the largest and most diversified. Production involves formation of webs by carding or air laying and subsequent subjection to either needlefelting, spunlacing or spraying with an aqueous resin. The five distinct processes involved in production of the needlefelt fabrics are blending,carding,web forming,needling and finishing.

<u>Spunbonded NWF</u>-In this case, a bulk thermoplastic polymer is melt-spun and the resulting fibres are bonded by heat, pressure or chemical activation.

It must be noted that in the commercial market fabrics are available which are made by a combination of needlefelting with either thermic bonding or spunbonding.

At present the principal NWF are either solely made of or a combination of Polypropylene,Polyester,Polyethylene,Poly-vinyl Chloride,Polyamide (Nylon-6 or 6.6) and Polystyrene fibres. Literature (42) suggests that in the commercial market Polypropylene dominates the non-woven fabrics field. Interpretation of physical,chemical and mechanical properties of the NWF vary largely due to the non-existence of standard testing procedures (40,41). Nevertheless, reviews of general properties of the fabrics/fibres have been given(42,43,48). Present experience in using NWF on SSF has proved that manufacturing processes influence fabric performance. For cases where fibres are bonded with binders, the stability and resilience of the binder upon exposure to adverse conditions is critical in fabric selection (43).In almost all cases, the mechanical strength of fabrics is an important factor in relation to handling during cleaning . This means excessively fragile NWF are not suitable for this purpose. Recent experiments have shown a big difference in filtration performance for two batches of fabrics needlefelted on scrims with different properties as observed with a scanning electron microscope. The results seem to suggest that the use of multifilament scrims is more favourable than monofilament scrims in NWF.

Some production processes involving calendering as a surface
finish for fabrics are already suspected to render NWF unsuitable
for use on SSF beds largely due to blinding. This observation
conforms to doubts expressed previously (42).   Experiments
carried out todate (44,45) point out that most spunbonded NWF
(and generally thin dense NWF) are not suitable for application
on SSF. The non-existence of in-depth filtration within this
type of fabric is a major disadvantage as dominance of cake
filtration results in very short filter run times (48). In the
study described in this paper both NWF used were made from
Polypropylene which is reported to have the following advantages
over the other polymers (37,46,47):

-High abrasion resistance.
-Blends well with other fibres during production.
-Since it is free from polar groups (like Amides in Nylon and
 Esters in Polyester) it has a good fibre stain resistance and
 hence can be expected to be easier to clean when dirty.
-Has an excellent resistance to most chemicals, acids, alkalis
 and oxidizing agents met in drinking water treatment.
-Has sufficient resistance to fungus and organic acids.
-Reasonably high resistance to dry heat or U.V. light degradation
 especially when stabilized.

In general polypropylene fibres offer most of the characteristics
of an ideal NWF at lower costs in comparison to other polymers.

PREVIOUS RESEARCH STUDIES

Initial research work in the use of NWF on SSF beds at Imperial
College, UK, started in 1982 (49). This work contributed to the
design of a disaster relief water treatment package which used an
arbitrarily chosen fabric (45). The performance of such packages
installed in refugee communities in Somalia in 1985/86 has been
reported (24,50). However, the inability of the selected fabric
to stop penetration of impurities into the sand bed prompted
further research work in the selection of a better fabric(s). A
comprehensive review of research work done in the use of NWF has
been presented previously (39). After carrying out a survey of
commercially available NWF in the U.K. market, a selected range
of fabric samples and one SSF sand sample were subjected to
hydraulic and filterability tests in the laboratory (44).
This work led to the selection of the fabrics used in this study.

The objectives of this study were to evaluate the performance of
pilot-scale SSF units fitted with a surface matting of particular
configurations of NWF layers. The potential benefits were
considered to be two-fold; firstly, to substantially decrease the
rate of headloss development, and secondly, to limit (if not
completely eliminate) the passage of impurities through the
fabric into sand such that routine maintenance involves cleaning
of the fabrics alone.

PILOT PLANT DESCRIPTION

Layout

The pilot plant was located at the Thames Water Ashford Common Treatment Works, Surrey. It consisted of six (U.V.- stabilized) medium density polythene circular SSF tanks, designated AFM1 to AFM6, with a nominal diameter of about 1.4 metres and a height of about 1.2 metres (Figs.1,2 & Plate 1). Each tank was prefabricated with a 5cm. wide rim at mid-height. The six SSF units were fed by one header tank placed at a height of approximately 3.0 metres above ground level. Water was pumped to the header tank from the influent to the main works SSF beds No.3 or 4 by two submersible pumps in the SSF supernatant water. Filtered water and influent overflow water from the SSF units were discharged to a drain. A separate header tank overflow was provided as indicated on the layout plan of the pilot plant in Fig.1 . A typical cross section through a SSF unit is shown in Fig.2 . Totalizing Kent meters were provided downstream of the filter beds in order to confirm the filtration rate. Each SSF unit was provided with a float valve rate controller downstream of the flow meters which operated as indicated in Figs.3.1 and 3.2 at the begining and end of the filter runs, respectively (32). Four of the SSF tanks were filled with 600 mm of sand for studying the performance of SSF with and without NWF protection. In order to evaluate directly the perceived benefits of a NWF layer, the two additional SSF units did not contain 600 mm of sand but were fitted to support a 25 mm. layer of either sand (Fig.4) or fabric.

Raw water source

Between June 1987 and May 1988, Ashford Common Treatment Works received raw water from the Queen Mary and Wraysbury reservoirs. The main source was Queen Mary but when its quality deteriorated additional water was drawn from one of the Staines group of reservoirs, and in this particular case Wraysbury reservoir. A survey of weekly identification of algae in the two reservoirs indicates that throughout the experimentation period, Blue green algae and Diatoms were predominant. While the Chlorophyta Volvox sp. were dominant only in August and September 1987,other green algae species were predominant as early as June 1987. The yellow green algae Tribonema were noticeable from September 1987 until January 1988 and also in May 1988. Some Dinoflagellates were predominant from July until October 1987. Occasional peaks of Cyanophyta (Anabaema and Aphanizomenon sp.) and Pennate diatoms were observed. Very high levels of diatomaceous algae were reported in the Queen Mary reservoir in October 1987.

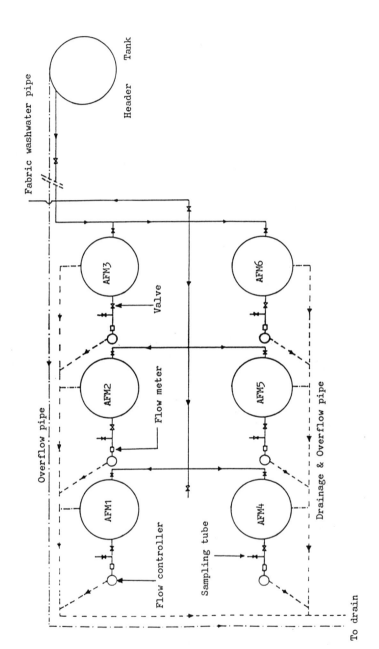

Fig.1 Layout Plan of Ashford Common Pilot Plant.

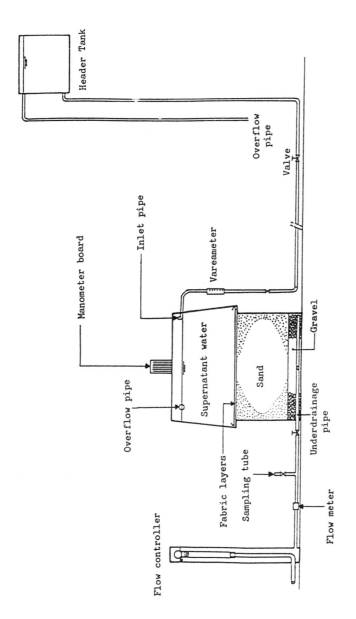

Fig.2 Typical cross section through a SSF unit.

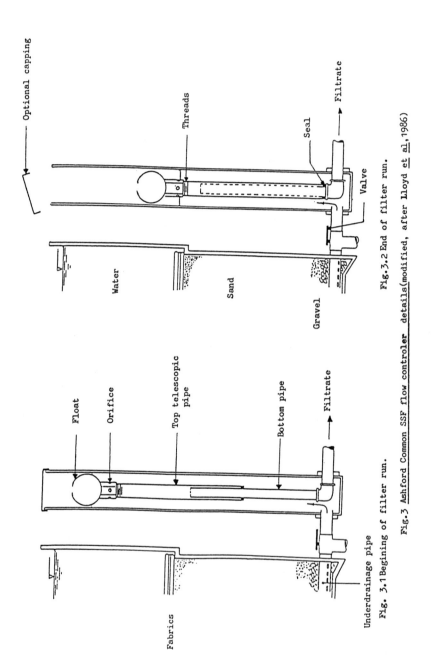

Fig. 3.1 Begining of filter run.

Fig. 3.2 End of filter run.

Fig. 3 Ashford Common SSF flow controler details(modified, after Lloyd et al,1986)

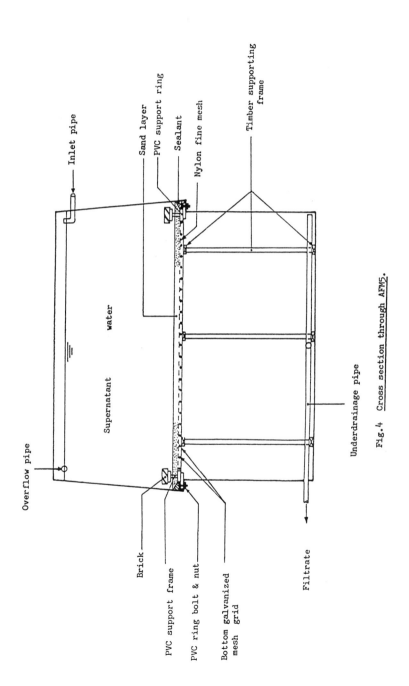

Fig.4 Cross section through AFM5.

Plate.1   Close-up of slow sand filter units
          at Ashford Common pilot plant.

With regard to the identification of the dominant species of
animals in the reservoirs, from June until November 1987 and
March to May 1988 Cladocerans and Copepods were predominant.
Howevever, between December 1987 and February 1988 Cladocerans
alone were predominant while Rotifera were observed at high
levels in April and May 1988. A high level of Daphnia was also
reported in June 1987.   The quality of reservoir water and that
leaving the main works primary filters (micro-strainers) was
continually monitored for Particulate Organic Carbon (POC) and
Chlorophyl-a (Chl-a) in order to establish the removal of algae
and also to quantify the organic load going into the SSF beds.
Figs. 5 and 6 show the variation of POC and Chl-a during the
experimentation period, respectively.   It should be noted that
during periods of high algal blooms in the reservoirs, the
micro-strainers became so much overloaded that a proportion of
the raw water was allowed to go unstrained to the SSF beds in
order to maintain the works flow.

Although it is worthwhile to have a knowledge of the biochemical
quality of reservoir water, experience (52,53,16) has shown that
algae predominance in the reservoir water subject to primary
filtration does not necessarily correspond to that in the

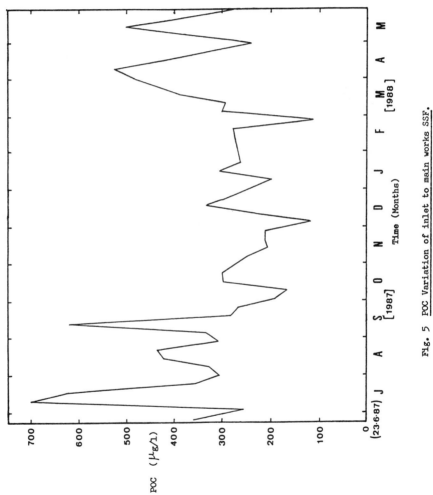

Fig. 5 POC Variation of inlet to main works SSF.

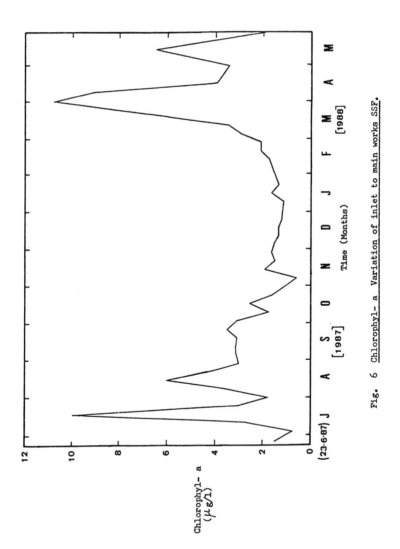

Fig. 6 Chlorophyl- a Variation of inlet to main works SSF.

subsequent SSF basins. This is because the algal populations form part of a complex ecosystem whose balance is largely dependent on the intensity of micro-organisms (bacteria,fungi,animal predators, etc) and the organic nutrient levels of the influent flow and water already in the SSF basin. High levels of algae in the SSF basins can be a result of accumulation during the course of filtration, especially in the case of smaller algae like Stephanodiscus astraea, Stephanodiscus hantzschii, Asterionella and Fragilaria which although do not reproduce rapidly in the SSF are predominant in the raw water reservoirs. Otherwise they can be an outcome of fast proliferation in the SSF basin after having passed the primary filters; e.g. diatomaceous form like Chlamydomonas spp.,Melosira varians, Cladophora and Nitzschia acicularis which are reported to reproduce rapidly in the SSF basins (16,53). Unfortunately in this study no identification was possible of either the raw water (main works supernatant water) or the pilot plant SSF schmutzdecke in order to establish the dominant algal species (apart from the filamentous algae observed).

FILTER MEDIA CHARACTERISTICS

The characteristics of the sand used as SSF media was as follows: an effective diameter (d10) of 0.30 mm and a uniformity coefficient (d60/d10) of 1.93. These are in conformity with generally accepted ranges of effective diameter and uniformity coefficient (20,34) . However, recent work (6,19,35) has showed that there are no major risks involved in using sand having specifications which slightly deviate from the conventional range.

Table.1 gives a summary of the physical and geometric characteristics of the two NWF used during these experiments. Prior to this study, field tests were undertaken which enabled a selection to be made of the most promising fabrics for filtration of Ashford Common raw water from among those found in the U.K. market (39,44). Fabric lab. No. 28 has properties that lie in the mid-range of the commercial NWF previously studied (44). Fabric lab. No. 44 represents a relatively dense, modest porosity material which has a very high filtration efficiency.

Table.2 gives the filter media composition of the SSF units during the period of experiments. AFM1 and AFM5 contained a thin layer (25 mm) of fabrics and sand, respectively. AFM3, 4 and 6 contained various fabric layers on top of 600 mm of sand.

Table.1 NWF Physical and Geometric Properties

Property	Fabric Lab. No. 28	Fabric Lab. No. 44
Fibre Polymer Comp.	Polypropylene	Polypropylene
Approx. fibre diam. (μm)	33	27.7/30.5/41.4
Average nominal fabric thickness (mm)	4.8	1.3
Fibre specific gravity	0.91	0.91
Average fabric bulk density(g/cc)	0.100	0.231
Calculated porosity (%)	89	75
Specific surface area (Sq.m/Cu.m)	13,266	33,238

Table.2   Filter   Media   Composition

SSF	23.06.87 - 11.02.88 (34 weeks)	11.02.88-17.03.88 (5 weeks)	17.03.88-17.05.88 (9 weeks)
AFM1	5 Layers lab. No. 28(24.0 mm thick) only	5 Layers lab. No. 28(24.0 mm thick) only	5 Layers lab. No. 28(24.0 mm thick) only
AFM2	No Fabric 600 mm sand only (Reference unit)	No Fabric 600 mm sand only (Reference unit)	No Fabric 600 mm sand only (Reference unit)
AFM3	4 Layers lab.No.28 over 1 layer lab. No.44(20.5mm thick) above 600 mm sand	No Fabric 600 mm sand (2nd Reference)	Unit stopped
AFM4	4 Layers lab.No.28 (19.2mm.thick) above 600 mm sand	4 Layers lab. No. 28(19.2 mm thick) above 600 mm sand	No Fabric 600 mm sand (2nd Reference)
AFM5	No Fabric (25 mm. sand only)	No Fabric (25 mm sand only)	No Fabric (25 mm sand only)
AFM6	6 Layers lab.No.28 (28.8 mm thick) on 600mm sand	6 Layers lab.No. 28(28.8mm thick) on 600mm sand	6 Layers lab.No. 28(28.8mm thick) on 600mm sand

PILOT PLANT TESTS

The pilot plant was operational from the end of June 1987 until
May 1988. However, the units with suspended layers of fabric No.
28 (AFM1) and sand (AFM5) presented many problems largely due to
excessive sagging of the supporting frame at the centre as a
result of being side-supported.   Either short-circuiting on the
edges or mesh splits had to be dealt with from time to time until
they were provided with new supporting frames at the end of
March/begining of April 1988.   After effecting these changes no
further operational problems were observed.   The SSF unit AFM3
had to be stopped on 01.03.88 during its second run as a second
reference as a result of excessive leakage around the flow
controller base.

All SSF units were operated at a constant filtration velocity of
0.15 m/h and filter runs were terminated when this flow velocity
could no longer be maintained.   This corresponded to a headloss
of approximately 300 mm . Ambient and SSF water temperatures were
measured with a mercury thermometer capable of registering
temperatures in the range of -10 to 40 degrees Centigrade.
Filter headloss was monitored by eight manometer tubes,
individually located at heights of 70 mm above the rim (No.1), at
the level of the rim ($\pm$ 0 mm, top sand bed level,No.2), 25 mm
(No.3), 55 mm (No.4), 80 mm (No.5), 110 mm (No.6) and 210 mm
(No.7) below the rim level.   The eighth tube was located within
the supporting gravel depth to register the maximum headloss
across the filter beds.

As expected, the deposition of impurities in the reference units
was largely confined to the top 10 - 15 mm.   These units were
cleaned by removing the top 40 mm of sand, to ensure the total
removal of all deposited impurities, and restoring the bed depth
with fresh sand. For the SSF units with fabric, filter run times
were so long that only a single filter run was possible in the
time available.

For all the units protected with fabrics on top of the sand, the
majority of the headloss was registered by manometer tube No.2 .
However, in some cases, sand beds compacted slightly while fixing
in position the metal hooks holding down the mesh grid placed
above the fabrics. This led to failure of manometer tube No.3 to
register headloss in the sand bed.   Instead, manometer tube No.4
acted as the measure of headloss in the top of the sand.   In
general, units AFM1 and AFM5 registered maximum headlosses in
manometer tube No.2 and headloss records helped to confirm short-
circuiting or mesh splits in view of their tendency to decrease
soon after its occurrence.   However, sometimes headloss recovery
occurred across the SSF beds especially during spring/summer as
observed in AFM6 and AFM2 and previously (51) .

The following water quality parameters were routinely monitored; Turbidity, Faecal Coliforms , Particulate Organic Carbon (POC) and Particle size analysis. In addition, water and ambient temperature, filter headloss and filtration rates were monitored continuously.

Turbidity-Raw water and SSF effluents were sampled and measured with a HACH model 2100A turbidimeter.

Faecal Coliforms-Raw water and SSF effluents were enumerated by the membrane filtration method using a Delagua field water testing kit, developed by the University of Surrey, U.K.. The experimental procedure is fully described elsewhere (54). Membrane Lauryl Sulphate was used as the nutrient and the processed membranes were incubated at 44±0.5 degrees Centigrade for between 20 - 24 hours prior to counting. For each sample, duplicate sets of membranes were prepared and average results were used.

Particulate Organic Carbon (POC)- Water samples were analyzed for POC once per week at the Thames Water Wraysbury laboratory according to a procedure described in detail elsewhere (45) based on the oxidation of organic Carbon with a Chromic acid solution. While usually the remaining Chromate would be titrated with Ferrous Ammonium Sulphate, to save time the titration is replaced with absorbance measurements which are directly related to the Chromate concentration. Due to financial limitations, this parameter was monitored only from July until September 1987.

Particle Size Analysis-Particle size analysis was carried out with a TAII Coulter Counter/PCA I .Both the raw water and SSF filtrates were analyzed with an 100 μm. orifice diameter tube. During spring the presence of Daphnia and Chironomid larvae in the raw water resulted in frequent blockages of the counter orifice. A microscopic observation of some Daphnia showed an average maximum chord diameter (55) of 200 μm. It should be noted that the Coulter volume diameter is not directly comparable to the maximum chord diameter. In general, particle count results could be substantially influenced by any hydraulic instability of the flow such as draining and tracking especially for the suspended media in AFM1 and AFM5.

DISCUSSION OF RESULTS

From 23.06.87 until 17.05.88, the ambient recorded temperature ranged from 3.0 to 33.0 degrees Centigrade and the SSF water temperature ranged from 3.5 to 24.1 degrees Centigrade.

Hydraulic Performance

Table.3 gives the filter run times for the reference and fabric protected SSF units. The average filter run length of 38 days for the reference filter (AFM2) compared well with the main works SSF units.

Table.3   Filter Run Times

SSF Unit	Filter Media Composition	Average Filter Run Length(days)	Improvement Factor
AFM2	Sand only (Reference)	38	1
AFM3	4 lab. No.28 on 1 lab.No.44	160*	4.2
AFM4	4 lab No. 28	178*	4.7
AFM5	25mm. Sand	32*	0.84
AFM6	6 lab. No.28	319*	8.4

*   Single filter run

In order to separate and quantify actual differences in filter run times because of the application of NWF from natural variations between two parallel units, two reference filters were operational for certain periods of the investigation (Table 2). The average run time for AFM4 as a reference unit was 40 days, which compares well with AFM2 . The AFM2 winter run time of 51 days is fairly comparable to 56 days recorded for AFM3 reference just a month later. On the basis of this evidence it was concluded that there were no significant behavioural differences in the sand media of AFM2, 3, 4 and 6. During filter runs, the reference filters registered most of the headloss in the manometer tube No.3 located at about 25 mm depth in the sand (Fig.7) . This together with the close similarity between the average filter run times of AFM5 and AFM2, clearly demonstrated the surface nature of the treatment process.

The magnitude of the improvement to the filter run time by the application of the NWF layers is clear from Table 3. Unfortunately, as a result of the very long run times of AFM3, 4 and 6 it was not possible to achieve successive filter runs after the first. It therefore remains an uncertainty what the long term average run times would be for these units. Evidence from previous work (44,45) and current studies are inconclusive as to whether successive filter runs have an average run time which is significantly shorter than the initial run. It is clear from the headloss profile data (e.g. Fig.8) that, generally, the

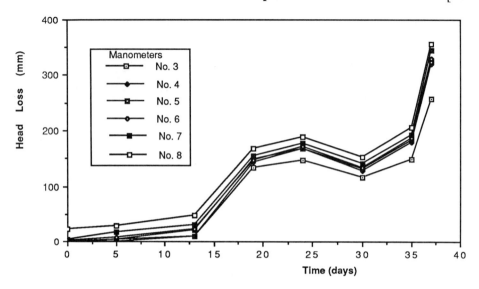

Fig. 7 AFM2 Typical Filter Head Loss

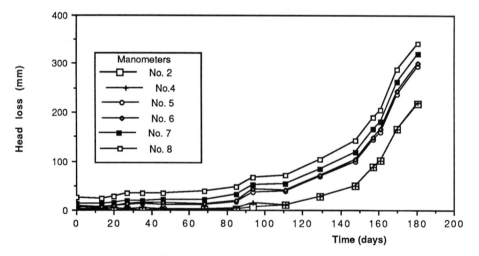

Fig.8   AFM4   Filter   Head Loss

deposition of influent water impurities occurs almost entirely
within the fabric layers. Because of the advantageous properties
of the fabrics the consequential rate of headloss development is
much slower than for the reference filter. This is evident from
the headloss data for the AFM1 and AFM5 units, where the latter
(25 mm sand layer) reached its operational limit after 32 days in
contrast to the same thickness of fabric lab. No. 28 which only
registered a headloss of 10 mm after 50 days.

For unit AFM3 with 4 layers lab.No.28 over one layer lab. No. 44,
impurities appeared to have penetrated fully into the top three
fabrics with intense biological activities confined to the top
two fabric layers. Partial penetration of the fourth lab.No. 28
fabric and also onto the bottom fabric was observed. In this
case it seems the blockage of pores in the bottom fabric (No.44)
might have been responsible for bringing about an earlier end of
the filter run time. Penetration of a few impurities through to
the sand was visible and edge penetration was apparent on the
sand bed upon removal of the fabrics. This observation rules out
the use of calendered NWF for protection of SSF sand beds;
previously (44) the washed fabrics were found to lose their
porosity progressively with time. AFM4 with 4 layers of fabric
lab No.28 showed a deposition pattern similar to AFM3 but visible
penetration of impurities onto the sand bed was more pronounced
and deeper (about 20 mm.); edge penetration on the sand bed was
also very apparent. This suggests that the thickness of the
fabrics provided was not sufficient, as suspected previously
(44).

In the case of AFM6 with six layers of fabric lab No.28,
inspection of the sand throughout the bed depth at the end of the
filter run showed no evidence of solids accumulation or impurity
deposition. After draining the supernatant water of AFM6 at the
end of the filter run , a large quantity of micro-organisms were
observed to be grazing in the heavy growth of filamentous algae
which was attached to the mesh grid and also onto the fabric
fibres. This attachment of algae prevented them from floating
during the day and hence causing operational problems related to
disturbance of the schmutzdecke, which has been previously
reported. The algae layer was full of silt indicating that it
acted as an additional filter medium.

In agreement with the observations of other workers (53),
Trichoptera, May flies (Ephemeroptera), Midges (Chironomids) and
some Crustaceans (Daphnia,Bosmina) were identified on the
schmutzdecke. Impurities penetrated extensively through the top
two layers of fabrics. The middle two fabric layers showed
considerable silt deposition but not as much as the top two
layers. The bottom fabrics were lightly tainted brown as a
result of intense edge penetration. The sand was lightly
tainted in the top 10mm with localized deposits on the edges
probably due to short circuiting after complete blockage of the
top fabrics. It is worthwhile to remember that these deposits
corresponded to 319 days of continuous filtration.

Treatment Performance-In terms of turbidity, the filtrate of AFM5 was always slightly higher than AFM1 such that their typical means were 0.66 and 0.50 NTU, respectively. The performance of the rest of the SSF units, whether fabric protected or not, showed no significant differences in terms of turbidity. Between 21.10.87 and 19.05.88,the influent water turbidity ranged from 0.95 to 5.5 NTU, and the effluent turbidities of AFM2,3,4, and 6 were usually in the range of 0.20-0.45 NTU. The corresponding average particle number concentrations for the influent water and the filtrate from AFM2 are shown in Table 4.

Table. 4 Average Particle Number Concentration

Average particle diamater (μ m)	Average Number Concentrations per ml.	
	Influent Water	AFM2 Filtrate
2.26	38361	14738
2.85	17395	6627
3.59	8385	3539
4.52	4767	2359
5.70	2401	1145
7.18	1343	524
9.04	756	247
11.39	500	145
14.35	167	73
18.08	97	55
22.78	55	2

From 08.07.87 until 23.09.87, the influent water POC ranged from 898 to 393 (μ g/l). Except during the initial fabric ripening period, the POC removal in AFM1 ranged from 52-89% showing that the fabrics were able to achieve a notable POC removal from raw water. The removal range for AFM5 was generally of equal magnitude to AFM1 during stable filter runs; however, after drainage or maintenance, the subsequent reduction in filtrate POC was a very gradual process for these units. The performance of the reference unit (AFM2) was predictably better than AFM1 and AFM5 with its POC removal being in the range of 65-93%. The fabric protected filter AFM3 performed very similarly to AFM2, but its variability was much less (i.e. removal range 82-93%). Overall, the best performance of all the units was AFM6 whose POC removals were consistently very high (87-95%).

Between 21.10.87 and 19.05.88, the influent water Faecal Coliform counts ranged from 0 to 105 per 100 ml. . There were no Faecal Coliforms detected during the whole month of April 1988. This phenomenon might have been a result of changes in the ecological

Table.5   Faecal Coliforms per 100 ml.

Date	Raw water	SSF Unit					
		AFM1	AFM2	AFM3	AFM4	AFM5	AFM6
21.10.87	33	1	0	1	0	-	0
25.10.87	40	1	0	0	0	-	0
30.10.87	34	2	0	1	0	-	0
06.11.87	39	2	0	2	0	24	1
10.11.87	35	2	0	0	0	21	0
18.11.87	28	2	0	1	0	16	1
22.11.87	28	3	1	1	0	27	1
25.11.87	39	3	1	0	0	32	1
29.11.87	28	2	1	1	1	30	1
06.12.87	27	3	0	2	1	27	1
13.12.87	42	3	0	-	0	38	2
21.12.87	29	4	1	5	0	-	0
06.01.88	58	39	0	-	0	-	0
02.02.88	80	48	-	0	-	-	0
10.02.88	105	63	-	0	-	-	0
21.02.88	28	7	1	1	2	7	0
28.02.88	34	10	0	3	0	5	0
06.03.88	15	14	0	-	0	7	0
20.03.88	14	-	0	-	1	-	0
27.03.88	8	-	-	-	0	0	0
14.04.88	0	0	0	-	0	0	0
24.04.88	0	0	0	-	0	0	0
01.05.88	0	0	0	-	0	0	0
04.05.88	1	0	0	-	0	1	0
08.05.88	0	0	0	-	0	0	0
19 05.88	1	0	0	-	0	0	0

regime in the SSF supernatant as a result of the spring algal blooms along with the high levels of Rotifera, acknowledged as bacterial predators (16,57) which were observed in the raw water reservoirs in April and May 1988. Table.5 gives a summary of the Faecal Coliform concentrations determined during the study period. In general, SSF units AFM2,3,4 and 6 showed very satisfactory removals(> 95%) all the time such that if filtrate Faecal Coliforms were not nil then they were in most cases not more than 2 per 100 ml. . The latter counts usually occurred a few days after cleaning the units or after some form of flow disturbance. However, the results for the suspended layers of fabrics (AFM1) and sand (AFM5) showed higher filtrate Faecal Coliforms during periods of operational problems and in both cases, after the units were rehabilitated no Faecal Coliforms were detected in the filtrate.

CONCLUSIONS

1. The application of an optimal thickness and type of NWF as a surface matting can dramatically increase SSF run times.

2. By the selection of an optimal thickness and type of NWF matting, particle deposition and biological accumulation can be contained within the fabric, thereby avoiding headloss development within the underlying sand.

3. The correct selection of thickness, type and configuration (if more than one fabric) of NWF matting is crucial to avoiding solids penetration into the sand and maximising filter run times. Experience suggests that calendered or thermic bonded NWF are not suitable in this application.

4. In general, the overall physical and biological treatment performance of conventional slow sand filters is very high so that application of NWF matting appears to make a negligible improvement to this.

ACKNOWLEDGEMENTS

The authors gratefully ackowledge the support and assistance of the Thames Water Authority throughout this project. The technical support of Clarke,B. and Fairlamb,M. (Kingston Polytechnic) in field test work is hereby acknowledged.

REFERENCES

1.Baker,M N , "The quest for pure water:The history of water purification from the earliest records to the twentieth century", American Water Works Association,2nd ed.,Vol.I,1981,USA .
2.Simpson,J ,"Filtration of Water at Chelsea Water Works", The Civil Engineer and Architects Journal,No.15,p.392,1838.
3.Slezak,A L and Sims,R C , "The Application and Effectiveness of Slow Sand Filtration in the United States",JAWWA, 76,12,p38,1984.
4.Cullen,T R and Letterman,R D , "The Effect of Slow Sand Filter Maintenance on Water Quality", JAWWA, 77,12, p.48,1985.
5.Bellamy,W D ,Silverman,G P, Hendriks,D W and Logsdon, G S, "Removing Giardia Cysts with Slow Sand Filtration",JAWWA, 77, 2,p.52,1985.
6.Bellamy,W D, Hendriks,D W and Logsdon,G S , "Slow Sand Filtration:Influences of Selected Process Variables", JAWWA, 77,12,p.62,1985.
7.Cleasby,J L,Hilmoe,D J and Dimitracopoulos,C J, "Slow Sand and Direct in-line Filtration of a Surface Water", JAWWA, 76,12,p.44,1984.
8.McNair,D R, Sims,R C , Sorensen,D L and Hulbert, M , "Schmutzdecke Characterization of Clinoptile-Ammended Slow Sand Filtration" , JAWWA,80,12,1987.
9.Fox,K R ,Miltner,R J , Logsdon,G S , Dicks,D L and Drolet,L F, "Pilot Plant Studies of Slow-rate Filtration", JAWWA, 76,12,p.62,1984.

10.Seelaus,T J , Hendricks,D W and Janonis,B A, "Design and Operation of a Slow Sand Filter", JAWWA, 78,12,p.35,1986.

11.Rachwal,A , Rodman,D , West,J and Zabel,T , "Uprating and Upgrading Slow Sand Filters by pre-Ozonization", Effl. and Wat. trmt. Jour.,26,5/6,p.143,1986.

12.Warden,J H and Craft,D G , "Waterworks Sludge Production and Disposal in the UK", Technical report TR 150,WRC, Medmenham,UK,1980.

13.Windle,T E ,"39th Report on the results of bacteriological, chemical and biological examination of the London waters for the years 1959-1960, Metropolitan Water Board(MWB), London,UK,1961.

14.Windle,T E ,"41st Report on the res. of bact.,chem. and biol. exam. of the London wat. for the yrs. 1963-1964",MWB, London,UK,1965.

15.Windle,T E , "42nd Report on the res. of bact.,chem. and biol. exam. of the London wat. for the yrs. 1965-1966", MWB,London,UK,1967.

16.Windle,T E , "43rd Report on the res. of bact.,chem.,and biol. exam. of the London wat. for the yrs. 1967-1968",MWB, London,UK, 1969.

17.Windle,T E , "44th Report on the res. of bact.,chem. and biol. exam. of the London wat. for the yrs. 1969-1970",MWB, London,UK, 1971.

18.Windle,T E , "45th Report on the res. of bact.,chem. and biol. exam. of the London wat. for the yrs. 1971-1973",MWB,London,UK,1974.

19.Mbwette,T S A, "Horizontal Flow Roughing Filters for Rural Water Treatment in Tanzania", M.Sc. Thesis, University of Dar-es-Salaam, Tanzania, 1983.

20.Visscher,J T ,Paramasivam,R, Raman,A and Heijnen,H A, "Slow Sand Filtration for Community Water Supply: Planning,Design,Construction, Operation and Maintenance", IRC TP.24, The Hague,The Netherlands,1987.

21.Ellis,K V , "Slow Sand Filtration",CRC Reviews in Environmental Control,15,Issue 4,p.315,1986.

22.Mbwette,TSA, "Experiences with Slow Sand Filters in Tanzania", Proc. 11th WEDC Conference on Water and Sanitation in Africa, p.E3.1,Dar-es-Salaam, Tanzania.

23.Wegelin,M , "Horizontal Flow Roughing Filtration: An Appropriate Pretreatment for Slow Sand Filters in Developing Countries", IRCWD Newsletter No.20, Zurich, Switzerland,1984.

24.Graham,NJD, "A Review of Water Facilities in Refugee Communities in Southern Somalia", Dept. of Civil Eng., Imperial College, London,1985.

25.El Basit,S and Brown,D., "Slow Sand Filter for Blue Nile Health Project", Waterlines, 5,1,p.29,1986.

26.Mhaisalkar,V.A., "Planning, Design and Construction of Slow Sand Filters", Paper presented at the International Seminar on SSF, Nairobi, Kenya,1983. (Unpublished).

27.Paramasivam,R , Mhaisalkar,V A and Berthouex,P M, "Slow Sand Filter Design and Construction in Developing Countries", JAWWA, 75,4,p.178,1981.

28.Pardon,M ,Wheeler,D and Lloyd,B , "Process aids for Slow Sand Filtration", Waterlines,2,2,p.24,1983.

29.Vigneswaran,S.   ,   "The   design   of   filters   with   low   cost
materials for small communities", Proc. 8th WEDC Conf. on Water
and Waste Engineering in Asia,p.51,Madras,India,1982.
30.Mbwette,TSA and Wegelin,M , "Field Experiences with HRF-SSF
systems in Treatment of Turbid Surface Waters in Tanzania", Water
Supply, 2, 3/4,1984.
31.Mbwette,TSA and Wegelin,M , "Engineering Considerations of HRF
preceeding SSF", Waterlines,1988. (In Press).
32.Lloyd,B J ,Pardon,M and Wheeler,D., "Final Report on the
development, evaluation and field trials of a small scale, multi-
stage, modular filtration system for the treatment of rural water
supplies" , University of Surrey, CEPIS, PAHO, WHO and Ministry
of Health Peru,1986.
33.Wheeler.D ,Pardon,M.,Lloyd,B J and Symonds,C N, "Aspects of
pre-filtration concerned with the application of small scale SSF
in rural communities",University of Surrey, March 1985.
34.Huisman,L and Wood,W E ,"Slow Sand Filtration", WHO, Geneva,
1974.
35.Van Dijk,J C and Oomen,JHCM , "Slow Sand Filtration for
Community Water Supply in Developing Countries", IRC TP11,The
Hague,The Netherlands, 1978.
36.McCann,B , Ashford Common Waterworks, Personal Communication,
1988.
37.Manchester   Textile   Trade   Press,   "Manual   of   Non-
Wovens",KRCMA,RADKO, W.R.C. Smith Publ.Co.,USA, 1971.
38.Blythe,A R , "Electrical Properties of Polymers".   Cambridge
solid state science series,p.2, Cambridge University, UK.
39.Mbwette,TSA and Graham,NJD ,"Improving the Efficiency of Slow
Sand Filtration with Non-Woven synthetic Fabrics", Filtr. and
Sep.,24,1,p.46,1987.
40.Sandstedt,H N ,"Non-Wovens in Filtration Applications", Proc.
2nd World Filtration Congress,London,1979.
41.Purdy,A T , "Structural Mechanics of Needlefelt Filter Media",
Proc. 2nd World Filtration Congress,London,1979.
42.Fletcher,E A ,   "Needled Filter Media for   Solid/Liquid
separation", Proc. 2nd World Filtration Congress,London,1979.
43.Ruddock,E A ,   "Fabrics and meshes in roads and other
pavements:A state of the art review", Technical Note 87,CIRIA,
London, 1978.
44.Mbwette,TSA and Graham, NJD ,"Investigations Concerning the
Use of Non-Woven Synthetic Fabrics with Slow Sand Filters", Final
Report for Thames Water Authority, 1987.
45.Bridges,G ," The Use of Fabrics in Slow Sand Filters", M.Sc.
Thesis, University of London, 1985.
46.Datye,K V and Vaidya,A A , "Chemical Processing of Synthetic
Fibres and Blends", John Wiley and Sons,New York:Chichester,1984.
47.Cannon,E W ,   "Fabrics in Civil Engineering",   Civil
Engineering, p.39,March 1976.
48.Gurtler,H G , "Filter Media for the Filtration of Liquids,
their Properties and Applications", Proc. Water Filtration
Symposium,p.6.1, Antwerp, 1982.
49.Liversidge,P W , "Potable Water Supply Scheme for Refugee
Camps - Evaluation of a Modular SSF, M.Sc. Thesis, University of
London,1982.

50.Graham,N. and Hartung, H , "Performance of Slow Sand Filters in Refugee Water Supplies in Somalia" <u>Waterlines</u>, <u>6</u>,3,p.19, 1988.
51.Clarke,B , "The Improvement of Slow Sand Filter Performance Using Synthetic Fabrics", M.Sc. Thesis, University of London,1988.
52.Ridley,J E ,"Experiences in the Use of Slow Sand Filtration, double sand filtration and Microstraining", Proc. <u>Soc.</u> <u>Wat.</u> <u>Treat. Exam.,16,</u>p170,1967.
53.Bellinger,E T ,   "Some Biological Aspects of Slow Sand Filters", <u>J. Inst. of Wat. Eng. and Sci.,33,</u>p.19,1979.
54.<u>Delagua</u> Water Testing Kit Users Manual, University of Surrey, Guildford,UK.
55.Svarovsky,L , " Characterization of Particles Suspended in Liquids",   Chapter.2   in   Solid-Liquid   Separation,   2nd ed.,p.8,Butterworths,UK,1981.
56.O'Melia,C R, "Particles Pretreatment and Performance in Water Filtration", <u>J.of Env. Eng., 111,</u>6,p.874,1985.
57.Lloyd,B ,"The Construction of a sand profile sampler:Its use in the study of Vorticella populations and the general intersticial microfauna of Slow Sand Filters",   <u>Water Research,</u>7,p963,1973.

5.2   ADVANCED TECHNIQUES FOR UPGRADING LARGE SCALE SLOW SAND FILTERS

A. J. Rachwal, M. J. Bauer and J. T. West — Thames Water Authority Research & Technical Development, Northumberland House, Mogden STW, Isleworth, Middlesex TW7 7LP

ABSTRACT

Thames Water operates eight slow sand filtration plants in London treating 2000 Mld^{-1} of reservoir stored lowland river water. A major programme of uprating and modernisation of the largest plants is in progress. Research at full scale showed that average filtration rates of 0.3 mh^{-1}, with peaks in excess of 0.4 mh^{-1} were sustainable with advanced filter management techniques. Filter downtime for cleaning and resanding was shown to be a key output limiting factor. A novel underwater sand skimming method has been developed aided by a laser guidance system and computer control. Work on shading of filters, with floating covers, to reduce in bed algal growth is reported. Pilot and full scale studies on the use of ozone with slow sand filtration showed improved filter performance and increased organics removal.

INTRODUCTION

The majority of London's water supply, some 2000 Mld^{-1}, is treated at nine works using the basically simple and well proven process of slow sand filtration. This process was developed in the 1800's as the first practicable water treatment means of controlling water borne disease. The first plants abstracted directly from rivers or used 1-2 day settlement basins prior to slow sand filtration at 0.1 mh^{-1}. They provided physical, biological and disinfection treatment in one stage.

In the River Thames and Lee basins this basic technology has been continually developed. Between 1900 and the 1980's the use of long term reservoir storage, to reduce river turbidity and pathogenic bacteria, together with the installation of primary rapid gravity sand filtration or microstrainers enabled slow sand filters to operate at 0.15 to 0.3 mh^{-1}. The last stage of treatment now comprises chlorination, followed by ammoniation, to provide a chloramine residual disinfection capability, protecting London's distribution system from both bacteria and the formation of potentially harmful chlorinated organics such as trihalomethanes (THM's).

Uprating

   To meet increasing water demand and improve cost effectiveness,
Thames Water carried out a programme of research and large scale trials
in the 1980's to uprate slow sand filter works by a further 30 to 50%,
operating up to 0.5 mh^{-1}. Successful completion of this work at an 500
Mld^{-1} plant in 1986, enabled Thames to adopt a strategy of uprating and
modernising its major slow sand treatment works and closing smaller,
less economic plants whilst still meeting increased demand (1).

Key Areas studied during this one year trial were:-

    - Effect of stored water quality

    - Primary treatment requirements

    - Slow sand filter run lengths, cleaning and sand handling
      requirements

    - Slow sand filtrate quality

    - Filter management and control needs

   In addition to the uprating programme, research efforts were also
directed towards techniques for improving efficiency and quality
performance of slow sand filtration works.

Efficiency

   Research on improving efficiency has concentrated on slow sand filter
cleaning and sand handling which represents the major operating cost.
Techniques investigated included;

 - reducing applied load by enhanced primary filtration
    treatment using ozone or chemical coagulant aids with
    improved media and backwash specifications.

 - reducing growth of algae within filters by the use of
    ozone or shading.

 - reducing filter cleaning downtime and costs by developing an
    automatic, underwater sand skimming system hydraulically
    linked to sand washing plant.

 - use of fabric layers on the sand filter bed to prolong filter
    runs and reduce sand handling costs.

Quality

   Slow sand filters achieve excellent removal of particulate organic
matter, turbidity, algae and bacteria. Removal efficiency for dissolved
organic material is lower with only 15-25% reduction of colour and total
organic carbon (TOC) reported (2). Thames Water, other Water authorities
(WA's) and WRc have been investigating additional unit treatment
processes, principally ozone and granular activated carbon (GAC) that
could enhance organics removal at slow sand filtration works.

This paper presents a review of research progress in Thames water on these aspects of advanced slow sand filtration.

SLOW SAND FILTRATION TRIALS 0.3 - 0.5 mh^{-1}

Previous research

Research work carried out by the Metropolitan Water Board (London) in the 1960's and early 1970's established that it was possible to operate a slow sand filter at filtration rates up to 0.5 mh^{-1} (3). The experimental filter treated reservoir stored and primary sand filtered, River Thames water. Filter run lengths were reduced but no significant deterioration in slow sand filtrate quality was reported.

Trial objectives

A trial to verify these results at works scale was initiated by Thames Water at Ashford Common in 1984. The aim was to achieve an annual mean works filtration rate of 0.3 mh^{-1} with a peak rate of 0.36 mh^{-1}. These filtration rates were to be calculated on the basis that on average 20% of the slow sand filters were out of service for draining, cleaning, resanding, maintenance etc, which was typical for filter downtime in London. To achieve these mean works filtration rates individual filters would need to operate at up to 0.5 mh^{-1} at the start of a filter run, declining to 0.1 - 0.2 mh^{-1} prior to cleaning. A full year of trials was undertaken to investigate seasonal influences on performance and establish filter cleaning needs at high rates.

Works scale trials

Ashford Common works was designed to treat 410 Mld^{-1} of reservoir stored water using primary microstrainers followed by 32 slow sand filters each of 3120m^2 area. The works can be divided into two parallel streams and was operated in this manner for the uprating trials. The East side (control) operated with 16 filters at an average works filtration rate of 0.2 mh^{-1}. The West side (uprated) operated with 11 filters at the target rate of 0.3 mh^{-1}.

Hydraulic analysis prior to the trials had established that the principal flow constraints were in the individual filter outlet pipework and venturi flow meters. Ultrasonic time-of-flight flow meters in straight pipe sections were installed to remove this hydraulic constraint. Differential pressure transducers were provided to measure filter headloss. A fibre optic telemetry system linked to a microcomputer gathered data from these and other works sensors to provide both real time and historic analysis of individual filter and overall works performance. The results of computer analysis and modelling of these data are reported elsewhere (4).

Results - Filtration rates and water quality

During the period April 1985 to March 1986 the 11 uprated filters achieved a mean filtration rate of 0.32 mh^{-1} compared to 0.21 mh^{-1} for the control filters (Table 1). A peak week rate of 0.38 mh^{-1} with individual filters operating up to 0.53 mh^{-1}, was maintained when reservoir and primary treated water was of good quality but could not be sustained under poor feed water quality. Good quality primary treated water was defined as < 5 ugl^{-1} chlorophyll a and < 500 ugl^{-1} particulate organic carbon (POC) (1). Poor quality water was defined as > 10 ugl^{-1} chlorophyll a and > 750 ugl^{-1} POC.

Maintenance of peak filtration rates was determined by the ability to meet filter cleaning needs and not by any concern over filtrate quality. No significant differences between control and uprated filter quality was found during the year's operation (Table 1).

Table 1. Uprated Slow Sand Filter Trial Results. Ashford Common
1st April 1985 - 31st March 1986

Heading		Control Filters	Uprated Filters
No of filters (each 3120m^2)		16	11
Mean total output	Mld^{-1}	203	209
Mean filtration rate	mh^{-1}	0.21	0.32
Peak week filtration rate	mh^{-1}	0.27	0.38
Individual filter peak	mh^{-1}	0.35	0.53
Mean run length	days	40	28
Volume filtered per run	m^{3}m^{-2}	190	195
No of filters cleaned		107	115
Mean skim depth	mm	36	41
Mean Chlorophyll a (algae)			
Primary filtrate	ugl^{-1}	4.7	4.9
Slow sand filtrate	ugl^{-1}	0.3	0.3
Mean Particulate organic carbon (POC)			
Primary filtrate	ugl^{-1}	460	490
Slow sand filtrate	ugl^{-1}	29	35

Results - Filter run length

Under similar load conditions an inverse relationship between filter run length and mean filtration rate was observed; doubling filtration rate halved filter run length. Mean run lengths of 40 days at 0.21 mh^{-1} and 28 days at 0.32 mh^{-1} were obtained with a range of 11 to 70 days for individual runs. The productivity or cumulative volume filtered per run remained essentially the same for both uprated and control filters, averaging 195 and 190 m^{3}m^{-2} filter area respectively (Table 1).

Results - Filter cleaning

Both control and uprated filters were cleaned using mechanical sand skimming machines operating in drained filter beds. Skim depth was controlled by the operator using visual judgement of the cut depth required to leave clean sand. Means of 36 mm and 41 mm were reported for the control and uprated filters. Core samples were taken before and after cleaning to determine algae and silt penetration. No significant difference between control and uprated filters was determined by this method. The results were scattered and not conclusive for either method of determining the effect of uprating on skim depth required. For planning purposes a 20% increase over current skim depth has been recommended for sizing sand handling plant and further trials initiated.

Uprating trial conclusions

The trials concluded that works scale operation at 0.3 mh^{-1} (peak 0.5 mh^{-1} for individual filters) was feasible provided that reservoirs and primary treatment could supply good quality water containing < 10 ugl^{-1} chlorophyll a (algae) and < 750 ugl^{-1} POC. Filter cleaning and sand handling would be the principal output constraints. Enhanced systems of plant management and control would be needed to meet the increased workload. Capital investment and further research was committed to these areas as part of the uprating programme for London's slow sand filtration works.

RESERVOIR MANAGEMENT

Thames Water operates a complex system of reservoirs which provide storage of river water (for security of supply in the event of drought or pollution) and preliminary treatment. Each treatment plant has access to more than one storage reservoir. The reservoirs can be managed to reduce the impact of high river turbidity and to control the effect of algal blooms. Retention times can typically vary from 10 to 100 days with some operating as standing reserves with a 1-2 year retention for biological nitrate reduction.

In the prevailing climate of southern England water bodies which have surface areas less than 1 km^2, or depths greater than 10 m, can be subject to thermal stratification. For London's reservoirs, this potential resource loss during stratification has been counteracted by various mixing techniques using pumped recirculation, jets or diffused air (5,6).

As part of the slow sand filter works uprating programme, further flexibility is being introduced into the reservoir system with additional stored water links and mixing systems. Reservoir and computer specialists are also developing computer based management decision support models to enable the system to be operated at lowest possible cost whilst still meeting water quality, supply security and amenity objectives.

PRIMARY FILTRATION

The principal task of London's primary rapid gravity sand filters and microstrainers is to reduce reservoir derived algal loading applied to the secondary slow sand filtration stage. It is possible to treat reservoir stored river water directly with slow sand filters but seasonal algal blooms and turbidity short circuiting result in rapid blocking of the filters unless low filtration rates, 0.1 mh^{-1}, are used. Existing rapid gravity filters, operating at 5-6 mh^{-1}, permit economic slow sand filter run lengths of 50-60 days at filtration rates of 0.15 to 0.25 mh^{-1}.

The uprating strategy in Thames Water requires an increase in the filtration rate of existing primary filters from 6 mh^{-1} to 10 mh^{-1} whilst maintaining or improving the removal of algae and turbidity. The subsequent slow sand filters are to be uprated to operate at a works average of 0.3 mh^{-1} peaking at 0.5 mh^{-1} on individual filters.

The principal primary filter modifications needed to achieve higher filtration rates and higher particulate removal efficiency were improved media specification and more effective backwashing. The major constraint with existing structures was hydraulic limitation particularly head available for clean bed headloss and subsequent headloss development due to filter blocking. In one case only 1.3 m total headloss was available.

The original media specifications for the rapid gravity filters provided a shallow 0.5 - 0.7 m bed of sand with an effective size (ES) of 0.75 mm and uniformity coefficient (UC) of 1.7. In practice 95% of the sand media was sized between 0.6 and 2.5 mm and overlayed 0.5 m of graded gravels. Average performance of these rapid gravity filters at 5 mh^{-1} was 50% removal of incoming algal and particulate organic material with a range of 30 to 80% depending for example on particle size and algal species (1,6).

The uprated primary filters are required to achieve an average of 67% removal of particulate matter in the 4-80 micron size range at 10 mh^{-1}. Pilot and full scale trials have shown that deeper beds, 0.7 - 0.9 m, of sand could achieve the required performance. Effective size could not be reduced significantly without compromising the head loss development limitations. At Hampton and Coppermills sand of ES of 0.7 mm and UC of 1.3 was specified for the uprated primary filters.

Provision of an effective combined backwash using air and water was found to be fundamental in ensuring the media was sufficiently cleaned after each run (6). Build up of silt and algae in poorly washed filters leads to the phenomenon of mud balling or build up of mud throughout the sand layer which will further reduce filtration efficiency. Optimisation of the primary filter backwash system including combined air and water has been a key uprating requirement.

DIRECT FILTRATION MODE FOR PRIMARY FILTERS

Further research on improving efficiency and water quality at the primary filtration stage has been carried out by Thames Water and WRc using a pilot plant at Coppermills (7).

The principle aim was to compare the effect of ozone or ferric sulphate on the subsequent performance of the primary filters. A target ozone residual of 0.2 mgl^{-1} was achieved with an applied ozone dose of 0.5 to 2 mgl^{-1}. A ferric sulphate dose of 1 to 3 mgl^{-1} as Fe was normally found effective for the direct filtration mode of operation.

Turbidity

Generally turbidity of the reservoir stored raw water was low, 1 to 3 FTU, with a peak of 8 FTU during winter filling of the reservoirs with high turbidity river water. Pre-ozonation had little effect on the removal of this mineral turbidity giving similar primary filtrate turbidities to the control stream 0.5 to 4 FTU (peak). The primary filter dosed with ferric sulphate always gave the best filtrate quality in terms of mineral turbidity, 0.2 to 2 FTU (peak), (Table 2).

Algae

Algal levels, as shown by chlorophyll a measurement, vary by up to two orders of magnitude in the 5 to 20 m deep raw water storage reservoirs supplying Coppermills and other London works. In winter chlorophyll a is usually less than 5 ugl^{-1} but in the spring and autumn algal blooms can give chlorophyll a concentrations exceeding 100 ugl^{-1}.

Without ozone or coagulation, chlorophyll a removal of 40 to 60% is generally achieved by London's primary filters. This can be insufficient during large algal blooms. In the spring pre-ozonation gave the best primary filter algal removal performance, 75 to 90%, with ferric sulphate addition giving 70 to 80% removal. The spring algal blooms are frequently dominated by small, 5 to 15 micron, filter penetrating algae, these algae were markedly affected by the addition of ozone or ferric sulphate, tending to clump together to form larger more filterable material.

In the autumn larger algae are more common and a better overall removal performance was achieved by all primary filter streams, pre-ozonation or ferric sulphate dosed filters achieving 80 to 90% chlorophyll a removal compared to 60 to 80% by the control filter, (Table 2).

Table 2 Comparison of Ozone and Ferric Dosing on Primary Filtration

Parameter	Reservoir Water	Control	Ozone	Ferric
		(Primary Filtrates)		
Turbidity (FTU)				
Mean	2	1	1	0.6
Peak (Winter)	8	4	4	2
Chlorophyll a				
'Normal' range	3 - 10	2 - 5	1 - 2	1 - 4
Spring peak	75	35	15	20
Autumn peak	70	15	7	8
Particulate Organic Carbon (POC)				
'Normal' level	600	300	130	200
Spring peak	2300	1000	600	700
Autumn peak	2200	600	300	300

Particulate Organic Carbon (POC)

It should be noted that ozone bleaches the chlorophyll pigment in algae and therefore chlorophyll a, determined colorimetrically, is not a true measure of algal removal performance by a pre-ozonated primary filter. Particulate organic carbon (POC) is a better measure of assessing both live algae and total organic particulate load. This technique was developed by the Thames Water Regional Laboratory for slow sand filter monitoring (8).

In the winter, reservoir stored raw water POC levels may be 300 ugl^{-1}, rising to over 3000 ugl^{-1} during algal blooms. Target primary filtrate POC concentrations of less than 500 ugl^{-1} average and 750 ugl^{-1} peak are recommended for successful operation of slow sand filters at filtration rates in excess of 0.3 mh^{-1} (1). Pre-ozonation or ferric sulphate addition both gave enhanced POC removal by the primary filters, as observed with chlorophyll a. Pre-ozonation was not significantly more effective than ferric sulphate during the spring algal bloom as had been the case when comparing chlorophyll a removal, (Table 2).

Headloss (Primary Filters)

The primary filters could develop 1.5 m headloss before backwashing was initiated. Under average loading conditions the control primary filter gave filter runs of 24 to 48 hours. At peak load, filter runs could reduce to less than 12 hours. Pre-ozonation had little effect on headloss and filter run length although a greater proportion of the applied load was being removed. The ferric sulphate dosed primary filter always gave the fastest rate of headloss development and shortest filter runs, reducing runs to less than 6 hours at peak loading. It should be noted that the filter media was not optimised for coagulant addition. For example, a dual media anthracite/sand combination would be expected to give a better performance in terms of run length than the 0.9 m of single media sand used in tests so far.

Headloss build-up (Slow Sand Filters)

The slow sand filter following the iron dosed primary filter regularly gave the lowest rate of headloss development. This was consistent with the primary filter results, which showed that iron gave the lowest turbidity primary filtrate. A cautionary note was that if incorrect iron dosing occurred then carry over into the slow filter resulted in rapid blocking.

The most recently constructed (1980) slow sand filter works in Thames Water is at Reading (Fobney). This works uses upflow sand filters for primary treatment of directly abstracted river water. Ferric sulphate is used at times of high turbidity loading to enhance filtration at rates up to 12 mh^{-1}. In general the primary filters have performed well but capital works are in hand to improve coagulant dosing control and provide additional filtration capacity. The subsequent slow sand filters are designed to operate at 0.25 mh^{-1} average but have been operated at up to 0.5 mh^{-1} for experimental trials.

The authors also note with interest recent results from Los Angeles (9) where a combination of ozone, ferric sulphate and polyelectrolytes enable a rapid gravity direct filtration system to achieve in excess of 30 mh^{-1} albeit with generally higher quality raw water than London's reservoirs could provide. Further pilot trials on the direct filtration mode of operation are planned for London's primary filters.

DEVELOPMENTS IN SLOW SAND FILTER CLEANING

Slow sand filter cleaning and sand handling represent 60-80% of the direct process operating costs and are the main reason for filter downtime. Current practice in London is to drain a filter when its output, headloss or filtrate quality is outside target values. A cleaning team of up to 6 men, using augur based skimming machines and dump trucks, remove a 15-50 mm deep layer of dirty sand after each run. A typical 3100m^2 filter is cleaned in 4-8 hours followed by slow filling and gradual increase in filtration rate to peak output. No water is filtered during the day of cleaning and reduced rates apply for 3-5 days. Resanding is needed every 12 - 18 filter runs and takes some 2 weeks followed by a 1-3 week filter conditioning period before full output can be achieved.

Uprating will increase the frequency of filter cleaning in London with filter runs reducing from more than 50 days at 0.15 - 0.25 mh^{-1} to less than 30 days at 0.3 mh^{-1} and higher filtration rates. Although filter production per run was not reduced during uprating trials and daily output per filter was higher, maintaining total works output also requires that filter downtime is minimised. This is of fixed duration regardless of filter run length or filtration rate. Thus filter downtime increases with filtration rate. A theoretical increase from 10-15% to

20% filter downtime has been allowed for in calculating uprated works output. Computer models have been developed to assist in the planning of filter cleaning enabling operation at near optimal output (4).

Thames Water also started developing an alternative wet cleaning sand handling system in 1986 based on underwater suction dredging with hydraulic transport of dirty sand to a washing plant. The objectives were to minimise filter downtime and increase works productivity in terms of reliable output and cost per unit treated. No draindown or refill would be needed and the possibility of 24 hour automatic operation, cleaning whilst filtering, was to be investigated. A conventional dredger was modified to operate automatically with a laser level controlling the depth of cut. Other sensors were linked to a computer for controlling output rate to match that of the remote sand washing plant (Figure 1). A patent has been filed on this new system.

Figure 1. Schematic of wet slow sand filter cleaning

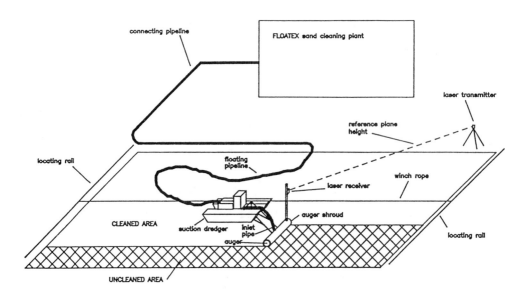

Wet cleaning is achieved by the use of a specially adapted auger head comprising an opposing screw auger and hydraulically adjustable shroud which enable bi-directional cleaning. The depth of sand cut and hence pumped to a remote sand cleaning plant is controlled by a laser transmitter and a receiver mounted above the auger. Wet cleaning has significantly reduced downtime by eradicating the need to drain down and recharge a filter as in dry cleaning operations. Also, as the biological community of the bed is not exposed to the detrimental effects of either solar radiation or frost, faster start up times should be achievable.

Adoption of the laser depth control strategy has resulted in filters being progressively planed flat during successive wet cleans. A dry resanded bed can have a differential of 150mm between the highest and lowest points (Figures 2 & 3).

Figure 2. Slow sand filter contour of a new resanded bed before wet
          cleaning. Bed 3. (Differential between highest and lowest
          point 280 mm)

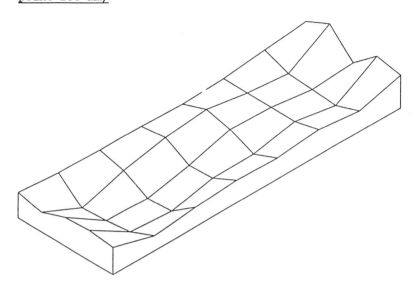

Figure 3. Slow sand filter contour after third wet cleaning. Bed 3.
(Differential between highest and lowest point 85 mm)

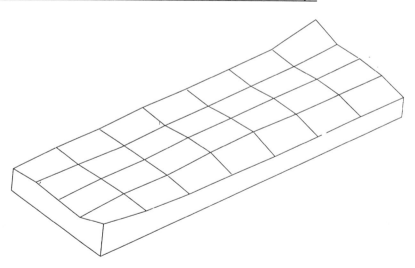

    Present dry skimming machinery is unable to plane out these contours
which result in certain areas of the bed at the end of their effective
life having sand depths greater or less than the minimum of 300 mm
allowable for quality reasons. By planing the bed relatively flat wet
skimming achieves better control of sand depth thereby maximising
effective bed life and possibly water quality.

Filtrate quality from the forty or so wet cleans to date has shown that quality is unaffected even whilst cleaning is proceeding. Although water is run to waste during wet cleaning there is scope for having the filter operational during this phase (zero downtime) albeit at a reduced rate. Analyses of sand cores taken whilst the filter was drained down have shown no deeper penetration of silt, chlorophyll a or POC. This is crucial as contamination of the deeper layers of sand would increase the clean bed headloss thereby shortening run lengths.

The logistics of wet cleaning have shown that one man and one machine can clean a filter (35 mm sand depth over 3100m^2) in an operational day (< 8 hours). With improved sand flow control, matching dredger output to the capacity of the sand cleaning plant and with development of the dredger location system the target cleaning period is < 6 hours.

Further developments include a wet resanding system and the feasability of mounting the underwater skimming head on a mobile bridge structure. The latter may be an advantage for works such as Ashford Common and Coppermills which have long rectangular filters. However a cable guided, floating system based on the current type of machine, will probably be the only option for works with irregular shaped filters.

Long term trials are continuing with a review of costs and effectiveness planned for 1989 followed, hopefully, by wider implementation of this promising underwater skimming system.

FILTER SHADING

Slow sand filter run length and filtered water production per filter run are determined by available headloss and rate of filter blocking. Under conditions of low light intensity and water temperature, or with turbid feed waters, the principal filter blocking load is provided in the incoming water. With high light intensity and water temperature together with low turbidity feed water, the principal filter blocking load can be that of algae growing in the filter bed. Large quantities of filamentous algae growing above filters also present handling problems for mechanical sand skimming machinery, adding time and labour to filter cleaning operations.

Techniques to reduce algal growth above slow sand filters include shading, pre-chlorination and pre-ozonation. The latter two techniques are discussed in the ozone section. Purpose designed enclosed slow sand filters have been installed in Europe and the USA both to minimise winter freezing problems and to reduce algal growth. Where large open slow sand filters exist the only economic option to provide a means of shading appears to be the use of floating covers.

In 1985 a 3120m^2 slow sand filter at Ashford Common was covered using material derived from swimming pool systems. The fibre reinforced plastic material was installed as 4 strips, 90m x 8.5 m, attached to stainless steel rollers for deployment and retrieval. Considerable materials handling problems were encountered particularly during deployment and retrieval under high wind conditions and eventually the system was abandoned.

Limited results were obtained during the period July - September 1985. These demonstrated complete elimination of filamentous algae or "blanket weed" that predominated on other, uncovered filters and required manual removal prior to cleaning. Filter run length was also improved by 50%.

Further design work on floating covers was carried out in 1986/87 and a more suitable buoyant cover material evaluated. However the costs of engineering a system, capable of deploying and retrieving in excess of 3000m^2 of cover under high wind conditions, were not considerd acceptable for works with 25 to 35 such filters. Cost-effective retrofitting of a shading system to large slow sand filters is considered desirable but unattainable at present.

FILTER MATS

The use of fabric layers on small slow sand filters to reduce the rate of blockage and increase filter run length has been reported for many years. A recent innovation has been to try geotextiles and non-woven synthetic fabrics as the top layer of a slow sand filter. Research into the use of fabrics, for improving the performance of small slow sand filters and the potential for application on large (3000m^2) filters, has been the subject of a joint project between Imperial College and Thames Water. The results for small filters have been encouraging and are reported elsewhere (10). However, like filter shading, engineering costs currently preclude economic use for large filters.

OZONE AND SLOW SAND FILTRATION

In 1980 Thames Water and WRc commenced pilot plant studies at Walton and in 1983, full scale at Fobney, on the effect of applying ozone between primary filtration and secondary slow sand filters. The use of ozone was part of the long term study investigating means of improving the performance of slow sand filtration treatment plants in terms of treated water quality, increased works output and overall process economics. The Water Research centre in England contributed to these studies particularly with respect to the use of ozone (2,11).

It was anticipated that if an ozone residual could be achieved in the water above a slow sand filter it would have a beneficial effect in terms of prolonging filter run length and reducing cleaning effort by controlling growth of floating and filter blocking algae. Ozonation was more acceptable than pre-chlorination in this role as the residual ozone would decay rapidly above the biologically active schmutzdecke layer unlike the more persistent chlorine which could harm the all important filter biology. Formation of potentially harmful chlorination by-products, would also be avoided with ozone.

The water supplied to the Walton plant was abstracted from the River Thames and stored for up to 50 days in raw water reservoirs. Treatment consisted of primary rapid gravity filtration (without coagulant addition) at a filtration rate of 6 mh^{-1}, followed by slow sand filtration at 0.12 mh^{-1}. The pilot plant was used to investigate the effect of pre-ozonation at various ozone doses on the performance of slow sand filters operating at up to 0.3 mh^{-1}, details of which are published elsewhere (2,7,11).

The full scale works at Fobney was designed to treat 45 Mld^{-1} and was completed and commissioned in 1983. This works abstracts directly from the river Kennet near Reading using ferric sulphate dosing with primary upflow filtration at 12 mh^{-1} and secondary slow sand filtration at 0.25 mh^{-1}. Facilties are available to dose the primary filtered water with ozone prior to slow sand filtration.

The effect of pre-ozonation at ozone doses of 0,1,3 and 5 mgl^{-1} on subsequent slow sand filter performance were studied at Walton. Results showed that pre-ozonation at 3 and 5 mgl^{-1} provided similar improvement in performance in terms of filter run length and filtrate quality whereas at 1 mgl^{-1} there was limited benefit, (Table 3).

Table 3. Effect of ozone dose on slow sand filter performance. (Walton)

Ozone Dose mgl^{-1}	Residual Ozone mgl^{-1}	Run Length days	TOC	254 nm (% reduction from) (primary filtrate)	400 nm	ATP value 10^{-1} ugl^{-1}
0	0	34-51	15	12	24	1.0
1	0	38-51	24	47	71	1.6
3	0.2	90	32	65	86	2.0
5	0.5	90	32	66	86	2.1

An ozone dose of 3 mgl^{-1} resulted in a residual of 0.2 mgl^{-1} above the pilot slow sand filter and a typical 50 to 100% improvement in filter run length during the high algal growth period between spring and autumn as reported in earlier papers (2,11).

On the full scale plant at Fobney it was not possible to apply more than 2 mgl^{-1} based on the design output of the ozonizer. As at Walton slow sand filtrate quality was improved with respect to TOC, absorbance at 400 nm (colour) and 254 nm (aromatic and aliphatic compounds with double bonds) (Table 4). The effect on filter run length was less pronounced on the full scale plant at Fobney than the pilot plants at Walton. This could be attributed to a number differences:-

- An ozone residual above the pilot slow sand filters was readily achieved at an applied dose of 3 mgl^{-1}. No ozone residual was detected above the full scale filters at Fobney even though there was a residual of 0.5 mgl^{-1} leaving the contact tank. Thus control of algal growth would be expected to be less at Fobney.

- Ozone was also applied intermittently allowing algal laden water that had not been ozonated to 'seed' the filters.

- Coagulant control was also not ideal and carry over of algal, ferric or turbidity flocs from the primary filters to the ozone contact tank and slow sand filters could occur. A thick scum of material built up in the ozone contact tank as a result of air flotation effects. This could be released to the slow sand filters as a result of level variations during primary filter backwashing.

The results are therefore not conclusive with pilot trials at Walton and at a site in Wessex WA (12), indicating considerable benefits to slow sand filter run length but with little benefit observed full scale at Fobney. Further full scale results are needed from other sites.

Table 4. Effect of pre-ozonation on slow sand filters (Fobney)

Ozone Dose mgl^{-1}	Residual contact tank	Filter Run Length Days	TOC % Reduction from filtrate	254 nm	400 nm primary
0	0	20–80	15	13	23
2	0.5	30–90	35	61	87

Effect of ozone on slow sand filter organics removal

The main effect of pre-ozonation was related to the reduction in absorbance at 254 and 400 nm which were reduced by up to 70 and 90% respectively, compared to reductions of only 15 and 25% respectively for slow sand filters with no pre-ozonation, (Tables 3 & 4). However most of the reduction was due to the pre-ozonation itself with only a small additional reduction of 10% occurring within the slow sand filter. Pre-ozonation also significantly increased the TOC removal by the slow sand filters by an average of 35% compared with 15% by conventional filters. Reduction of TOC was more pronounced in warm summer months than cooler winter periods, indicating biological mechanisms were possibly responsible.

Analysis by gas chromatography/mass spectrophotometry (GC/MS) showed large increases in peaks corresponding to aliphatic aldehydes, ketones and acids after ozonation which were absent or much reduced after subsequent slow sand filtration. These compounds are readily biodegradable.

WRc carried out adenosine triphosphate (ATP) measurements to determine the effect of ozone and subsequent treatment on the amount of biodegradable material present in the water. Ozonation significantly increased the ATP value of primary filtered water, increasing its biodegradability. Slow sand filtration after pre-ozonation reduced the ATP value to essentially the same as for non-ozonated slow sand filters.

The ATP value of the Walton pilot plant primary filtered water, dosed
with 1 mgl^{-1} ozone, was lower than that for the water dosed with
3 mgl^{-1}. There was no difference between the waters ozonated with 3 and
5 mgl^{-1} of ozone (Table 3). This may explain why pre-ozonation at 3 & 5
mgl^{-1} resulted in similar TOC removal, headloss development and filter
run length for these slow sand filters (2,7,11).

SUMMARY AND CONCLUSIONS

Slow sand filtration is not an example of a stagnating 150 year old
water treatment technology. It has been continually developed so that it
is appropriate for not only small communities in developing areas of the
world but can also be used as part of a sophisticated treatment process
stream in some of the largest treatment works in the world.

Modifications are in progress to uprate 3 slow sand filter works to
produce 2000 Mld^{-1} for London at filtration rates of 0.3 - 0.5 mh^{-1}. The
works will be highly automated, with computer monitoring and modelling
of reservoir, primary and slow sand filtration processes, reliably
producing water of a high quality.

For the future, the use of ozone appears to offer slow sand
filtration a potential role to play in the biological reduction of
organics, thereby reducing the level to which chlorinated organics
potentially may form during chlorine disinfection. Research is
progressing rapidly in this area. Pre-ozonated primary filters and
granular activated carbon (GAC) are also part of current slow sand
filtration research.

Research into novel cleaning methods, fabric replacement of sand,
modelling and understanding of slow sand bio-physical mechanisms are
also continuing.

It therefore seems reasonable to expect to see the slow sand
filtration process continue as part of effective water treatment systems
well into the 21st Century.

ACKNOWLEDGEMENTS

The authors would like to thank Mr G.A.Thomas, Head of Technology and
Development and Thames Water for permission to prepare this paper. The
views expressed are those of the authors.

REFERENCES

1. Uprating Ashford Common Slow Sand Filters. Rachwal, A.J., Internal
    Thames Water Report, (May 1986).

2. Rachwal, A.J., Rodman, D.J., West, J.T. and Zabel, T.F., Effluent
    and Water Treatment Journal, (May 1986).

3. Windle Taylor, E., <u>44th Reports on Results Bacteriological, Chemical</u> <u>and Biological Examination of London's Waters for the years 1971-73.</u> Metropolitan Water Board, London.

4. Woodward, C.A and Ta, C.T., Developments in Modelling Slow Sand Filtration, Slow Sand Filter Conference, Imperial College, London, (November 1988)

5. Steel, J.A., The Management of Thames Valley Reservoirs. <u>Proceedings</u> <u>of a Symposium, The Effects of Storage on Water Quality,</u> Reading University, (March 1975), Organized by WRc Medmenham.

6. Toms, I.P., <u>Arch. Hydrobiol. Beith. Ergebn. Limnol.</u> 28, pp. 149-167, Stuttgart, (August 1987).

7. Rachwal, A.J., Bauer, M.J. and Chipps, M.J., Ozone's Role in Biological Filtration Processes. <u>Presented at 2nd International</u> <u>Conference on The Role of Ozone in Water and Wastewater Treatment.</u> Edmonton, Alberta. (April 1987).

8. Steel, J.A. and Toms, I.P., Particulate Organic Carbon <u>Determination., Internal Thames Water Report.</u>

9. McBride, D.G. and Stolarik, P.E., Pilot to Full-Scale Ozone and Deep Bed Filtration at the Los Angeles Aqueduct Filtration Plant. <u>Presented at American Water Works Association National Conference,</u> <u>Kansas City,</u> (June 1987).

10. Mbwette, T.S and Graham, N.J.D., <u>Proceedings of the Filtration</u> <u>Society.</u> (January- February 1987).

11. Hyde, R.A., Rodman, D.J. and Zabel, T.F., <u>Journal of the Institute</u> <u>of Water Engineers and Scientists,</u> 38, No 1, (Feb 1984).

12. Knight, M.S. and Tuckwell, S.B., <u>Wessex Water Authority Report on</u> <u>Ozone and Slow Sand Filtration.</u> Published in discussion section of WRc Proceedings of Seminar on Ozone in UK Water Treatment Practice The Future, London, (1984).

5.3 DEVELOPMENTS IN MODELLING SLOW SAND FILTRATION

C. A. Woodward and C.T.Ta — Thames Water Authority, Research and
Development, Northumberland House, Mogden Works, Mogden Lane,
Isleworth, Middlesex TW7 7LP

ABSTRACT

   A set of equations are derived relating normalized headloss,
time, flow rate, input water quality and primary productivity.
   Results are presented showing the performance of the model in
predicting daily average flow rates given a table of daily average
headlosses and initial flow data.
   The model is employed to forcast flows and headlosses through a
network of hydraulically linked slow sand filters, each at a
different point in its operating cycle.

I.   INTRODUCTION :

      Thames Water's Authorities strategy for water supply and
distribution into the 21st century provides for closure of a number of
smaller, less cost effective works and uprating of a number of existing
slow sand filtration works. The outputs required of the uprated works
will be equivalent to a filtration rate of 0.36 mh^{-1} at peak. When
operating   at   this   filtration   rate,   filter   run   lengths   may   be
considerably   shorter than those experienced to date (at filtration
rates in the order of $0.2mh^{-1}$).
      Filter cleaning, trenching and resanding times are unaffected by
uprating. This means that the non-productive time will become a higher
proportion of the life of a bed. This in turn means that the filtration
rates required when the bed is in service are disproportionately greater
than the average. As higher average filtration rates are attempted, this
effect is accelerated.
      The flow rate through a filter is, in practice, hydraulically
limited - even with perfectly clean sand. There must be an optimum time
for cleaning a filter which will depend on the shape of the normalized
headloss development curve and the actual average filtration rate
required. At low filtration rates and with good quality raw water, the
optimum position may be ill defined with the average output capability
and operating costs being more or less constant over a range of cleaning
times. At higher filtration rates, the optimum cleaning time is much
more   sharply defined. Indeed, the range of cleaning times which will
allow the target output to be achieved may be very small.
      This   paper   descibes   some   of   the   early   work   done   towards
development of mathematical models to assist in the management and
operation of uprated slow sand filtration works.
      In section II, a set of equations relating normalized headloss
and   flow   rate   are   derived.   A   procedure   to   obtain   the   relevant
parameters is also discussed.   In section III, we compare predicted
and measured values for filter beds at Ashford Common Water Treatment
Works (WTW) in the Thames Valley.

II.   THEORY :

      The nature of fluid flow is characterised by the Reynolds number,
which is the ratio between inertia and viscous forces (c.f.(1)) in the
Navier-Stoke equation.   For a slow sand filter, the effective size of
the sand grains is about 0.3mm so that if the   filtration rate is
8.3×10^{-5} ms^{-1} (or $0.3mh^{-1}$) the Reynolds number would be about 0.025. A
laminar flow can therefore be assumed and Darcy's law for fluid flow
through a vertically homogeneous porous medium can be applied. Headloss
(H) across the bed is therefore directly proportional to flow rate (Q)
and inversely proportional to the cross-sectional clean sand area (A).

$$H = k \frac{Q}{A} \qquad\qquad\qquad\qquad (1)$$

or$\qquad\qquad\quad$ $$L = k \frac{Qs}{A} = H \frac{Qs}{Q} \qquad\qquad\qquad (2)$$

where k is a constant and L, referred to as the normalized headloss, is
the headloss across the bed under standard flow (Qs) conditions. It is
conventional in practice to choose a standard flow equivalent to a
filtration rate of 5.6×10^{-5} ms^{-1} ($0.2mh^{-1}$).
      The proportionality constant k, referred to as the coefficient of
permeability or the hydraulic conductivity, depends on the current
characteristics of the filter bed. More explicitly, k is a   function

of the effective size and shape of the sand grains and hence the porosity of the filter bed. It also depends on the viscosity (u) and the density (p) of the influent water, which are temperature dependent.

In the capillary model, a filter bed is assumed to be made up of a number of capillary tubes of radius r. Flow through the filter bed is therefore of Hagen-Poisseuille type and the value of k is determined as

$$k = \frac{(L\ 8\ u)}{N\ \pi\ r^4 p\ g} \qquad (3)$$

where g is acceleration due to gravity, L and N respectively are the height and number density (per unit area) of capillary tubes.

We now briefly review some of the mechanisms which are responsible for the clogging process.

Water passing through the slow sand filters undergoes both physical filtration and purification through microbiological processes. Filtration principally involves straining, sedimentation and adsorption.

The effectiveness of the straining process is determined by the sizes of the pores and of the suspended particles. For clean sand of effective size 0.3mm, a triangular constriction would retain particles larger than about 25µm and straining is therefore relatively unimportant for removing small particles such as bacteria (<15µm) and colloidal materials (0.001 to 0.1µm). Later in the filter run, when a filter skin or schmutzdecke has formed, the straining process becomes of major importance.

In the sedimentation process, materials with high density travel vertically downward under gravity, hence they are able to cross the streamlines and come adjacent to the grain surface.

Smaller particles near to the grain surface may be removed by adsorption which involves either electrostatic attraction, London-Van der Waals forces or adhesion to the zoogloeal film on the surface of the sand grains (2).

Microbiological processes involved in purification are complex. Inorganic and organic matter enters the filter continuously in the raw water. Photosynthesis yields a further input of particulate and dissolved organic material. Soluable matter entering the sand is utilised by bacteria and other microorganisms. The extracellular slime produced by these organisms further entraps small particles, dead cells and part-oxidised material. Zooplankton grazing occurs and respiration of the entire biomass is continuous.

The benthic algae constituent may be of unattached or attached forms (3), the proliferation of which may ultimately clog the filter. Certain larger filamentous species cause problems operationally in the cleaning process, apparently without directly contributing to the clogging.

While the mechanisms of slow sand filtration are reasonably well understood, quantitative theories meet difficulties because of the complexities and the difficulty of direct measurement of the relevant state variables.

For a clean sand filter, a simplification can be made if the clogging is assumed to be uniform. In the capillary model, materials deposited or growing in filter cause a reduction of void space and hence decrease the diameter of the capillaries. A filter bed can also be considered to consist of particles, characterised by a mean diameter. The flow is then governed by Stokes law and the clogging kinetics can be studied in terms of that mean diameter. Semi-empirical methods utilising Kozeny–Carman model study the clogging process in terms of porosity, which decreases as more and more materials deposit on the sand grains.

For a partially clogged filter, where particles removed may have comparable size to the pores, non-uniform clogging theory is necessary. In this case, it is clear that the clean sand area effectively decreases because of deposited materials. In the capillary model, this is equivalent to a reduction of the number (N) of free capillaries (c.f. equation (3)).

In the present work, we will employ the non-uniform clogging model to study the hydraulic behaviour of slow sand filters. Normalized headloss (L) is of particular interest as it provides a useful tool to investigate the behaviour of the filter bed in day to day operation. Figure 1 shows typical behaviour of normalized headloss with time. It can be observed that during day 5–15 normalized headloss increases slowly from an initial value. From day 15 onward, it increases rapidly. High values of normalized headloss mean that the filter bed is clogged and cleaning should be carried out.

As mentioned above, because of deposited materials, the clean sand area effectively decreases. We assume this reduction to be directly proportional to the amount of deposited materials (M) thus

$$L = \frac{Lo}{1 - K M} \qquad (4)$$

where Lo is initial normalized headloss and K is constant.

It is clear from the discussion above that materials from the influent water and those produced by biological activity both contribute to the clogging process. If we assume that the effect is the same whether materials were derived from the influent water or from photosynthesis, we can write:

$$\frac{dM}{dt} = g \cdot M \left(1 - \frac{M}{a} \right) + m Q \qquad (5)$$

where   g : growth rate
         a : constant
         t : time
and   m : concentration of materials removed from the influent water.

The first term on the right hand side of equation (5) represents a logistic curve which corresponds to a simple density dependent population growth. Without the term mQ, equation (5) is the well-known Verhulst-Pearl logistic equation (c.f.(6)). Use of this expression means that the population is assumed to increase exponentially at first but to approach asymtotically a steady state in the later stages.

This expression cannot be used to describe the initial build up of the population in a clean bed but does describe the expected behaviour of operational filters which are never truly clean. It is noted further that since the different classes of materials have not been distinguished, the value of g represents only an overall average growth rate which may be significantly smaller than that expected for microbial growth. The second term (mQ) clearly represents the amount of matter added to the filter bed during filtration. Since almost all materials are removed during filtration, the quantity m is in fact directly related to the quality of input water.

From equations (4) and (5), it can be easily shown that

$$\frac{d}{dt}\left(\frac{Lo}{L}\right) - g\left(\frac{Lo}{L}\right) - \frac{g}{aK}\left(1 - \frac{Lo}{L}\right)^2 = -(mQ + g) \qquad (6)$$

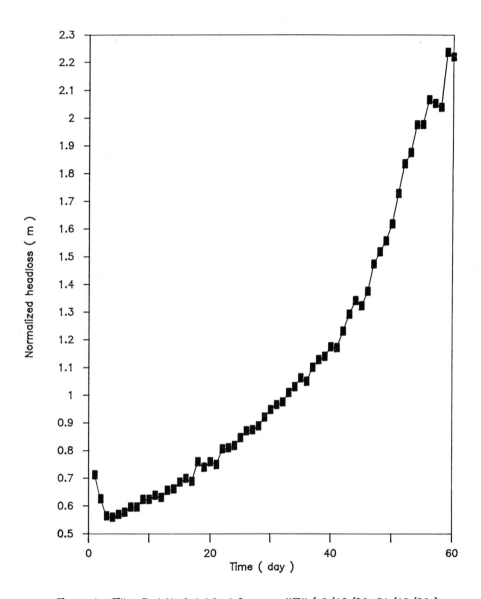

Figure 1 : Filter Bed No.9 Ashford Common WTW (2/10/86–31/12/86)

We now further assume that flow rate should decrease smoothly to zero so that dQ/dt → 0 as Q → 0.

Equation (6) gives

$$aK = 1 \qquad (7)$$

Using equations (6) and (7), we have

$$\frac{d}{dt}\left(\frac{Lo}{L}\right) + g\left(\frac{Lo}{L}\right) = -mQ \qquad (8)$$

where we have ignored the $(Lo/L)^2$ term since, after the first few days of a filter run, normalized headloss increases generally with time. In order to check the validity of this simplification, checks have been performed by using equation (6) directly (c.f. section III.1), the results obtained show no significant improvement over equation (8). During the first 2-4 days, normalized headloss falls. This behaviour cannot be explained within the scope of the present model. A possible explanation for this feature is that air trapped in the sand media during the cleaning operations is gradually released or dissolved from the filter bed.

Equation (8) can then be integrated to obtain

$$L = \frac{Lo \exp(gt)}{yo - Km \int_o^t Q \exp(gt)\, dt} \qquad (9)$$

where yo is the ratio between the actual starting and calculated clean bed normalized headloss. If the quality of the inflow water varies during a filter run, it is necessary to leave the term Km inside the integral. In subsequent discussion, Km is assumed to be constant for simplicity.

If the growth rate is sufficiently small, the clogging of the filter bed depends on the amount of material removed from the influent water. When the amount of deposited material approaches the value 1/K, the normalized headloss increases rapidly.

If flow rate is constant (Q'), then

$$L = \frac{Lo}{1 - Km\, Q't} \qquad (10)$$

where yo is taken to be unity.

To compare this result with an exponential model for straining mechanism given by Boucher (6)

$$L = Lo \exp(bt) \qquad (11)$$

we expand

$$L = \frac{Lo}{1 - KmQ't} = Lo\,(1 + KmQ't + (KmQ't)^2 + \cdots)$$

and

$$L = Lo \exp(bt) = Lo\left(1 + bt + \frac{(bt)^2}{2!} + \cdots\right)$$

If b is equal to KmQ', then, at the beginning of a filter run equations (10) and (11) should agree (error~Lo(KmQ't)^2). The normalized headloss obtained using the exponential model however becomes lower at the end of a filter run. Figures 2 and 3 compare the results calculated using equations (10) and (11) together with measurements from a slow sand filter at Ashford Common Water Treatment Works. The values of b, Lo and KmQ' are obtained by regression analysis on day 5-9 data for filter bed no.4.

It is observed that equation (10) apparently gives better agreement with the measurements. The exponential model tends to be too low at the end of the filter run. If the exponential model is to be fitted with data from the end of a filter run, the exponent b may be more than twice as large as its orginal value (c.f. also (7)). The

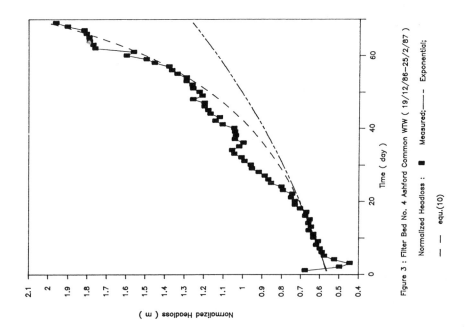

Figure 3 : Filter Bed No. 4 Ashford Common WTW (19/12/86−25/2/87)

Normalized Headloss : ■ Measured;──── Exponential;

— — equ.(10)

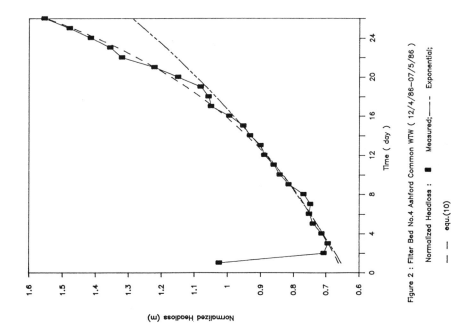

Figure 2 : Filter Bed No.4 Ashford Common WTW (12/4/86−07/5/86)

Normalized Headloss : ■ Measured;──── Exponential;

— — equ.(10)

exponential model therefore does not provide reliable information on the run length of slow sand filters. Boucher's empirical 'law' can in fact be derived from the capillary model for constant flow (8). In the derivations, however, an assumption that L/Lo should be less than 2 is necessary so that, again, the exponential model is only adequate at initial stages of filtration.

Although equation (10) does give reasonable agreement with the measurements, it is sensitive to a change in either Lo or KmQ'. Accuracy can only be obtained when sufficient data are available. Furthermore, this model does not take into account any variation of flow rate during filtration or any biological growth. However, application in practice (section III) shows that the denominator on the right hand side of equation (9) in fact determines the behaviour of the normalized headloss. The rapid increase of the normalized headloss at the end of a filter run occurs when the value of integral $m\int Q \exp(gt) \, dt$ attains the value yo/K. If, on the other hand, the product Km is small, headloss does increase exponentially with time.

In order to apply equation (9), four parameters (Lo, g, yo and Km) are required. In the present work, these are obtained from the headloss and flow rate data by the following procedure.

-Initial normalized headloss (Lo) is estimated by using the exponential model.
-For a given value of g, values of Km and yo are obtained by regression analysis.
-The resultant parameter set is fitted to the known data and the deviations are calculated.
-The value of g is varied to obtain the smallest possible deviation between the predicted flow rates and the measurements.

This procedure has the advantage that the initial headloss and flow rate data, can be used directly as the inputs of the model. Furthermore, the parameters can be easily updated on-line if required.

                    III. APPLICATION :

Ashford Common WTW in Thames Valley is one of nine major treatment works of Thames Water's Authorities. The works supplies about 20% of London's water and employs 32 slow sand filters for secondary treatment of microstrained reservoir water. From March to September 1985, part of the west side of the works was uprated in order to achieve a target mean filtration rate of $0.3mh^{-1}$ to meet the London Water Treatment Strategy requirements.

In order to maintain the output of the works, efficient day to day management was essential. In this section we discuss how the model developed in section II can be employed to assist the operational planning of the works.

In section III.1, we study the behaviour of an individual slow sand filter bed. Flow rates through a slow sand filter bed are forecast on the basis of known headloss data. In section III.2, equation (9) is incorporated in a hydraulic model to study the slow sand filter bed network. Flow rates and headlosses for all filters are forecast on the basis of known total flow data. Comparisons between the calculated values and measurements will also be presented.

Measured data presented in this section were obtained by using a fibre optic based telemetry and sensor system linked to a microcomputer based data acquisition, control and data analysis system (9). The flow rates and headlosses, respectively, are measured by ultrasonic time of flight flow meters and differential pressure transducers.

III.1 <u>Single slow sand filter bed</u> :

The parameters required for the slow sand filter model were obtained by the procedure discussed in section II.   Flow rate and headloss data at days 6 to 10 from the start of a filter run were used. The flow rates were then evaluated iteratively from headloss data (H) by using equations (2) and (9):

$$ Q = \frac{H \ Qs}{(Lo \ exp(gt))} \ (yo - Km \int_{o}^{t} Q \ exp(gt) \ dt) \hspace{2cm} (12) $$

A convergence threshold of 1 m^3day^{-1} (or $0.001Mlday^{-1}$) on the flow rate was used. Simpson's and trapezoidal rules are employed to integrate the right hand side of equation (12).

As a check, a second method has also been employed to obtain the flow rates.   Equations (2), (4) and (5) are substituted into equation (1), a differential equation for flow rate is obtained

$$ \frac{dQ}{dt} = \frac{Q}{H} \frac{dH}{dt} - gQ + g \frac{Lo}{HQs} Q^2 - Km \left(\frac{Qs \ H}{Lo} \right) Q $$

and is integrated using Runge-Kutta method. Since headloss is now required as a continuous and differentiable function, cubic spline interpolation has been employed on daily average headloss data.

In both methods, a time step of 1 day has been employed. Table 1 compares the flow rates evaluated by these two methods and the measured data. It is observed that the results evaluated by the two different methods are in good agreement. In subsequent evaluations, we therefore simply employ equation (9).

<u>Table 1</u>   Comparison between calculated and measured flow rates. Data for filter Bed number 4 at Ashford Common WTW (14/10 -31/10/87)   (Calculated values by A:differential and B: integral method; C: measured data)

Day No.	Flow rates (Ml/day)		
	A	B	C
7	17.38	17.61	16.01
8	15.84	16.06	15.44
9	14.74	14.95	15.12
10	13.52	13.72	13.39
11	13.68	13.88	13.59
12	14.34	14.55	14.17
13	16.57	16.82	16.24
14	16.18	16.42	15.73
15	13.91	14.12	13.84
16	12.85	13.06	12.82
17	11.78	11.99	13.40
18	14.07	14.32	14.61
19	14.02	14.27	14.87
20	14.10	14.36	15.05
21	13.46	13.71	14.63
22	12.63	12.87	13.99
23	10.75	10.57	11.73

    Figures 4 and 5 show typical results. For a winter run (fig.5),
because biological activity both in the inflow water and in the filter
is low, the run length of the filter is significantly longer than that
in the summer period (fig.4). In both cases, however, good agreement
(error < 5%) is observed over a long period. It should be noted that
the first five data are those employed to obtain the required parameters
of the model. The rather erratic behaviour of the flow rate (fig.5) is
due to variations in the headloss across the filter bed, caused in turn
by changes in the works total output.
    Tests for other filter runs at Ashford Common WTW have also been
performed. In general, the predicted flow rates are in reasonable
agreement with measurements (deviation < 15%). Tables 2a and 2b show
typical values of g and Km obtained for a series of runs on filter beds
nos. 13 and 5. The average errors (E) between the evaluated values and
the   measured data have also been included. It is observed that g
apparently reaches a maximum value during the March to July period while
the value of Km is highest from March to May. This behaviour possibly
corresponds   to the high biological activity in the influent water (Km)
and in the filter bed (g).

Table 2a : Values of g (day^{-1}) and Km over a one year period for
           Filter bed 13 at Ashford Common. E (Mlday^{-1}) :
           average deviation of predicted flow rates from
           measurements.

No.	Period		g	Km	E
1	Nov.86	Jan.87	0.02421	0.000084	0.6
2	Jan.87	Mar.87	0.02499	0.000300	3.3
3	Mar.87	May.87	0.03077	0.002272	1.2
4	May.87	Jul.87	0.05646	0.000019	5.8
5	Jul.87	Sep.87	0.02761	0.000218	5.2
6	Sep.87	Nov.87	-	-	-

Table 2b : Values of g (day^{-1}) and Km over a one year period for
           Filter bed 5 at Ashford Common. E (Mlday^{-1}) : average
           deviation of predicted flow rates from measurements.

No.	Period		g	Km	E
1	Nov.86	Jan.87	0.02514	0.000152	1.1
2	Jan.87	Mar.87	-	-	-
3	Mar.87	May.87	0.03418	0.000261	0.4
4	May.87	Jul.87	0.02201	0.000191	0.9
5	Jul.87	Sep.87	0.06814	0.000112	1.2
6	Sep.87	Nov.87	0.02185	0.000146	0.9

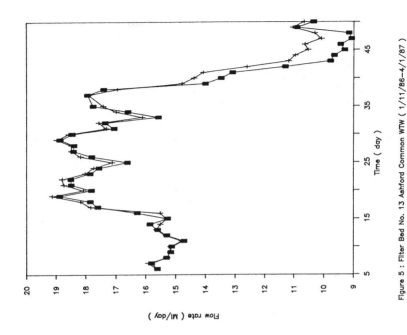

Figure 5 : Filter Bed No. 13 Ashford Common WTW (1/11/86—4/1/87)
Flow rates : ■ Measured ; + Calculated .

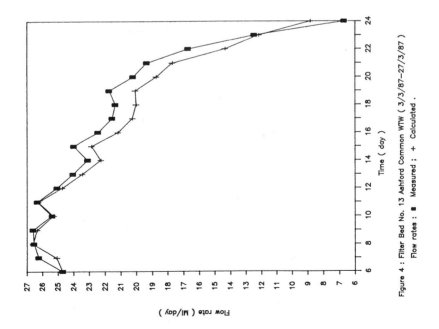

Figure 4 : Filter Bed No. 13 Ashford Common WTW (3/3/87—27/3/87)
Flow rates : ■ Measured ; + Calculated .

III.2 Filter Bed Network :

Figure 6 shows schematically the West side of Ashford Common water works. Only the slow sand filter network has been shown. The arrows indicate the direction of the flow. The input and output of the network are at nodes A and B respectively. Pipe coefficients and valve characteristics for this network are known (10). In the subsequent analysis, we shall assume full pipe flow. A computer program, which we refer to as SANNET, has been set up to study this system. The flow diagram is as shown in figure 7.

Initial average daily flow rate and headloss data are employed to determine the model parameters for each filter bed (c.f.section II). These average data are obtained directly from those logged hourly at Ashford Common. When a filter bed is out of service or there is a lack of valid data (failure of meters or the logging system), average values of the parameters obtained for the other filter beds are employed. For the results presented subsequently in this section, a set of five daily average pairs of flow rate and headloss data for each bed have been used.

Table 3 displays a typical output from SANNET immediately after parameter estimation. For filter beds nearing the end of their useful life, high values of Lo are observed (e.g. beds 1, 3, 9 and 16). Average values have been employed for beds 12 and 13, both of which were out of service for part of the observation period. Negative values of g are obtained for filter beds 1, 7 and 10 and imply negative net growth of living organisms in the filter.

The parameters which are determined at this stage are used until the filter bed is assumed to have been cleaned. When SANNET is employed for long term forcasts, average values are employed for filters that are assumed to have been returned to service after cleaning.

Table 3 : Typical output of SANNET program in determining the parameters for slow sand filter model. E (Mlday^{-1}): estimated deviation of flow rates from measurement, Qo (Mlday^{-1}) initial flow rate data, other symbols are the same as in section II.

Bed	Lo	yo	g	Km	Qo	E
1	1.014	1.013	-0.0491	0.00090	6.39	2.41
2	0.622	1.003	0.0381	0.00165	16.32	0.25
3	1.270	1.003	0.0816	0.00082	10.56	0.51
4	0.621	1.001	0.0441	0.00013	16.33	0.15
5	0.542	1.001	0.0306	0.00027	17.93	0.06
6	0.683	1.006	0.0909	0.00014	17.87	0.23
7	0.409	0.969	-0.1571	0.01057	22.52	2.39
8	0.797	1.002	0.0539	0.00120	15.32	0.72
9	0.954	1.002	0.0283	0.00167	12.48	0.14
10	0.499	1.013	-0.0645	0.01035	13.39	1.31
11	0.483	1.001	0.0217	0.00015	17.73	0.04
12	0.737	1.002	0.0420	0.00100	0.00	*
13	0.737	1.002	0.0420	0.00100	0.00	*
14	0.729	1.000	-0.0083	0.00146	11.41	0.15
15	0.372	1.002	0.0534	0.00003	20.11	0.15
16	1.034	1.005	0.0280	0.00347	11.82	0.63

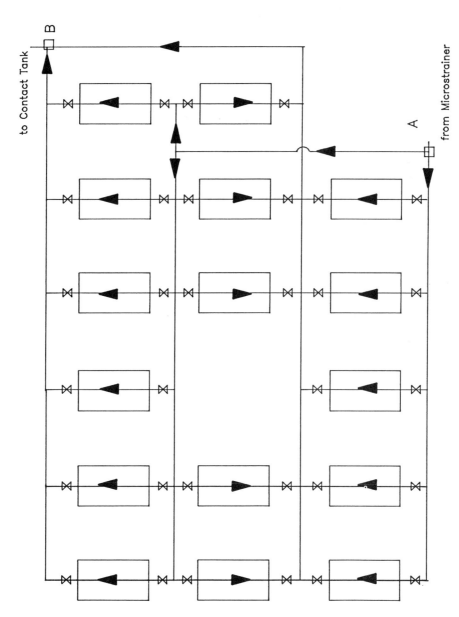

Figure 6 : Ashford Common Water Treatment Works — West side
Schematic diagram.⋈ Valve ☐ Filter bed ◀ flow direction

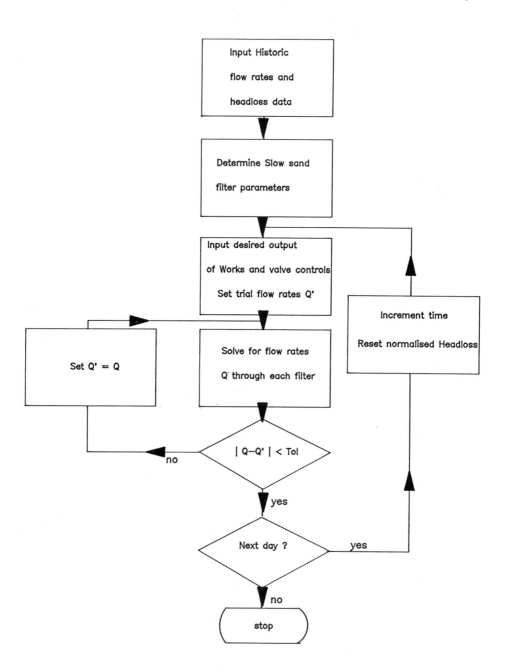

Figure 7 : SANNET flow diagram

When parameter estimation is complete, a set of headloss equations and the continuity equation at node B (figure 6) are set up. In this approach, the program requires the total daily outputs of the network for the period of the forcast, as input data. Since the outlet valves of the filters are controlled as part of the normal works operations, information about valve positions is also required for realistic forecasting.

The headloss and continuity equations are solved iteratively for the flow rates through each slow sand filter. In each iteration step, solutions of 16 simultaneous linear equations are obtained by Crout's factorisation method (11). A convergence threshold of $0.001 Mlday^{-1}$ on each flow rate has been employed. Convergence can normally be achieved after 10 to 15 iterations.

The accuracy of the SANNET program depends on various factors. It is clear that if the slow sand filter parameters are to be determined with sufficient accuracy, reliable and precise initial data are required. The deviations evaluated in the fitting procedure (c.f. section II) can be used to access the reliabilities.

When an outlet valve is restricted, the headloss developed across it is one of the major contributions. Adequate calibration of headloss against proportional openings and accurate data on valve positions are therefore essential.

Figures 8 and 9 respectively show typical predicted flow rates and headlosses across a filter bed. In general, fair agreement with the measured data is obtained (average error < $5 Mlday^{-1}$). Results, however, should be improved when more accurate input data, particularly information on the outlet valve controls, are available.

When flow rates are known, the total headloss across the slow sand filter network (from nodes A to B in figure 6) can be readily obtained. The calculated values for the period 14/10/87 to 1/1/88 are plotted in figure 10 together with the measured data. It can be observed a general agreement is obtained.

The SANNET program can be employed to assist the operational planning of a slow sand filter works principally in two ways. Firstly, if target outputs of the works are given, SANNET can be used to estimate possible values of the flow rate and headloss through each filter bed. The information about the flow rates assists effective planning of the cleaning schedules and allows the effect of various control options to be assessed. In this mode headloss information indicates when works capacity is approached or the required headlosses approach the maximum achievable value.

Secondly, if the aim is to determine the highest possible output from the works, the total headloss, rather than the output flow, of the network is known. Trival modifications of SANNET have been done by replacing the continuity equation at node B (figure 6) by a headloss equation. The flow rates and headlosses for each filter bed and hence the total output of the works (figure 11) can therefore be estimated.

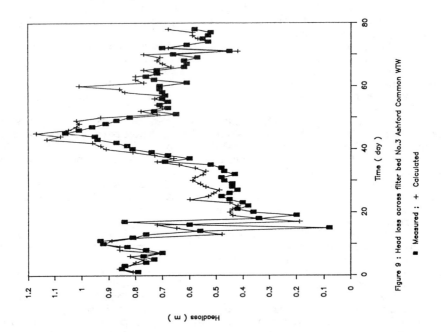

Figure 9 : Head loss across filter bed No.3 Ashford Common WTW

■ Measured ; + Calculated

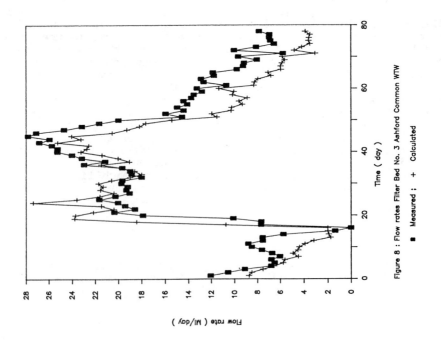

Figure 8 : Flow rates Filter Bed No. 3 Ashford Common WTW

■ Measured ; + Calculated

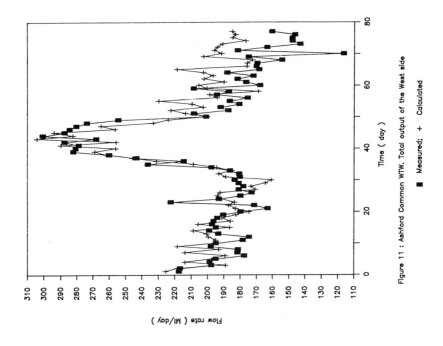

Figure 11 : Ashford Common WTW. Total output of the West side

■ Measured; + Calculated

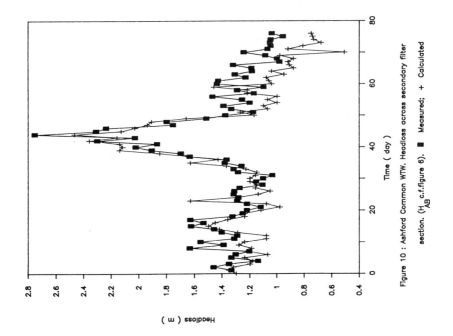

Figure 10 : Ashford Common WTW. Headloss across secondary filter

section. (H_{AB} c.f.figure 6). ■ Measured; + Calculated

IV.   CONCLUSION :

A mathematical model for the hydraulic behaviour of a slow sand filter has been developed. For a single filter, the model predicts the flow rates for given headloss data within 15% of the observed values.

This model has been incorporated in a FORTRAN computer program (SANNET) to study the slow sand filter network at Ashford Common WTW. From the total output of the network, the flow rates and headlosses for all filters can be estimated within 5Ml/day and 0.2m respectively. A microcomputer version of SANNET running on 20MHz 80386 processor requires less than 15 seconds for each forecast.

The program will form the heart of a Filter Bed Management system, to be implemented at three or more of the Authorities works in the near future.

ACKNOWLEDGEMENT

The authors would like to thank K Aston (Thames Water) for providing valuable historic data.

REFERENCES

1. Tritton D J Physical Fluid Dynamics Van Nostrand Reinhold p 82 (1982)
2. Huisman L and Wood W E Slow Sand Filtration World Health
        Organisation (1974)
3. Bellinger E G I.W.E.S 33 19 (1979)
4. Lloyd B Water Research 7 , 963 (1973)
5. Pielou E.C Mathematical Ecology p20-40 John Wiley & Sons (1977)
6. Boucher P.L. J.Int.Wat.Engrs.   p561 (1951)
        J. New England Wat Wks. Assoc. 70 1 (1956)
        Municipal Utidities Mag 95 , 59 and 95 , 22 (1957)
7. Toms I.P Developements in London's water supply system, Arch.
        Hydrobiol. Beih, 28 p149.(1987)
8. Ives K.J Proc. Inst. Civ. Engrs. 17 , 333 (1960).
9. CAMM (Computer Aided Manufacturing Management) Technology, Series
        2000
10. Halcrow (1987) Ashford Common Uprating Hydraulic Model, Interim
        report (1988) Annexe report (1988) Stage 2 report (draft)
11. NAG Fortran Workstation Library (1986) Implementation F04ATF.
        Numerical Algorithms Group.

5.4 THE APPLICATION OF POLYURETHANE TO IMPROVE SLOW SAND FILTERS

P. Vochten, J. Liessens and W. Verstraete — Laboratory of Microbial Ecology, State University of Gent, Coupure L653, B-9000 Gent, Belgium

ABSTRACT

The potential use of reticulated polyurethane prefilters in slow sand filtration was studied on small pilot-scale. Two alternative types of prefilters were studied. In a first type a 10 cm layer of highly porous polyurethane foam was installed on top of the filter bed. By this modification, the biological activity was distributed across the prefilter surface, creating a kind of expanded "Schmutzdecke". This resulted in a significant increase of oxygen consumption over the filter indicating an increased microbial activity. The installation of the polyurethane matrix prevented clogging and decreased the head - loss across the sand filter resulting in longer filtration cycles. In a second modification, a 75 cm prefilter of polyurethane was aerated at the bottom. This submerged aerated matrix, with high microbial activity, enhanced the slow sand filter performance especially for low quality water containing substantial amounts of NH_4^+ and biodegradable organic matter. The removal of BOD and NOD over the filter was significantly better compared to the reference filter. A better bacteriological quality of the filtrate was observed with regard to the number of *Escherichia coli* and *Streptococcus faecalis*. The aerated prefilter also prolonged the length of filter runs in a substantial way.

INTRODUCTION

Slow sand filtration is an effective low cost water purification process. The method was introduced at the beginning of the 19th century and is still used to prepare drinking water from surface waters with low loads of pollution and turbidity [1]. Its simple operation makes it very appropriate for application in developing countries. Slow sand filtration is a highly effective means of removing bacteria and other microorganisms of intestinal origin from the water. It is reported that between 99 and 99,9% of pathogenic bacteria are removed during filtration [2]. The filter performance can be strongly influenced by parameters such as temperature, filtration rate, sand bed depth, size of the sand and frequency of filter maintenance [3,4,5]. The removal of organics is significant but depends on the circumstances of the study. COD/TOC reduction percentages between 15 and 74% are reported by Ellis (6), while Joshi et al. (3) observed an average removal of 50%.

The purification achieved in a slow sand filter may be considered to be the result of the combination of different processes, such as : mechanical straining, sedimentation, inertial and centrifugal forces, diffusion, mass-attraction and electrostatic attraction, but most importantly biochemical and microbial actions. The major part of this removal takes place in the top layer of the filter bed, where the filter skin or *Schmutzdecke* develops. This intensely biologically active layer is composed of algae, bacteria, fungi, actinomycetes, an other microorganisms, including protozoa and rotifera (6).

The biological activity in the filter bed is of an aerobic nature. When the water becomes anoxic, a number of undesirable phenomena can occur. Anaerobic activity leads to the production of unwanted end products which create taste and odour (e.g. H_2S). In addition, manganese and iron which are oxidized under normal aerobic conditions prior to filtration and subsequently precipitated on the filter surface, will be reduced and redissolved under anoxic conditions. This can lead to subsequent precipitation in the water supply system (6). From bacteriological point of view, anaerobic conditions also have an adverse effect (3,6).

In a normal slow sand filter, the zone with the highest microbial activity - the *Schmutzdecke* - is concentrated in the top few millimeters of the sand bed. It is postulated that a layer of very fine reticulated polyurethane can, because of its higher porosity (sand has only a porosity of some 50%), expand this layer over a larger depth. In this way the residence time of the water in this Expanded Schmutzdecke (ES) will be longer, and thus also the time it is subjected to the high microbial activity. The prefilter is installed in the hydraulic water head on top of the sand bed. Reticulated polyurethane was chosen as matrix, because this material - already in practice in anaerobic and aerobic waste water treatment - has a high porosity (96-97%) and a relatively high specific surface (2000-4000 m^2/m^3 depending on its pore size). It is furthermore readily colonized by micro-organisms (7,8).

Under conditions where the COD of the raw water exceeds 20 mg/l, the bacteriological and physical removal efficiency is reported to be insufficient without preliminary treatment (3). To avoid deterioration of the filtrate, due to high loads of organic matter or nitrogenous compounds, a matrix of reticulated polyurethane is placed on top of an air diffusor in the otherwise "iddle" supernatant water of the filter surface. On this Submerged Aerated Prefilter (SAP), microorganisms, that aerobically biodegrade these pollutants, attach and grow.

Both types of prefilters were tested under laboratory conditions.

MATERIALS AND METHODS

The experiments were carried out in three tests (Table 1). In test 1 the ES concept was studied with a moderate quality raw water susceptable to precipitation of $CaCO_3$. In the first half of the test (test 1 (a)) the water used was hard (≥ 4.0 meq Ca^{2+}/l), in the second half (test 1 (b)) soft water (≤ 3.0 meq Ca^{2+}/l) was used. In test 2, the incoming water quality was adjusted so that no further $CaCO_3$ deposition occurred. In test 3, the SAP concept was studied with a low quality raw water not susceptable to $CaCO_3$ precipitation.

Table 1. Overview of the different test runs.

Test	Quality of water	Duration (days)	Height of the sand bed (cm)			
			Experimental filter	Reference filter		
			Run 1	Run 1	Run 2	Run 3
1(a)	Hard	49	⎫ 73.5	⎫ 72.0	⎫ 69.5	⎫ 67.5
(b)	Soft	41	⎭	⎭	⎭	⎭
2	Soft	90	69.5	66.5	-	-
3	Soft	90	73.5	73.5	-	-

The lab-size filters (Fig. 1) were made of plexiglass tubes measuring 2 m of height and with an internal diameter of 50 mm. The filter bed consisted of a layer of ca. 70 cm of sand with an effective size of 0.22 mm and a uniformity coefficient of 1.9. The filter bed was separated from a layer of fine gravel by a thin layer of glass-wool, replacing the normal gradation in gravel size. The height of the supernatant water was kept constant by means of an inlet weir. The rate of filtration was kept constant at 0.1 $m^3/m^2.h$ by daily adjusting the height of the outlet weir. The difference between these two weirs was a measure of the head-loss developed by the filter. If after a certain time of operation the head-loss developed by the filter exceeded 100 cm, the water level was lowered below the sand surface and the top centimeters of the filter bed were carefully scraped of. The two filters used in test 1 and test 2 had been in use for 4 months prior to the start of the experiment and were fully colonized. The ones used in test 3 were started up as the experiment commenced.

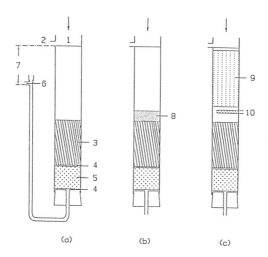

Fig. 1. Set-up of the filters (a) Reference filter (b) ES filter (c) SAP filter
1. raw water 2. raw water weir 3. filter bed 4. glass-wool 5. gravel 6. filtrate weir 7. head-loss 8. ES (expanded *Schmutzdecke*) 9. SAP submerged aerated prefilter) 10. air diffusor

Two types of reticulated polyurethane foam (Recticel N.V., Wetteren, Belgium) were used. For the experiments with the ES filter, a 10 cm thick block of reticulated polyurethane with a porosity of 80 pores per inch (80 ppi) was placed on top of the filter bed (Fig. 1(b)). For the experiment with the SAP, mats of reticulated polyurethane with 30 pores per inch (30 ppi) were installed vertically on top of the air diffusor, 15 centimetres above the *Schmutzdecke* of the filterbed. The latter was not physically disturbed by the aeration. The diffusor produced medium size (1 to 3 mm) air bubbles and assured a flow rate of 0.18 m^3 air per m^2 surface per minute.

The raw water used in the experiment was tap water, artificially polluted. Two levels of water pollution were used. For the moderate quality raw water the following amendments were made per liter : 6.0 mg COD (as NaCH$_3$COO); 0.5 mg NH$_4$$^+$-N (as NH$_4$Cl); 5.0 mg NO$_3$$^-$-N (as NaNO$_3$); *Escherichia coli* (grown in a nutrient solution) and *Streptococcus faecalis* (idem). For the low quality raw water, the following amendments were made per liter : 20 mg COD (as NaCH$_3$COO), 2.0 mg NH$_4$$^+$-N (as NH$_4$Cl), 5.0 mg NO$_3$$^-$-N (as NaNO$_3$), 25 ml presettled domestic sewage (containing some 180 mg BOD$_5$^{20} l^{-1}, 150 mg NH$_4$$^+$-N l^{-1}); *Escherichia coli* (grown in a nutrient solution) and *Streptococcus faecalis* (idem). In the first half of test 1 (test 1 (a)) the raw water had a Ca^{2+} content of 80 to 120 mg Ca^{2+} l^{-1} by diluting the tap water with decalcified water. Yet this raw water was still susceptable to precipitation of CaCO$_3$ because of its high pH. In the following two tests (2 and 3) the pH of the soft raw water was adjusted by a pH control unit to 7.0.

The follwing parameters were monitored : the head-loss of the filters; pH; dissolved oxygen (DO); concentration of NH$_4$$^-$-N ,NO$_3$$^-$-N, NO$_2$$^-$-N and BOD$_5$^{20}; number of *Escherichia coli* and *Streptococcus faecalis*. The dissolved oxygen was measured with a Beckman Monitor System II. The nitrate concentration was determined spectrophotometrically by means of the Nessler-reagent and the method of Montgomery and Dymock (9) respectively. The total number of *E. coli* and *S. faecalis* were counted by plating out on MacConkey and Streptococcus KF agar (Difco) and incubated at 43 °C for 24 hours and 37 °C for 48 hours, respectively. The BOD$_5$^{20} was determined as the decrease in DO after 5 days of incubation in the dark at room tempurature in a Winkler bottle of a (diluted when necessary) sample inoculated with 1 ml/l inoculum and supplemented with 2.0 mg/l nitrapyrin as nitrification inhibitor.

RESULTS

The expanded *Schmutzdecke* concept (ES)

Test 1 : Moderate quality water with CaCO$_3$ precipitation

In the first 49 days (test 1(a)) hard water with a high level of Ca^{2+} (80 to 120 mg Ca^{2+} l^{-1}) was used. In the second half (test 1(b)) softer water (40 to 60 mg Ca^{2+} l^{-1}) was used. As shown by Fig. 2, the ES filter showed hardly any change in head-loss over the 90 days of operation. The reference filter on the other hand displayed a rapid increasing head-loss over the first two filter runs. The increase of its head-loss was less, but still significant important, when during the second half of this test (test 1(b)) soft water was used.

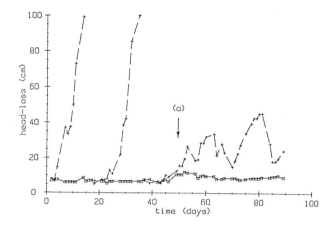

Fig. 2. Head-loss of ES and reference filter during test 1 (□——□ ES filter,
+——+   reference filter).
(a) Ca^{2+} content of the water was lowered from 80-120 mg Ca^{2+} l^{-1} to 40 -
60 mg Ca^{2+} l^{-1}.

        The DO of the filtrates of both filters was about equal and some four
units lower than that of the raw water (Fig. 3 and Table 2). From about
the 40th till the 60th day the DO of the raw water and the filtrates
increased temporarily. This was due to a temperature drop because of
severe winter circumstances. No difference could be noticed between the
performances of the filters in subtests 1(a) and 1 (b). Both filters gave
a BOD_5^{20} removal between 40· and 50%. The NOD removal (NOD = 4.33 x NH_4^+-N)
averaged 80%. The pH of the filtrates of both filters dropped from 7.9 to
7.6. Fig. 4 indicates that both filters gave a very substantial reduction
of the faecal bacteria *Escherichia coli* and *Streptococcus faecalis*; the
filtrate of the reference filter was slightly better in this respect.

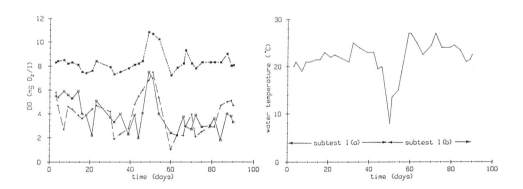

Fig. 3. DO of the raw water and the filtrates of the ES and the reference
    filter during test 1, in relation to the water temperature (•——·——•
    raw water, □——□ filtrate ES filter, +——+ filtrate reference filter,
    ———— water temperature).

Table 2. Average values and standard deviations for selected parameters of
raw water and filtrates of ES and reference filter during test 1.

Parameter	Unit	Raw water		Filtrate ES filter		Filtrate ref. filter	
		$\bar{x}$	s.d.	$\bar{x}$	s.d.	$\bar{x}$	s.d.
NO_3^--N	mg/l	9.7	0.7	9.5	0.8	9.7	0.9
NO_2^--N	mg/l	-		0.08	0.10	0.09	0.17
NH_4^+-N	mg/l	0.68	0.13	0.12	0.08	0.14	0.10
BOD_5^{20}	mg/l	4.1	0.6	2.4	0.7	2.0	0.6
DO	mg O_2/l	8.3	0.9	4.0	1.4	4.0	1.5
pH		7.9	0.4	7.6	0.3	7.6	0.3
Temp.	°C	21.0	5.3	-		-	

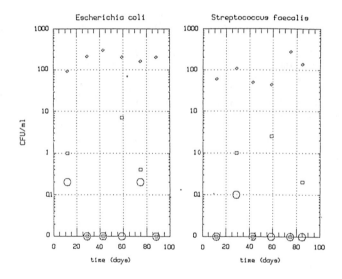

Fig. 4. *Escherichia coli* and *Streptococcus faecalis* in the raw water and
the filtrates of the ES and the reference filter during test 1 (◇ raw
water, □ filtrate ES filter, ○ filtrate reference filter).

Test 2 : Moderate quality water without CaCO₃ precipitation

In this second test, soft raw water with a Ca^{2+} content of 40 to 60
mg Ca^{2+} l^{-1} was used. The ES filter showed an almost constant head over
the whole period (Fig. 5). The reference filter on the other hand
displayed a gradually increasing head-loss from the 35th day on.

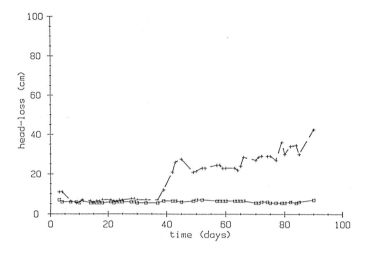

Fig. 5. Head-loss of ES and reference filter during test 2 (□———□ ES filter, +— —+   reference filter).

Fig. 6 gives the DO of the raw water and the filtrates. In the second part of the test, the DO of the ES filter is clearly lower than that of the reference, indicating an increased biological activity. Beyond 45 days, the filtrate of the ES filter has an average DO of 2.7 mg O_2 l^{-1} versus 5.3 mg O_2 l^{-1} for the reference. This difference is significant at $\alpha = 0.001$. The BOD_5^{20} removal also increased to between 60 and 70% in this second test (Table 3). Both filters gave excellent removal of the faecal bacteria (Fig. 7).

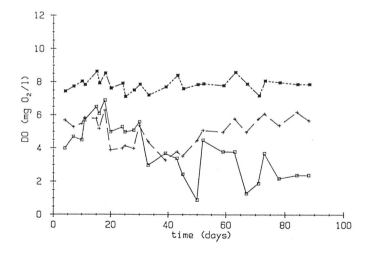

Fig. 6. DO of the raw water and filtrates of the ES and the reference filter during test 2 (■— ···—■ raw water, □———□ filtrate ES filter, +— —+   filtrate reference filter).

Table 3. Average values and standard deviations for selected parameters of raw water and filtrates of ES and reference filter during test 2.

Parameter	Unit	Raw water		Filtrate ES filter		Filtrate ref. filter	
		$\bar{x}$	s.d.	$\bar{x}$	s.d.	$\bar{x}$	s.d.
NO_3^--N	mg/l	11.1	0.7	10.9	1.0	10.7	0.9
NO_2^--N	mg/l	-		0.03	0.03	0.01	0.01
NH_4^+-N	mg/l	0.84	0.24	0.11	0.08	0.07	0.05
BOD_5^{20}	mg/l	4.7	0.8	1.5	0.5	1.9	0.6
DO	mg O_2/l	7.8	0.6	3.6	1.4	5.0	0.9
pH		6.9	0.2	7.5	0.2	7.5	0.9
Temp.	°C	22.5	1.4	-		-	

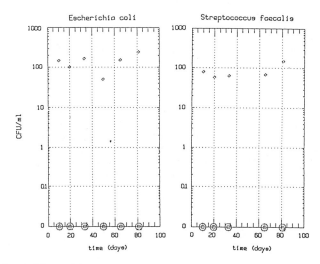

Fig. 7. *Escherichia coli* and *Streptococcus faecalis* in the raw water and the filtrates of the ES and the reference filter during test 2 (◇ raw water, ◻ filtrate ES filter, ◯ filtrate reference filter).

The   submerged aerated prefilter concept (SAP)

In a third series of experiments, a different approach was tried out, i.e. the installation of a submerged aerated matrix (Fig. 1 (c)). Low quality soft raw water was used. Fig. 8 shows that the head-loss of the SAP filter stayed more or less constant over the whole test period. The head-loss of the reference filter developed similarly untill approximately day 60 when it started to increase.

Measurements of the DO of the raw water, the effluent of the prefilter and the final filtrate showed a large decrease of the DO over the sand filter bed (Fig. 9). In the first 35 days, the DO of the water over the reference filter decreased only slightly. From then on, the reference filter approached anoxic conditions. The first 35 days can be considered as a start-up period, because these filters were not colonized at the onset of this test.

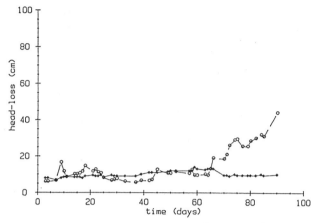

Fig. 8. Head-loss of SAP and reference filter during test 3(●———● SAP
filter, ○— —○ reference filter).

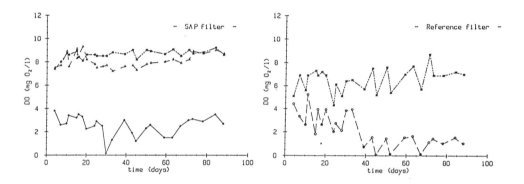

Fig. 9. DO of the raw water, the water leaving the prefilter and the final
filtrates during test 3 (■— --—■ raw water, ▲— —▲ effluent prefilter,
●———● filtrate SAP filter, ○— —○ filtrate reference filter).

The results from day 35 on, when both filters were fully operational,
are presented in Table 4. The BOD_5^{20} of the final SAP and reference
filrate amounted 1.5 and 5.2 mg l^{-1} respectively for an average influent
BOD_5^{20} of 16.9 mg l^{-1}. This difference is significant at $\alpha = 0.001$. The
NH_4^+-N content of the raw water dropped from 8.69 mg NH_4^+-N l^{-1} to 0.06 mg
NH_4^+-N l^{-1} in the filtrate of the SAP filter and to 0.45 mg NH_4^+-N l^{-1} in
the reference filtrate, which is significantly different at $\alpha = 0.001$. The
average amount of NO_2^--N in the SAP filtrate - 0.01 mg NO_2^--N l^{-1} was
significantly lower ($\alpha = 0.001$) compared to the filtrate of the reference
filter - 0.28 mg NO_2^--N l^{-1}. The modified SAP filter also gave an
excellent removal of both *Escherichia coli* and *Streptococcus faecalis*
(Fig. 10). For the reference filter, this removal was rather poor. The
results of the O_2 removal over the filters are difficult to compare, since
the DO of the influent of both filters differed slightly. This was due to
the fact that the raw water DO had to be measured at the heighth of the
raw water weirs (= the top of the supernatant water). Because of the
aeration of the prefilter, the DO of the raw water of the SAP filter was
obviously higher compared to the reference filter (Fig. 9).

Table 4. Average values and standard deviations for selected parameters of raw water, prefilter effluent and filtrates of SAP and reference filter during test 3 from day 35 on.

Paramater	Unit	Raw water		Effluent SAP		Filtrate SAP filter		Filtrate ref. filter	
		$\bar{x}$	s.d.	$\bar{x}$	s.d.	$\bar{x}$	s.d.	$\bar{x}$	s.d.
NO_3^--N	mg/l	10.4	1.9	14.2	1.8	15.3	0.9	13.8	2.5
NO_2^--N	mg/l	-		0.15	0.20	0.01	0.02	0.28	0.26
NH_4^+-N	mg/l	8.69	1.65	0.43	0.39	0.06	0.07	0.45	0.22
BOD_5^{20}	mg/l	16.9	2.8	2.5	0.4	1.5	0.8	5.2	1.0
DO	mg O_2/l	8.7 [1] 0.3 6.8 [2] 1.0		8.1	0.5	2.4	0.7	1.0	0.6
pH		7.0 [1] 0.1 7.6 [2] 0.2		7.9	0.2	7.5	0.1	7.4	0.3
Temp.	°C	22.3	1.2						

[1] Top supernatant water SAP filter
[2] Top supernatant water reference filter

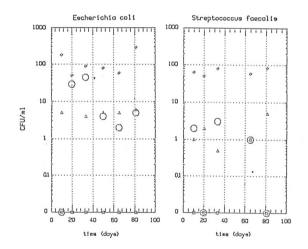

Fig.10. *Escherichia coli* and *Streptococcus faecalis* in the raw water, the prefilter effluent and the final filtrates (◇ raw water, △ effluent prefilter , □ filtrate SAP filter, ○ filtrate reference filter).

DISCUSSION

Fig. 2 and Fig. 5 show very clearly that the application of a layer of reticulated polyurethane on top of a slow sand filter prevents premature clogging of the filter. In the case of $CaCO_3$ saturation, it precipitated on the polyurethane before the water entered the sand bed. In this way, premature clogging of the pores between the sand grains by the $CaCO_3$ - as happened in the reference filter - was avoided. In the case where no $CaCO_3$ precipitated, the reticulated polyurethane provided, due to its high porosity (96-97%), a larger depth over which a high microbial activity could develop, resulting in an expansion of the *Schmutzdecke*. This higher porosity also prevented rapid biological clogging of the filter bed, since much thicker biofilms than on the sand grains had to develop, before the pores between the polyurethane matrix would get clogged. Barnes and Mampitiyarachichi (10) also advocated the use of highly porous material. They found that for raw water, in which inorganic suspended solids were present, rice hull ash (a material with a high porosity) instead of sand, could be used advantageously because thus the head-loss was spread out over a larger depth.

In the second part of test 2, the decrease in dissolved oxygen over the ES filter increased substantially from day 45 on (Fig. 6). For this period, the oxygen consumption over the ES and the reference filter was significantly different at $\alpha = 0.001$. Since the oxygen consumption correlates with the aerobic microbial activity on the filter, this is an indication of a beneficial effect by this modification on the operation of the slow sand filter. Test 1 and test 2 succeeded each other in time. The $CaCO_3$ that had been deposited on the polyurethane matrix during the first test, redissolved probably in the softer water with low pH during the first period of test 2. The support for the microorganisms to grow on was instabile, untill only the polyurethane matrix remained. From then on, a more stable microbial community could develop and the oxygen consumption increased concomitantly.

For the third test, the first two weeks of operation have to be considered as a start-up period. This may explain the slowly increasing head-loss of the reference filter (Fig. 8) and the fact that DO of the filtrate - an indication of the filter activity - only decreased after 35 days (Fig. 9). The main cause of filter clogging was microbial growth since the raw water was polluted with mainly soluble organic pollutants and ammonium.

The BOD_5^{20} removal by the SAP filter and the reference filter was significantly different at $\alpha = 0.001$ for the period of steady state operation. Yet, it is surprising that so much BOD_5^{20} was removed by the reference filter in test 3, in view of the relatively low reduction of the DO over this filter (Table 4). Apart from the aerobic removal, some of the BOD_5^{20} could have been dissimilated anaerobically. This is not unlikely, due to the almost anoxic status of the filtrate on the one hand (Table 4) and the substantial amount of nitrogen loss in the system. The N-input is 10.4 mg NO_3^--N l^{-1} and 8.69 mg NH_4^+-N l^{-1}, or a total of 19 mg N l^{-1}. The N-output is 13.8 mg NO_3^--N l^{-1}, 0.28 mg NO_2^--N l^{-1} and 0.45 NO_4^--N l^{-1}, with a total of 14.5 mg N l^{-1}. The deficit is probably due to denitrification of NO_3^- to N_2 gas. This is possible since the average DO of the filtrate is 1.0 mg O_2 l^{-1}, which is lower than the 2.0 mg O_2 l^{-1} set by Focht and Verstraete (11) as indicative for denitrification to take place. Oxygen limitation also seemed to have an influence on nitrification, as NO_2^- appeared in the filtrate in a relatively high concentration (Table 4). The semi-anoxic conditions also had an adverse effect (Fig. 10) on the removal of *Escherichia coli* and *Streptococcus faecalis*. Similar phenomena have been reported before (3,6). The better bacteriological removal efficiency can be due to a higher protozoal activity in the prefilter. However, this aspect was not studied and warrants further research.

CONCLUSIONS

Lab-scale experiments indicated that a layer of reticulated polyurethane, installed just above the sand bed, created an expanded

Schmutzdecke. This modification showed a very effective resistance of the filter to clogging by precipitation of $CaCO_3$. Under circumstances where no $CaCO_3$ precipitated, filter runs were also substantially prolonged. The biological activity of the filter increased significantly as indicated by an almost doubling of the dissolved oxygen consumption. No differences were detected for the pH or the amount of NH_4^+-N, NO_3^--N, NO_2^--N or faecal bacteria in the filtrate, between the modified and the reference filter.

    The slow sand filter equipped with a submerged aerated matrix of reticulated polyurethane also prolonged the lenghth of the filter runs. The removal of biodegradable organic matter and NH_4^+-N were significantly better due to an increased aerobic microbial activity. The filtrate was also of a much higher quality with regard to the number of residual *Escherichia coli* and *Streptococcus faecalis* present.

REFERENCES

1. L. Huisman and W.E. Wood in La Filtration lente sur Sable, World Health Organization, Genève, 133 p (1975).
2. J.M.G. Van Damme and J.T. Visscher, H₂O, 17, 464-469 (1984).
3. N.S. Joshi, P.S. Kelner, S.S. Dhage, R. Paramasivam and S.K. Gadkari, Indian Journal of Environmental Health, 24, 261-276 (1982).
4. W.D. Bellamy, D.W. Hendricks and G.S. Logsdon, Journal of American Water Works Association, 77, 62-66 (1985).
5. T.R. Cullen and R.D. Letterman, Journal of American Water Works Association, 77, 48-55 (1985).
6. K.V. Ellis, CRC Critical Reviews in Environmental Control, 15, 315-354 (1985).
7. P. Huysman, P. Van Meenen, P. Van Assche and W. Verstraete, Biotechnology Letters, 5, 643-648 (1983).
8. S. Deboosere in Het gebruik van een circulerend sponsbed voor de nitrifikatie van afvalwater, Engineer Thesis State University of Gent, 80 p (1985).
9. H.A.C. Montgomery and J.F. Dymock, Analyst, 86, 414-416 (1961).
10. D. Barnes and T.R. Mampitiyarachichi, Waterlines, 2, 21-23 (1983).
11. D.D. Focht and W. Verstraete, Advances in Microbial Ecology, 1, 135-214 (1977).

6

Developing country case studies

6.1 DUAL MEDIA FILTRATION FOR THE REHABILITATION OF AN EXISTING SLOW SAND FILTER IN ZIMBABWE

T. F. Ryan — Department of Civil Engineering, University of Zimbabwe

ABSTRACT

Studies were carried out aimed at improving the performance of an existing slow sand filter unit which suffered from extremely short filter runs. Investigations showed the main cause of the short runs was too small a sand grain size. Length of filter run could be substantially increased by replacing the existing sand by a coarser sand but at considerable cost. The solution proposed was a dual media filter whereby the existing fine sand was superimposed by a layer of coarser sand. This had the benefits of improved filter run, efficient coliform removal and importantly low cost implementation.

INTRODUCTION

The advantages of slow sand filtration as an appropriate form of water treatment for developing countries are continually stated. Visscher et al (1) have described slow sand filtration as "one of the most effective, simplest and least expensive treatment processes and is therefore particularly suitable for rural areas in developing countries". Slow sand filtration is a technology that is sustainable by most developing countries i.e. there are generally no foreign currency requirements in the construction of the unit, no (imported) chemicals or mechanical plant are required, and the filters can be maintained by semi-skilled operators.

Yet despite the apparent enthusiasm for slow sand filtration amongst many (predominantly first world) engineers and scientists successful examples of this technology are comparatively rare in many developing countries. The reasons for this are many and varied and could include the fact that:

- The training of many local and expatriate engineers is Western orientated with sufficient emphasis often not given to developing world conditions.

- The conservative nature of the profession results often in a reluctance to try new or 'different' technologies - for which no proven track record exists.

- Cost comparison between slow sand filtration and rapid gravity sand filtration often favour the latter, particularly if a pretreatment unit is required. The major cost of slow sand filtration over the design life of the scheme is the initial capital cost of construction, therefore any major savings must lie in finding cheaper alternatives to reinforced concrete structures.

Many slow sand filtration units in developing countries fail to perform adequately due to lack of operator training.

The tendency to continually promote slow sand filtration on the grounds that it is a simple (i.e. appropriate) technology well within the capabilities of most developing countries, could also have adverse effects. Because slow sand filtration is supposedly 'simple' it may not get the degree of attention it deserves in terms of design and operation. The design may be relegated to more junior staff with little input from senior, more experienced staff. As slow sand filters can be operated by semi-skilled staff (again cited as an advantage of slow sand filtration) there is the danger these operators will suffer from lower status and reward and hence receive less training than their more skilled counterparts. In developing countries where resources are limited there is an obvious tendency to put more resources into the more 'difficult' and challenging projects with the result that they are often successful. The 'simple' projects conversely, often fail due to lack of resources. In Zimbabwe, as an example, the major towns and cities have water and sewage facilities comparable to those in any Western country, successfully operated to high standards. However, in many parts of the rural areas there are numerous examples of 'simple' technologies (e.g. pit latrines, hand pumps), which have failed. Slow sand filtration is in danger of falling into this category of 'simple', hence neglected technology.

OVERVIEW OF SLOW SAND FILTRATION IN ZIMBABWE

Slow sand filtration is not extensively used in Zimbabwe and hence experience in this area is lacking. It is being used on a small scale in certain remote areas, e.g. farms, mission stations etc. However this experience is not being made available for the benefit of others. Experience within the relevant Government Ministries and Municipalities is largely with the conventional chemical coagulation/rapid sand filtration type treatment plants. However foreign currency constraints and the severe shortage of technical skills coupled with the post-independence emphasis in Zimbabwe on the development of communal (rural) areas has caused decision makers to consider other forms of water treatment - which are less demanding on the limited resources of the country. This has led to the design and construction of a number of slow sand filtration plants in Zimbabwe within the last few years. Unfortunately, this experience has not done much to promote slow sand filtration in Zimbabwe, as in many cases the plants failed to perform satisfactorily due to poor design, insufficient operator training and lack of previous experience with slow sand filtration. This recent experience has highlighted the lack of slow sand filtration expertise within Zimbabwe and the need for adequate research and pilot plant studies prior to the introduction of any 'new' technology such as slow sand filtration.

It has further indicated that this technology, though relatively 'simple' is often not well understood. The lack of success of slow sand filtration in Zimbabwe, in recent years, is in danger of prejudicing local engineers and decision makers against this form of treatment when it has potentially much to offer. What is required is properly planned programmes of research and training, backed up by experience gained through the monitoring of existing slow sand filter plants within Zimbabwe.

This paper deals with experience gained by the Department of Civil
Engineering at the University of Zimbabwe in attempting to resolve some of
the problems of poor performance at a recently constructed slow sand filter
installation at the Howard Institute in the Chiweshe Communal Lands, 100
kilometers North of Harare, in Zimbabwe. The investigation was based on
undergraduate student project work carried out on site, and pilot plant
investigations carried out at the Department of Civil Engineering in Harare.

HOWARD INSTITUTE SLOW SAND FILTERS

The slow sand filters at the Howard Institute were designed to provide
450 cubic meters of potable water per day for the use of a mission hospital,
school and associated community. The raw water is pumped from the nearby
Sawi River to an intermediate reservoir situated on high ground. From here
the water gravitates to the water treatment works consisting of two identical
slow sand filtration units, a clear water tank and elevated storage. The
water then enters the Institute's existing reticulation system. Chlorination
was added shortly afterwards to ensure the supply of good quality water for
the hospital's operating theatre. In the original design no provision was
made for any form of pretreatment - more by omission than intent, however
as will be discussed later, raw water turbidity levels were fortunately not
a problem.

Initial attempts at commissioning the Howard filters were not entirely
successful and exposed several weaknesses in the design. An example was
the lack of provision for bottom filling the sandbeds. Initial attempts
at filling and operating the filters were impeded by the presence of pockets
of air within the sandbed. This problem was subsequently overcome by modifying
the interconnecting pipework to permit bottom filling.

The major problem with the Howard filters however, was the extremely
short filter runs of 5 to 7 days which were consistently achieved. This
placed an additional burden on the part-time plant operators and led to
significant water wastage due to the very frequent ripening operations.
Design flows were rarely achieved for any length of time and the net effect
on the community was severe water restrictions.

In an attempt to determine the probable cause or causes of the short
filter runs, university civil engineering students were employed to undertake
the investigative work. It was assumed that the short filter runs must be
due to either one, or a combination of, the following parameters:

(1) Sand grading
(2) Turbidity
(3) Algae

and this formed the basis of our further investigations.

Sand Grading

Sieve analysis showed that the sand used in the Howard slow sand filters
had an effective size of 0,1mm and a uniformity coefficient of 3,44. Because
sufficient quantities of river sand could not be located in the vicinity
of the Howard Institute, sand had to be imported from Harare, at great expense.

Unfortunately the sand delivered was finer than that specified. However this
fact only came to light once the sand was in place. This aspect highlights
another general assumption of slow sand filtration, i.e. that suitable sand
is generally available in the vicinity of the proposed construction. If this
is not the case, as with the Howard slow sand filters, then suitable sand must
be imported, often at considerable expense. It was unfortunate in this case,
that having resorted to this expensive option, sand finer than that required
was provided. Sources generally agree that the sand specification for slow
sand filters is an effective size (ES) in the range 0,15 - 0,3mm and a uniformity
coefficient (UC) of less than 3,0. The Howard sand did not meet this specification.

Turbidity

     Though the lack of any pretreatment stage was an omission in the design it
fortunately proved not to be a liability as turbidity levels in the river were
not excessive. Turbidity measurements taken over a 2 month period at the height
of the 1986/87 rainy season (a lower than average rainfall season) showed an
average turbidity over this period of 16 turbidity units, with a maximum of 30
turbidity units. Though raw water turbidity levels were acceptably low, some
suspended solids load was entering the system at the river intake necessitating
modifications.

     The submersible pumps and pipework were modified to permit the raising and
lowering of the pumps to match the river stage so that water could always be
drawn off from near the river surface. This improved on the previous situation
where the pumps situated at depth were stirring up and resuspending bottom
deposits which could then enter the pipe system.

     The intermediate reservoir situated on high ground acted as a break pressure
tank, allowing the raw water to gravitate to the slow sand filter units at the
Howard Institute. Water entered this tank via a float valve and exited via an
outlet pipe situated at the tank floor level. Retention times in this tank were
sufficient to permit sedimentation such that after a period of time silt accumulated on the tank floor and was drawn into the outlet pipe to the slow sand
filter. Though not the original intention, this tank acted as a sedimentation
basin, albeit an inefficient one. Modifications to the inlet and outlet together
with the addition of a scour pipe improved the performance of this tank enabling
it to act more efficiently as a sedimentation basin. The outlet pipework was
modified to act as a floating outlet, drawing off water at just below surface
level. The inlet pipe was baffled to minimise disturbance of settled deposits
and a scour pipe permitted occasional cleaning of the tank.

     The above improvements reduced the amount of suspended solids entering the
pipe system and permitted better operation and control of the water supply.

Algae

     The possibility of the presence of algae being a contributory cause of the
short filter runs proved more difficult to investigate. Algal blooms were not
generally visible in the supernatent above the sand filter. However an analysis
of this water did show the presence of the following algal species:
ankistrodemus , *diatoma, chlamydamonas, microcysts, haematocбccus* and *selenastrum*.
Diatoma and *microcysts* are potential filter blockers but whether they were present

in sufficient numbers to adversely affect the length of filter run was
difficult to ascertain. Had it been possible to cover one of the sand filters
then any marked improvement in length of filter run would have indicated an
algae problem in the uncovered filter. However this was not possible.

Initial investigations indicated that the fineness of the filter sand was
the most likely cause of the short filter runs. To prove this more conclusively,
and more importantly to find a solution to the problem of short filter runs, it
was decided to carry out tests using pilot filters to simulate the Howard slow
sand filter, as well as a range of other sand configurations. The pilot plant
investigations were carried out at the Department of Civil Engineering to
facilitate the monitoring of filter performance.

PILOT PLANT INVESTIGATION

The objectives of the pilot plant were to:

(a) Investigate the poor performance of the Howard filters, and
(b) Find a relatively simple and cheap remedy to improve the Howard filter
     performance.

The investigation was carried out in two stages. Stage 1 compared the
performance of three sand configurations: filter No 1 containing existing Howard
filter sand, filter No 2 containing a typical washed river sand, and filter No
3 containing a dual-media combination of existing Howard sand topped by a layer
of washed river sand. Stage 2 attempted to simulate more realistically the
proposed remedy for the Howard slow sand filters and was similar to the filter
No 2 case in Stage 1. However instead of starting with clean sand, Howard sand
which had been through a number of filter cycles was used, thereby ensuring some
deep bed penetration of biological activity had taken place. Figure 1 shows
the sand profiles used in Stage 1 and 2 pilot investigations.

The pilot filters were fabricated from galvanised steel sheeting to a
diameter of 285mm and an overall height of 2635mm. The raw water feed tank was
at a higher elevation permitting gravity supply to the pilot filters and the top
water level in the filters was controlled by ball valves. The flow was controlled
at the outlet by gate valves. No attempt was made to simulate the raw water
quality at the Howard Institute. However, the raw water feed tank was seeded
with settled sewage to give an intended total coliform count in the range of
500 - 1500 per 100 mℓ. The head loss through the pilot filter sandbeds was
monitored by recording the water levels in piezometer tubes inserted into the
sandbed at various depths (figure 1).

Stage 1
_____

Three pilot filters were set up as shown in figure 1 for direct comparison
of performance. Filter No 1 contained sand taken from the Howard sand filters
(ES = 0,10mm UC 3,44) and was intended to represent conditions there. Filter
No 2 contained river sand typical of the sort found in most rural areas. A
grading analysis carried out on the washed river sand showed an effective size
of 0,20mm and a uniformity coefficient 5,5. Though the uniformity coefficient
was larger than the values normally recommended for slow sand filters it is
representative of most river sands found in Zimbabwe and for this reason it was
decided to carry out the investigations using this sand. Filter No 3 contained

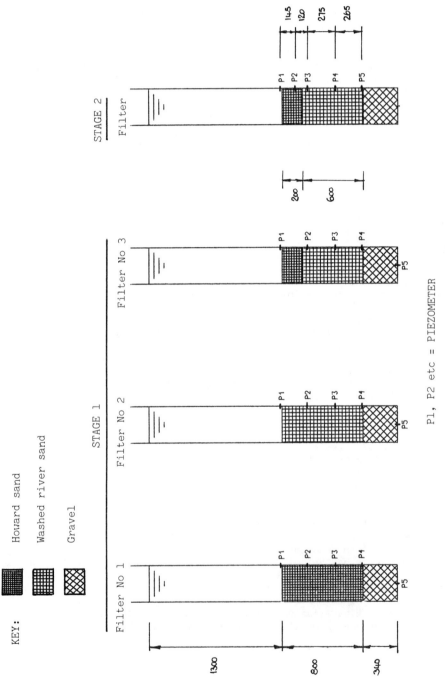

P1, P2 etc = PIEZOMETER

FIGURE 1   PILOT PLANT FILTER PROFILES

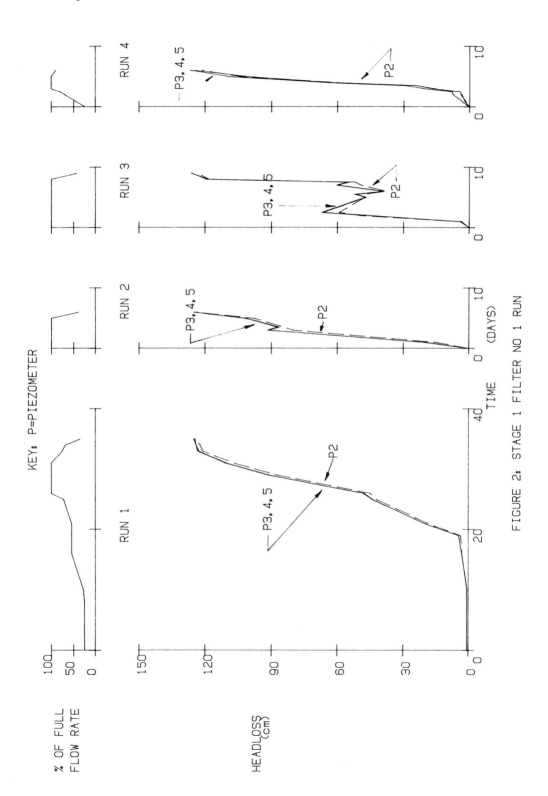

FIGURE 2: STAGE 1 FILTER NO 1 RUN

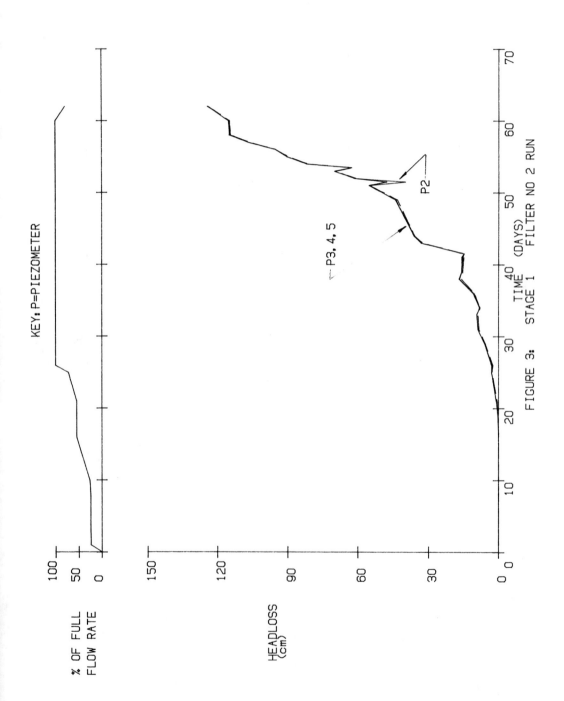

FIGURE 3:   FILTER NO 2 RUN
STAGE 1

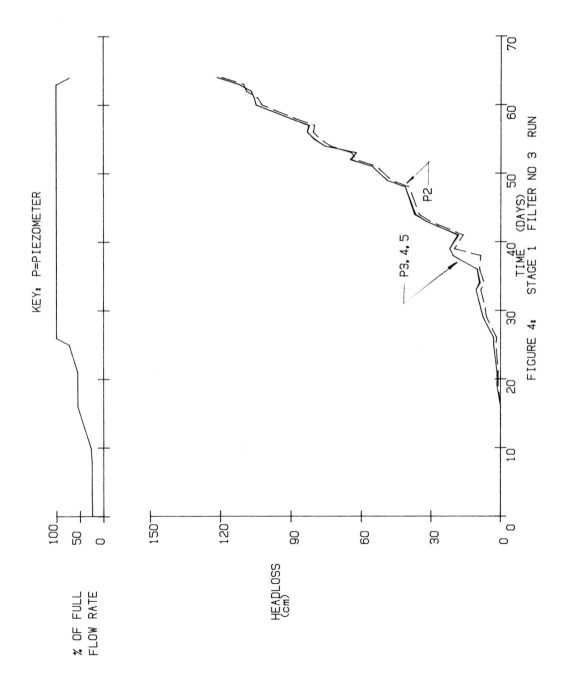

FIGURE 4: STAGE 1 FILTER NO 3 RUN

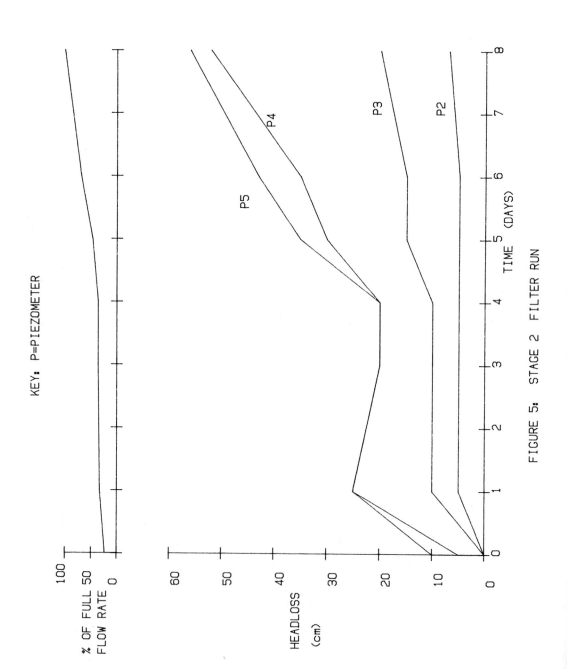

FIGURE 5: STAGE 2 FILTER RUN

the Howard sand (as in Filter No 1). However in this case the top 200mm of
Howard sand was removed and replaced with the coarser, washed river sand.
Thus two types of sand were used in the Stage 1 investigation. Filter No 3
examined whether filter runs could be lengthened without loss of effluent
water quality. It also provided a direct comparison with Filter No 2 in
performance.

The results obtained from the Stage 1 filter runs are shown graphically
in Figures 2-4. With Filter No 1, containing the Howard sand, four filter
runs were possible of duration 10, 5, 8 and 5 days respectively. The initial
ripening period for Filter No 1 was taken as 21 days based on the bacteriological quality of the filtered water. This period coincided well with the
increase in filter resistance due to the formation of the schmutzdecke
layer, as indicated by the piezometer water levels. In Figure 2, filter
run 3 shows a recovery in piezometer level commencing on day 2 of the
filter run and ending on day 6. This is as a result of filter short-circuiting due to ground vibrations caused by adjacent building operations. This
effects a longer than normal filter run. However lengths of filter run
achieved with pilot filter No 1 do show similarity to the actual filter runs
achieved with the Howard Institute slow sand filters.

Filter No 2 containing washed river sand showed a filter run duration
of 40 days after an initial ripening period of 21 days. Filter No 2 containing the dual-media i.e. a top layer of washed river sand overlying Howard
filter sand, gave very similar results to Filter No 2, with a filter run
duration of 42 days and an initial ripening period of 21 days.

Stage 2

In the Stage 2 investigation only one pilot filter - Filter No 1 - was
utilised. As this filter had been operated for four filter runs (in Stage 1)
some deep bed penetration of biological activity had therefore taken place.
The top layer of Howard sand was removed and replaced by a 200mm layer of
fresh washed river sand. Unfortunately only a limited operation period of
10 days was possible under Stage 2. This was insufficient time to enable a
full filter run to be carried out. However some general trends are discernable
from this limited filter operation period.

Full filter ripening was accomplished in a period of 7 days, substantially
shorter than the 21 day ripening period in Stage 1 (Figure 5). This indicates
the presence of residual biological activity - as a result of Stage 1 filter
operation - being present in the Howard sand layer at the commencement of the
Stage 2 operation. Further evidence of deep bed penetration is provided in
Figure 3 by the head loss recordings registered by the piezometer tubes.
After 10 days' filter operation, only 7mm head loss was recorded by piezometer
No 2, representing the top 120mm sand layer (Figure 1). The next 120mm sand
layer, which included 80mm of washed river sand and 40mm of the biologically
active Howard sand, recorded 18mm head loss over the same period (piezometer
No 3). Piezometer No 4 recorded a 30mm head loss in the next 275mm of Howard
sand, giving a total head loss of 55mm. The bottom 265mm sand layer recorded
only 10mm of head loss (piezometer No 5), to give a total head loss for the
filter sand bed, after 10 days' operation, of 60mm.

Thus at the commencement of ripening the initial head loss was due to existing biological activity within the top Howard sand layer, indicating deep bed penetration had taken place down to a depth of 300mm to 500mm in the Stage 1 Filter No 1 operation. With the further development of the schmutzdecke layer the majority of the head loss would be expected to occur at the surface of a sand bed. Although it was not possible to complete the filter run, the piezometer readings show some similarity with Stage 1 Filter No 3 - the overall head loss at the end of the ripening period is similar in both cases - hence a similar length of filter run could be expected. However further deep bed penetration into the finer Howard sand layer would add to the overall head loss and in the long term lead to shorter filter runs.

Bacteriological Investigation

Periodic total coliform counts were performed on the influent and effluent to monitor the bacteriological performance of the pilot filters, particularly during filter ripening. Due to the variable nature of the settled sewage used in the raw water mixture, the influent total coliform counts showed some variation. The average percentage coliform removal for each filter (fully ripened) are shown in Table 1. As could be expected, Filter No 3, the dual-media filter, gave a similar performance to Filter No 1 containing Howard sand, and performed better than Filter No 2 containing the washed river sand. As the slow sand filtration installation at the Howard Institute had a post-filtration chlorination stage, the bacteriological performance of the pilot filters was not our primary consideration. However the results do show satisfactory bacteriological performance.

Table 1 :   Percentage total coliform removal during Stage 1
(filtration rate $0,1mh^{-1}$)

Filter No 1	Filter No 2	Filter No 3
97,8%	96,5%	98,2%

DISCUSSION OF RESULTS

Pilot Filter No 1 containing Howard sand gave similar results to the slow sand filters at the Howard Institute. This investigation further showed that the cause of the short filter runs at the Howard Institute was due to the fineness of the filter sand alone. Turbidity levels in the raw water source at the Howard Institute were well within the acceptable range for successful slow sand filter operation, and therefore were considered as not contributing significantly to the short filter runs. Algae, another possible cause of short filter runs in slow sand filters, was in this case eliminated as a possible cause by covering the pilot filters. The fact that the (covered) pilot filters showed similar lengths of filter run to the Howard Institute slow sand filters implied algae was not a significant contributary cause of the short filter runs. If algae had been present in sufficient numbers to affect the length of filter run at Howard, then the pilot filters would be expected to give significantly longer filter runs. The main cause of the short filter runs at the Howard Institute, therefore, could be directly attributed to the fineness of the filter sand used.

One obvious remedy considered was to replace the whole volume of sand in the slow sand filters with more suitable sand e.g. washed river sand of the type used in the pilot investigations. However this would involve considerable expenditure given that a large proportion of the sand required would have to be imported from some distance away. The proposed remedy, which was the subject of the pilot plant investigations, appears to offer a solution at a greatly reduced cost. By replacing only a small proportion (approximately 20%) of the Howard sand with the washed river sand, the benefit in terms of increased length of filter run is substantial, at no sacrifice in bacteriological performance.

In a slow sand filter the majority of the head loss occurs across the schmutzdecke layer i.e. in the top 20mm or so of the sand bed. It is important therefore that this sand layer is of suitable sand (particularly in terms of effective size) as this has a major influence on the length of filter run achieved. The process of deep bed penetration is also of importance in slow sand filters where biological activity can penetrate to a depth of 400mm in the sand bed (2). In the case of the proposed dual-media filter, this would result in some biological activity penetrating the lower fine Howard sand layer resulting in some contribution to the overall head loss. In the long term this could lead to deep bed clogging and result in shortened filter runs. The following operational guidelines (3) should be adopted to overcome this problem with dual media filters:

At the end of each filter run the schmutzdecke layer is removed and discarded and when, after several filter runs, the upper coarser sand layer has reached a minimum thickness of 50mm, it too is removed.

The major portion of the lower finer sand layer is removed, the majority (over 50%) of this should be washed and replaced at the bottom of the filter bed covered by the unwashed portion.

The upper coarser sand is then placed on top, first the washed sand covered by the unwashed (50mm layer) coarse sand.

This operational procedure should realise good filter performance and facilitate the ripening of dual-media filters.

Increasing the thickness of the upper coarser sand layer results in a greater number of filter runs between resanding operations and a reduction in the effect of deep bed penetration - confining most of the biological activity to the upper coarser sand layer giving approximately 10 filter runs which should provide over one year's operation before resanding is required.

CONCLUSION

A cost effective solution to the problem of short filter runs experienced by the slow sand filters at the Howard Institute was found in the form of a dual-media filter bed. By largely retaining the original fine Howard sand and placing on top a coarser layer of sand greatly improved filter runs were achieved, while still retaining the good coliform removal characteristics of the finer sand. Dual-media filtration can therefore be considered as an option for the rehabilitation of slow sand filters suffering from short filter runs due to poor sand characteristics. It could also be considered

at the outset, in the design of slow sand filters where sufficient quantities
of suitable filter sand are not readily available. Importing only a relatively
small proportion of the total volume of sand required will ensure the important
top layer of the sand bed is of suitable sand. The remainder of the sand
bed, which is less subject to biological activity and therefore has less
influence on length of filter run, can be of other less suitable sand. Finally
further research is required into the long term effect of deep bed clogging
on the operation of dual media slow sand filters.

ACKNOWLEDGEMENT

Mr K V Tidy (BSc Civil Engineering, University of Zimbabwe) whose final
year undergraduate project formed the basis of the results contained in this
paper.

REFERENCES

1 Visscher J T, Paramasivam R, Raman A, and Heijnen H A.
  Slow sand filtration for community water supply: planning, design,
  construction, operation and maintenance (Technical paper No 24). The
  Hague, The Netherlands, IRC. (1987)

2 Ellis K V,Slow sand filtration CRC Critical Reviews in Environmental
  Control, 15, 4, 315-354.(1984)

3 Smet J, Personal communication. IRC, The Hague, The Netherlands. (1988)

6.2   THE PERFORMANCE OF SLOW SAND FILTERS IN PERU

B. Lloyd, M. Pardon and D. Wheeler — Del Agua, Environmental Health Unit, Robens Institute, University of Surrey, Guildford, Surrey GU2 5XH, U.K.

ABSTRACT

After 25 years experience of construction of rural water treatment systems in Peru recent diagnostic surveillance studies have demonstrated that slow sand filtration plants have uniformly failed to reduce contamination of surface water sources to provide a safe water supply. The reasons for the failure of slow sand filters are identified and strategies for improvement of water quality are proposed. Rehabilitation of treatment plants incorporating flow control at the point of abstraction, gravel prefiltration and protected slow sand filtration is described. Improvements in water quality are demonstrated and evaluated.

INTRODUCTION

The Peruvian National Plan for Rural Water Supply commenced in 1962 and during the subsequent 25 years over 1,300 small water supply systems were constructed for population sizes ranging from 500 to several thousands of people. Although the great majority were simple gravity systems about a quarter were treatment plants for surface water sources. These were built from a common design (Figure 1) typically incorporating the following sequence of components: screened intake, a grit chamber, a settler/sedimenter, a pair of slow sand filters and a reservoir leading to a distribution network.

A preliminary evaluation of 40 treatment plants in Peru (1) by the Panamerican Center for Sanitary Engineering and Environmental Sciences (CEPIS) in 1979 indicated that sedimenters and slow sand filters were often poorly maintained (Table 1). Subsequently a more intensive diagnostic survey of 60 systems, including 18 treatment systems, summarised here, was reported by DelAgua in 1985 (2). This demonstrated that the majority of system components had fundamental operational and design problems. These findings led to the development of a strategy for rehabilitation and improvement of treatment plants.

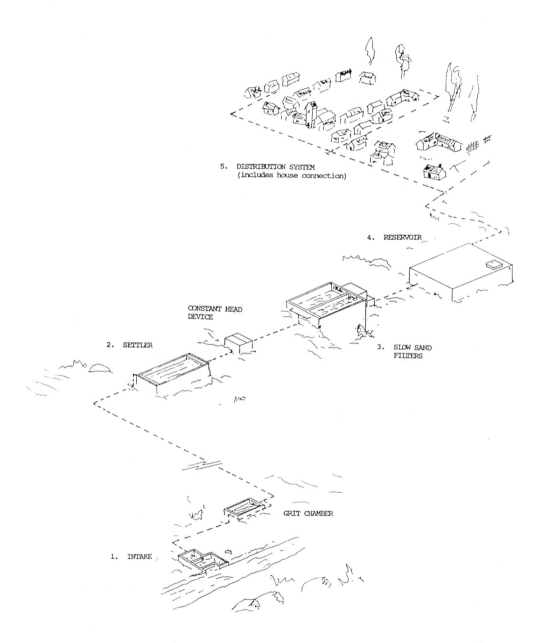

5. DISTRIBUTION SYSTEM
    (includes house connection)

4. RESERVOIR

CONSTANT HEAD
DEVICE

2. SETTLER

3. SLOW SAND
    FILTERS

GRIT CHAMBER

1. INTAKE

Figure 1. Layout of a rural water treatment plant in the peruvian highlands

TABLE 1

Preliminary Evaluation of Water Treatment in Rural Communities of
Peru (1979)

System Component & State of Unit	Nº	(%)
Settlers with sedimentation problems	15	(37)
never been cleaned	6	(15)
Slow Sand Filters bad state or not operating	13	(33)
never been cleaned	13	(33)
Disinfection not operated or maintained	35	(88)
Survey Size:	40	(100)

Source: Cánepa, L (1982) Investigación No 3, Filtros de arena en acueductos rurales,
Proyecto DTIAPA– CEPIS/ OPS (1).

TABLE 2

Summary of the diagnostic survey of Rural Water Treatment Systems in
the Central Highlands and High Jungle of Peru, 1985

System Component	Sites surveyed	Systems presenting major deficiencies and problems	
		Nº	%
Abstraction	18	16	89
Settlers	18	11	61
Slow sand filters	16	16	100
Disinfection	18	18	100

NOTE: The Ministry of Health reports a total of 28 rural treatment systems in the Region.

Source: Lloyd et al., (1986) "Developing Regional Water Surveillance in Health Region
XIII–Peru" DelAgua; ODA Phase One Report; Peruvian Water Surveillance Programme (2).

THE PROBLEMS OF RURAL TREATMENT SYSTEMS IN PERU

The results of a preliminary survey of slow sand filters carried out by CEPIS (Table 1) suggested that 33% of slow sand filters and 37% of sedimenters presented operating problems. The situation with respect to disinfection was worse, with 88% of systems not subject to disinfection control. It should be noted however that these results were based on sanitary inspection only and did not incorporate bacteriological or chemical analysis. The seriousness of the situation was not therefore fully revealed until a detailed diagnostic survey based on World Health Organisation guidelines (6), incorporating both water quality analysis and sanitary inspection, was carried out in 1985 in the Peruvian highlands and high jungle (2).

The 1985 survey included 18 treatment plants in 2 departments, Huancavelica and Junin. Two of the plants had rapid sand filters which were not working. The results for all 18 filtration plants are summarised in Table 2 and reveal that all slow sand filters and disinfection units had major deficiencies and operating problems. The reasons for this have been examined in detail and a clear pattern of common problems has emerged. They can be grouped in two main categories, a) administrative and b) technical.

a) Administrative problems of operation & maintenance.

The construction programme for the National Plan for rural water supply is largely the responsibility of the Rural Sanitation Division of the Ministry of Health. Systems within the National Plan are constructed jointly with the community which provides the work force. The community is required to establish an Administrative Committee (JAAP) responsible for operation, maintenance and recovery of operating costs. The administration and populations served by these schemes in the study area are summarised in Table 3a. This shows that the majority of rural schemes fall into the JAAP administrative category and therein lay a major problem in the supply of a safe, continuous supply of water. The JAAP were not properly trained, neither were there any incentives to provide a professional service. Until recently the committees almost never received professional supervisory support from the Ministry of Health after supplies were commissioned. We have therefore set up a pilot project to establish a training and a supervisory infrastructure through (DITESA) the national surveillance institution of the Ministry of Health (3).

b) Technical problems.

Athough some of the problems of treatment relate to faulty construction, the more fundamental problems were those associated with the raw water source quality and flows for which the standard designs were inadequate. With grossly contaminated source water it is essential to build a series of barriers to prevent contamination entering into distribution. This process of protection and treatment should begin with source water selection and abstraction. It is clear from Ministry of Health records that although a chemical analysis was carried out on intended source waters the more important turbidimetric and sanitary analysis was neglected and no adequate design provision was made to cope with high turbidity and faecal contamination.

Thus the problems of performance of slow sand filters cannot be considered in isolation from the other components of the system. In order to provide a better understanding of the whole, the technical problems identified are summarised in Tables 3b & 4 and described here in sequence from source water abstraction to terminal disinfection:

Abstraction

Water treatment processes are usually designed to work at a controlled flow rate. The standard design of the systems in Peru relies on flow control at the point of abstraction. This includes a 90º V-notch weir which is not capable of adequately controlling flows of less than $70m^3$/day (0.8 l/sec). The diagnostic survey revealed the following:

-In 16 out of 17 rural plants the V-notch was not installed.

-In 17 out of 17 plants the intake was poorly constructed or unprotected, either lacking a screen, or with a screen which was too coarse or broken.

-In 15 out of 17 systems there was no weir in the source water canal to ensure that there was a continuous minimum head, consequently the majority of systems suffer regular discontinuity of flow. This was often aggravated by neighbouring communities cutting the flow to the plant in favour of their perceived irrigation requirements. It was not well understood that the proportion of water required for potable supplies is insignificant compared to that required for agriculture.

TABLE 3a
Summary diagnostic of eighteen treatment systems
(Admin. & Population) Peruvian Water Surveillance Programme (1985)

Community	Admin. Authority	Total Population	Population Served	House Connections
Huancavelica	SENAPA	17,452	11,892	2,500
Yauli	JAAP	1,868	1,230	800
Palian	JAAP	1,334	693	228
Cocharcas	JAAP	1,195	355	100
San Martín de P	JAAP	Periurban	229	50
Tres de Diciembre	JAAP	1,002	248	50
San Agustín de Cajas	JAAP	4,246	1,265	230
Chaquicocha	JAAP	667	479	87
San José de Quero	JAAP	1,307	644	137
Huayao	JAAP	655	480	80
Churcampa	Council	1,859	715	140
Hualhuas	JAAP	1,751	1,375	233
Saños Grande	JAAP	5,000	1,750	350
El Mantaro	JAAP	3,016	2,280	420
Julcan	JAAP	2,126	1,167	200
Sacsamarca	JAAP	2,205	845	222
Tarmatambo	JAAP	2,042	1,370	234
Pichinaki	Council	1,890	1,364	600

SENAPA: National Water and Sewerage Authority
JAAP: Community Drinking Water Administrative Committee

TABLE 3b
Summary diagnostic of eighteen treatment systems
(Technical problems) Peruvian Water Surveillance Programme (1985)

	Structural & Maintenance Problems Identified						Water Quality (Res + Dist)		
Community:	Abst:	Line:	Sedi:	Filt:	Res:	Dist:	Clor: (mg/l)	Turb: (T.U)	Bact (Class)
Huancavelica(Urban)			+ (coagulation)	+ +	+		0.3	<5	A
Yauli			+ (coagulation)	+		+ +	00	15-32	D
Palian	+ + +		+	+ + + +	+		00	15->30	D
Cocharcas	+ + +		+ +	+ + + +			00	<5->25	D
San Martín de Porras	+ +		+ + (coagulation)	+ +	+ +		00	10-20	D
Tres de Diciembre	+ +		+	+	+		00	<5	B
San Agustín de Cajas	+ + +	+	+ +	+ + +	+	+ +	00	15-50	D
Chaquicocha	+ +		+ +	+ +		+	00	<5	C
San José de Quero	+		+ +	+ +		+	00	<5	B
Huayao	+ + +		+ + +	+ +	+	+	00	<5	C
Churcampa	+ +		+ + +	+ +		+	00	<5-9	C
Hualhuas	+ + + +		+ + +	+ + + +	+	+ +	00	15->50	D
Saños Grande	+ + + +		+ + +	+ + + +	+ +		00	18-500	D
El Mantaro			+	+ + +	+ +		00	10-60	C
Julcan	+ +	+		+	+	+	00	<5	C
Sacsamarca	+ + +		+ +	+ + +	+	+ +	00	6->50	D
Tarmatambo	+ + +		+ +	+	+ +	+	00	<5	C
Pichinaki	+	+	+	+			00	<5	C

+ = Nº of problems identified at each stage
Line= Conduction line. Res= Reservoir. Dist= Distribution system.
Sedi= Sedimentation tank. Abst= Abstraction. Clor=disinfection.Turb= Turbidity.Filt= sand
filters. Bact. class: A = 0, B = 1-10, C=11-50, D=> 50 Faecal coli/100ml

Settlers/sedimenters

These were generally subject to short circuiting through inadequate capacity and lack of baffling. Spot checks using salt tracers revealed minimum retention times of 10-20 minutes.
- In 16 out of 18 systems there was no form of baffling.
- Settler efficiency was further reduced by lack of routine cleaning.
Their efficiency in reducing turbidity was rarely more than 30% and consequently they failed to protect the slow sand filters which quickly become blocked. In evaluating raw water quality prior to construction of plants no turbidity or suspended solids analyses have been made during rainy periods in order to correctly size the sedimenters.

Slow sand filters

The fundamental problems suffered by slow sand filtration plants in many countries are their inability to cope with:
a) high turbidities
b) flow variation
c) incorrect media

In the Peruvian diagnostic survey it was repeatedly observed that the absence of flow control at the point of abstraction caused filters to be operated at an unstable or intermittent rate. It is well known that this results in sub optimal performance e.g., low bacteriological removal efficiency. More than half the plants had marginal or no effect in reducing turbidity and bacterial contamination. Table 3b demonstrates that a majority of treatment systems produced class D water, routinely allowing greater than 50 faecal coliforms/ 100ml into supply. Furthermore almost all of the systems with slow sand filters regularly supplied faecally contaminated water; only 3 out of 16 slow sand filters significantly reduced faecal coliform count to less than 10/100ml (Class A & B).

Table 4 highlights the serious neglect of basic operation and maintenance in slow sand filters.
-In 16 out of 16 plants sand bed depth was below the recommended 60cm minimum and two filters had no sand at all.
-In 14 out of 16 plants there was no reserve stock of sand
-In 16 out of 16 plants there was no installation or designated area for sand cleaning and storage.
None of the filters had any means of measuring the amount of water leaving the filter.

TABLE 4

Operational problems of sixteen rural sand filters.
Peruvian Water Surveillance Programme (1985)

Code: Community:•	Depth of sand bed (cm)	Volume of sand reserve (m³)	Distance to source of sand (Km)	Actions required
Palian	20	00	25	P
Cocharcas	00	00	5	P
San Martín de Porras	30	00	10	R
Tres de Diciembre	25	200	0.12	R,C,I
San Ag de Cajas	55	00	3	P
Chaquicocha	10	00	10	R,C,I
San José de Quero	40	00	12	R,C,I
Huayao	50	00	1	R,C
Churcampa	00	00	?	P
Hualhuas	35	00	18	P
Saños Grande	20	00	25	P
El Mantaro	10	00	?	P
Julcan	40	00	15	C,I
Sacsamarca	20	00	50	P
Tarmatambo	35	4	55	R,C,I
Pichinaki	40	00	?	I

•N.B Minimum sand depth recommended is 60 cm. P= complete rehabilitation
R = complete sand bed replacement   C = cleaning I= increase depth

Disinfection

A diffusion hypochlorinator is recommended by the National Plan which is designed for installation in rural supplies with a 1l/sec flow. This device was occasionally found installed on a retaining wire under the inspection cover inside reservoirs.

In 7 out of 18 systems the hypochlorinator was found hanging in the reservoir at points of convenient access but in none of these was it located in a flow of water.

On no occasion was a measurable chlorine residual detected at the outlet of any rural reservoir or in distribution in any of the rural systems. There is clearly an urgent need to optimise the design and reassess the point of installation of this device. This is the subject of a separate development project sponsored by the Overseas Development Administration of the British Government.

IMPROVEMENT & REHABILITATION STRATEGIES

It was concluded from the diagnostic survey (2) that the communities at greatest risk from waterborne disease were those with no option other than surface water sources since the existing treatment systems had failed to reduce critical contaminants to safe levels. As a result of the survey the Ministry of Health imposed a moratorium on the construction of new slow sand filter installations until pilot demonstration projects could demonstrate improved performance.

It was clear that in those supplies where the raw water turbidity overloaded the filters, rehabilitation and new construction incorporating prefiltration was required. The supplies graded as faecal coliform class D, required priority attention to improve their quality and the continuity of supply often required improvement as well.

TABLE 5

Improvement and rehabilitation strategies in rural water treatment plants in Peru*

Year of intervention	Location	Department	Type of intervention	Pre filtration	Enhanced Slow Sand Filt	Training of community	Follow-up evaluation
1984	Carhua	Lima	I		•		•
1984	San Buenaventura	Lima	I		•		•
1985	Azpitia	Lima	N	•	•	•	•
1985	Iscozacin	C.de Pasco	I		•	•	•
1986	Espachin	Lima	N	•	•		
1986	Cocharcas	Junin	R	•	•	•	•
1987	La Cuesta	L.Libertad	R	•	•	•	•
1987	Compin	L.Libertad	R	•	•	•	•
1988	Palian	Junin	R,C	•			
1988	Viccos	Junin	N,C	•	•		
1988	Collambay	L.Libertad	N,C	•	•		
1988	Cayanchal	L.Libertad	N,C	•	•		
1988	Simbal	L.Libertad	N,C	•	•		

I= Improved slow sand filters   N= New Scheme
R= Rehabilitation of existing scheme   C= In construction
*Information upto July 1988/Rural communities,<2000 hab.

The first pilot demonstration project to incorporate gravel prefiltration in Peru was a completely new supply constructed for the community of Azpitia (4). Prior to this only two slow sand filter plants had been improved (Table 5). These supplies, at Carhua and San Buenaventura, did not require gravel prefilters, since their raw water turbidities rarely registered more than 20TU even in periods of heavy rainfall. These two schemes were improved by sand replacement using appropriately graded filter sand and, in the case of San Buenaventura, by using synthetic matting to enhance performance and protect auxilliary packaged slow sand filters.

The Azpitia treatment plant has a small settler integrated with a three stage **vertical downflow** gravel prefilter in one structure and four parallel slow sand filters downstream. The scheme was completed in early 1985 and performance evaluation was carried out prior to the commencement of the first rehabilitation project, incorporating gravel prefiltration, of an existing slow sand filtration plant at Cocharcas.

The Cocharcas rehabilitation project was completed in 1986 and incorporated the first three stage **horizontal** flow gravel prefilters constructed in Peru. Figure 2 demonstrates schematically the principal points of intervention in the Cocharcas system. The pilot rehabilitation project for Cocharcas (5) incorporated the following components:

1) Administrative: Joint project planning with the Community Water Administration Committee and public meetings to ensure majority support and community participation in the proposed interventions.
2) Reconstruction of abstraction box: to ensure minimum flow, flow control and source protection.
3) Construction of 3 stage horizontal flow gravel prefilters: to protect slow sand filters from overload by turbidity and suspended solids, and provide multiple barrier treatment.
4) Rehabilitation of slow sand filters: to provide resanded beds, graded gravel underdrainage and flow control to optimise performance.
5) Terminal hypochlorite disinfection.
6) Water committee training: to ensure proper operation and maintenance after commissioning.
7) Operational evaluation: to establish sustained improvement in the water supply service.

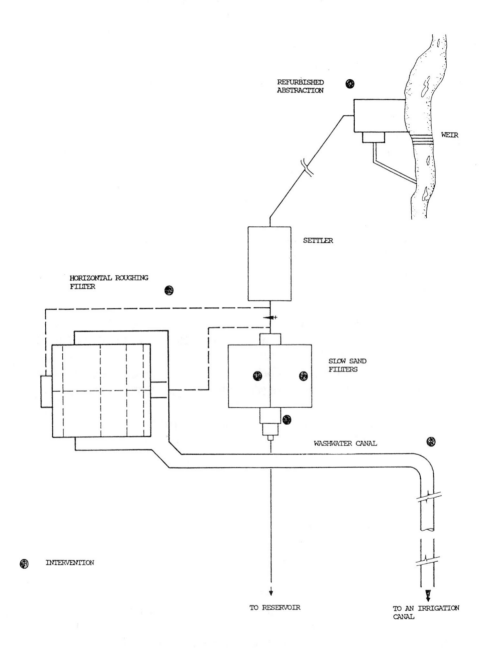

Figure 2. Schematic drawing of the rehabilitated water treatment plant of Cocharcas

PERFORMANCE OF NEW & REHABILITATED TREATMENT PLANTS AT AZPITIA AND COCHARCAS

Figure 3 and Table 6 demonstrate that at Azpitia, in the rainy season, the raw water turbidity peaks are so high (>5,000) that the plant frequently failed to maintain final filtrate quality below the WHO guideline maximum of 5TU in spite of a high level of commitment by the plant operator. By contrast the Cocharcas plant always produced acceptable turbidity, <5TU, because raw water peaks never exceeded 300TU. The Cocharcas prefilters have not thus far been rigorously tested. It may be argued that Azpitia is undersized and Cocharcas is oversized for their respective turbidity loadings. However it should be noted that few conventional treatment systems could cope with the high turbidity levels encountered in rivers on the Pacific coast of Peru.

Figure 4 demonstrates the loss of head of three parallel slow sand filters compared with the raw water turbidities which would , without intervening prefilters, cause very rapid blockage. It is important to note that the prefilters effectively protected the filters; the latter only required cleaning three times in the whole rainy season. The slow sand filtration units sustained acceptable **constant** filtration rates for at least 15 days for each filter run. Equally importantly the slow sand filters were designed to be cleaned rapidly by removing and washing the top layers of synthetic fabric, which overlay the sand bed, without draining the sand bed. This maintains the integrity of the biological community and facilitates an
immediate return to useful filtration. It should also be noted that the prefilters only required cleaning once during the 2.5 month period. The prefilter cleaning procedure was developed by the authors, using a rapid drainage gate valve leading to a wash water canal (Figure 2).

It is of interest to compare the bacteriological quality of the raw waters at Azpitia and Cocharcas (Table 6) in the context of removal efficiency. The Cocharcas raw water is an order of magnitude more contaminated; nonetheless examination of faecal coliform removal demonstrates that the final quality of slow sand filtered water was marginally better at Cocharcas and thus overall efficiency was significantly higher at 99.9% compared with 96%. Furthermore each of the treatment stages (sedimentation, gravel prefiltration and slow sand filtration) was more effective at Cocharcas.

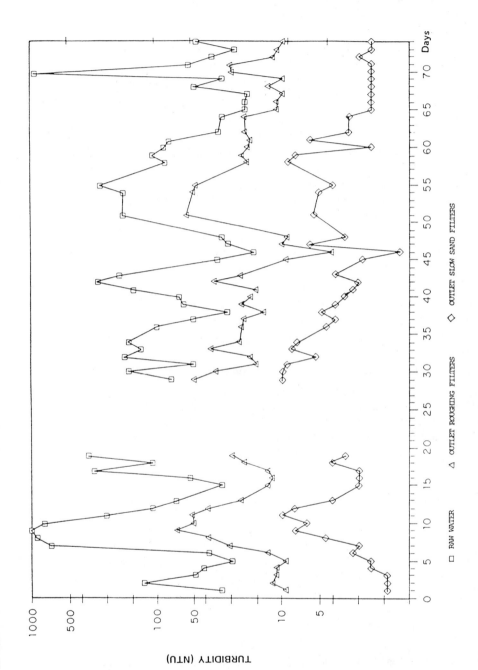

Figure 3. Turbidity levels at different treatment stages of a water supply scheme in the community of Azpitia. Evaluation period January - March 1986

TABLE 6

Performance of new and rehabilitated rural treatment plants in Peru

Location	Azpitia		Cocharcas	
Evaluation period	Jan'85 –	March'86	Feb'87 –	Apr.'87
Water quality parameter.	Turbidity TU	Bacteriology FC/100ml	Turbidity TU	Bacteriology FC/100ml
Nº of Samples	III	51	41	24
Raw Water				
Mean value	197	860	20	20,400
Max.	>5000	>2000	300	60,200
Min.	<5	150	<5	400
Std Dev.	294	837	45	–
Treatment Stage				
Sedimentation	143 (27.3)*	754 (12.3)	8 (60)	14,500 (28.9)
Prefiltration	28 (85.7)	246 (71.6)	<5 (>75)	1,200 (94.1)
Slow sand filters	8 (96.1)	26 (97.0)	<5 (>75)	20 (99.9)
Application rates				
Plant,$Q(m^3)/D$	35		103	
Sed,$V_s(m^3/m^2/D)$	10		8.6	
Gravel prefilt,$V_f(m/h)$	0.20		0.6	
Slow sand filt,$V_f(m/h)$	0.15		0.2	

NOTES: 1. Turbidity Units (TU). Bacteriological, Faecal Coliform count/100ml
2. Rainfall periods in areas under study: Jan–March= Wet season
: April–December=Dry season
* ( ) = (cumulative percentage reduction) alongside
mean values at outlet of each treatment stage

Sedimentation and prefiltration together produced a 71.6% reduction in faecal coliform counts at Azpitia compared with a 94% reduction at Cocharcas. Similarly the packaged plant sand filters at Azpitia perform less well than the conventional slow sand filters at Cocharcas. At this stage it is difficult to provide a reasonable explanation for these discrepancies since the lower flow rates at Azpitia, both through the gravel filters and slow sand filters, should produce better results. Figure 5 shows that bacteriological improvement at each stage of treatment at Azpitia was relatively consistently maintained in spite of the simultaneous major variation in turbidity during that period. In spite of the considerable improvement in water quality at both plants the final water quality only improved

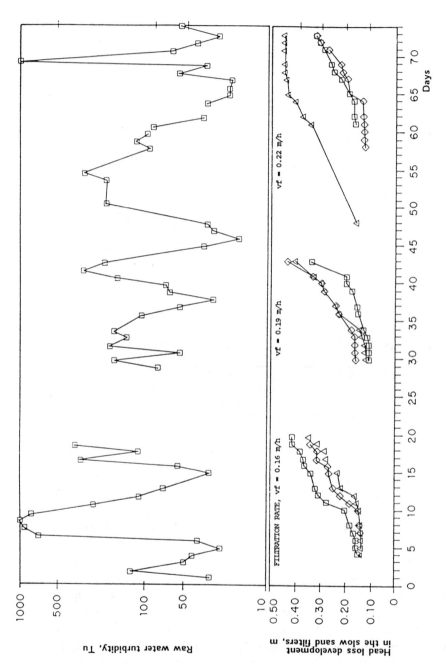

Figure 4. Comparative levels of raw water turbidity and development of head loss in three protected slow sand filters. Water supply scheme of Azpitia. Evaluation period January – March 1986

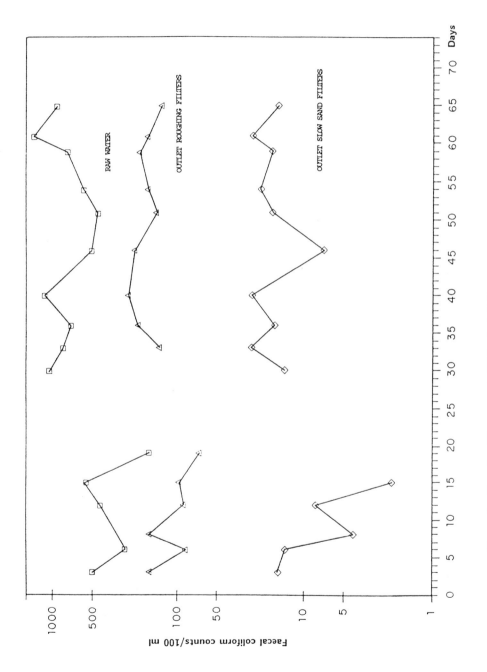

Figure 5.   Faecal coliform counts/100 ml at different treatment stages of a
water supply scheme in the community of Azpitia. Evaluation period
January – March 1986

from class D to C on the criteria presented in Table 3b. On these criteria Cocharcas is not oversized and it was necessary to install pot chlorinators at both plants, after the slow sand filters, to remove the remaining faecal coliform contamination and consistently achieve class A water quality.

CONCLUSIONS

A fundamental lesson learnt from the demonstration pilot projects at Azpitia and Cocharcas was the importance of matching the newly developed prefiltration technology to both the raw water source quality and the operational capacity of the village water administrative committee.

Many Peruvian slow sand filtration plants have been abandoned by their administrative committees because the operators were not trained to deal with the problem of blocked filters and because in some cases the plants were inadequately designed to treat grossly contaminated raw water. Nevertheless important conclusions may be drawn from the Peruvian experience including:
a) Filter maintenance is a procedure which can be easily carried out by the operator.
b) Gravel prefiltration may be effectively integrated with settlement and slow sand filtration to enhance overall plant performance and reduce the required frequency of cleaning.

We are confident that, at the technical level, filtration technologies can perform satisfactorily; the more difficult task ahead is to develop appropriate training and administrative infrastructure which will ensure an adequate level of operation and maintenance of these systems.

ACKNOWLEDGEMENTS

This work was carried out under contract to the British Overseas Development Administration and through co-operative funding from the IRCWD in Switzerland. The views expressed are the responsibility of the authors and not necessarily those of the supporting organisations.

REFERENCES

1.   Cánepa, L (1982) DTIAPA Report № 3 CEPIS/ WHO, Lima, Peru
2.   Lloyd, B.J., Pardon, M., Wedgewood, K and Bartram, J DelAgua
        (1986) Phase One ODA Report; Peruvian Water Surveillance
        Programme.
3.   Lloyd, B.J., Pardon, M and Bartram, J (1986) Proc. Int. Conference
        for Resource Mobilisation for Drinking Water and Sanitation in
        Developing Nations; Am Soc Civ Eng (1987) 640-652.
4.   Lloyd, B.J., Pardon, M and Wheeler, D.C (1986) Final report ODA
        Research project R3760, DelAgua-Robens Institute, July.
5.   DelAgua - Public Health Consultants (1986) ODA Pilot project
        report, Peruvian Water Surveillance Programme, July.
6.   WHO, Geneva (1984) Vol III Guidelines for drinking water quality.

412

Index

accidental pollution, 64
actinomycetes, 368
activated carbon, 265–280
Aelosoma hemprichi, 168
Aeromonas, 258
  A. hydrophilia, 197
algae
  blue-green, 13, 15, 309
  filamentous green, 25, 27
  at Leiduin and Weesperkarspel, 259
  removal of, 119–120
  yellow-green, 309
aluminium oxide, 41
ammonia, 202, 272
Amsterdam, 253
AMW (apparent molecular weight), 282
Anabaena sp., 306, 309
Ankistrodemus, 382
anthracite, 41, 282, 297
AOC, 162, 256–257, 262
AOM, 281–282, 291
Aphanizomenon sp., 309
aquatic organic matter (AOM), 281–282, 285, 291
aquifer, 92, 93, 108
artificial ground water recharge, 91–92, 100, 107, 108
Asellus aquaticus, 13
Ashford Common (UK), 163, 164, 170, 172, 175, 309–314, 333, 350–356
Ashland (USA), 283, 297, 299
Asian Institute of Technology (AIT), 117
Aspidisca costata, 169
assimilable organic carbon (AOC), 162, 256–257, 262
Asterionella, 306, 317
ATP, 270, 276, 280, 345
Austria, 107
automation, 52
Azpitia (Peru), 113, 116, 405–410

Bacillus licheniformis, 212, 218, 221
bacteriophage, 208, 211–212, 218–221, 226, 227
bank filtration, 94–98
Bdellovibrios, 169

benzoate, 285
bioassay, 191–205
biofilm, 225, 242, 248, 250
blanket weed, 343
blue-green algae, 13, 15, 309
Blue Nile Health Project, 119, 120
Bosmina, 323
Brazil, 124
break-point chlorination, 48
Bremen (W. Germany), 163, 170
Burma, 120

cake filtration, 297
Cali (Columbia), 116, 121
Canthocamptus staphylinus, 168
Catenula lemmna, 168
CFU, 258
chemical oxygen demand, 274, 275
Chilodonella sp., 168, 169, 172
chironomids, 323
Chlamydomonas sp., 317, 382
chlorination, 39
  break-point, 48
chlorine dioxide, 56–57
chlorophyll a, 13, 27, 33, 35, 163, 164, 170, 314, 316, 335
Chlorophyta sp., 120, 309
Chorro de Plata (Columbia), 112
Chromadorita leukarti, 168
ciliates, 164, 259
Cinetochilum sp., 168, 169
Cladocera, 259, 314
Cladophora sp., 13, 25, 26, 317
clarification, 50
classification of filters, 108–110
cleaning, 9, 112, 114, 120, 255, 282, 301, 335, 339–342
  methods, 286
  of Municipal Filter, 286–291
  of Pebble Matrix filter, 144
  standards, 23, 25
  wet, 340–342
clinoptilolite, 41, 282, 296–297, 306
Clostridium perfringens, 169

Index

Cocharces (Peru), 405–410
COD, 274, 275
coliforms, 30, 33, 35–37, 40, 137, 169, 208, 222, 231, 247
  faecal, 215, 320, 324–325
coliphages, 208, 220
colony forming units (CFUs), 258
colour
  apparent, 138
  removal, 153–162
  and turbidity, 247–250
Columbia, 111, 112, 114, 116
conditioning period, 15
conductivity, 199
contact coagulation, 69–82, 86
Copepoda, 259, 314
Coppermills (UK), 148, 336, 337
costs, 6–7
  of construction, 7, 299–301, 302
  maintenance, 7
  operation, 7
covered slow sand filtration, 253–264
Cryptophyceae, 181
Cryptosporidium, 41, 53
Cyanophyta, *see* blue-green algae
Cyclidium sp., 168, 169
Cyclotella sp., 120, 181, 182, 185–187, 189
  C. comta, 183, 184, 185
    C. c. fo praetermissa, 184
  C. glomerata, 188

Daphnia sp., 13, 314, 320, 323
deep-skimming, 16, 20
design considerations, 5–6
design criteria, 5, 148–152
design period, 5
diatoms, 13, 15, 25, 181–190, 259, 309, 382
  fibrilate discoid, 182
  pennate, 182
dinoflagellates, 309
direct filtration, 337–339
dissolved organic carbon (DOC), 262, 284
DOC, 262, 284
Dortmund (Germany), 107, 163
double filtration, *see* dual-media filters
down-flow roughing filter, 105, 113
downstream improvements, 56–57, 82–86
drainage cycle, 132–133
dual-media filters, 379–392
dynamic filter, 109, 112, 116

EBCT, 267, 275
echovirus, 208, 226
effective size (ES), 336
El Retiro (Columbia), 114, 116
empty bed contact time (EBCT), 267, 275
Enchytraeus bucholzi, 163, 168, 169, 170, 171
Enterobacter cloacea, 197, 212, 218, 220
enterovirus, 208, 226
Ephemeroptera, 323
Erwinia carotovora, 212, 218, 219
Escherichia coli, 14, 15, 25, 27, 169, 197, 208, 209, 212, 218, 232, 236, 237, 370, 375, 377, 378
Euplotes, 169
expanded schmutzdecke (ES), 368, 370–374

fabrics, 305–329
faecal coliforms, 215, 320, 324–325
faecal indicator organisms, 15
faecal streptococci (*see also Streptococci faecalis*), 211, 215, 222
ferric sulphate, 337
filamentous green algae, 25, 27
filter bed, 4, 6, 9
filter loading rates, 302
filter media, 119, 301–302
  size of, 13
filter modelling
filtration rate, 5–6
  and algae, 13
  in horizontal-flow prefilter units, 128
  low temperature limits and, 15
  seasonal, 13–14
flagellates, 164
flatworms, 169–170, 209
  microturbellarian, 163
flow control, 4–5
flow indicator, 5
fluorescence, 275
Fobney (UK), 339, 344
Fragilaria sp., 317
fulvic acid, 282
fungi, 368

GAC, 41, 82, 86, 266, 274, 275, 280, 282, 292–6, 332
germs, 199
Ghana, 115, 119
Giardia cysts, 29–31, 35, 37, 39–41, 44, 48, 53, 69, 281, 291
giardiasis, 30–31, 39
Glaucoma spp., 169
glucose, 202
Gomphosphaeria sp., 120
granular activated carbon (GAC), 41, 82, 86, 266, 274, 275, 280, 282, 292–296, 332
gravel prefiltration, 128, 130
green filters, 182, 187–188, 189
ground water, 91–101
  recharge, artificial, 91–92, 100, 107, 108

Haematococcus, 382
Hamburg (W. Germany), 47
Hampton (UK), 163, 169, 214, 336
harpacticoid copepods, 163, 170
head-loss, 16–17, 236–247, 322, 338–339
  normalised (NSH), 17, 20, 25
helminths, *see* flatworms
Hemiophrys sp., 168, 169
Hengsen (W. Germany), 93, 94–97
high-carbon feedwaters, 231–252
high-coliform feedwaters, 231–252
holotrichs, 169

Index

horizontal-flow roughing filtration (HRF), 109, 115, 116, 117–120
Howard Institute (Zimbabwe), 381–392
HRF, 109, 115, 116, 117–120
humic acid, 234, 282
humic compounds, 247, 248–249
hydraulic conductivity, 236–242

initial commissioning, 9
injuk, 119
inlet control, 4
inlet structure, 2
intake filter, 105, 109, 111, 116
intermediate drainage, 129, 132–133
International Reference Centre (IRC), 121
  for Waste Disposal (IRWCD), 119, 121
Invercannie (UK), 154, 155, 158, 162
iron, 94, 134–136, 138, 202
iron bacteria, 13
Ivry (France), 60–87

'jacking-up', 17

kaolin simulation, 119, 128
Klebsiella pneumoniae, 197
Kozeny–Carman model, 352

'Labo' ozoniser, 154
Leiduin (Netherlands), 253–255, 256, 258
Litonotus sp., 168, 169
Llanforda (UK), 153, 155
London, 11, 47
Lound (UK), 182, 187

Mafi Kumase (Ghana), 115
maintenance, 8–10
Mallomonas, 181
manganese, 94, 97, 134–136, 138, 202–203
mats, 343
mayflies, 323
medium, *see* filter medium
meiofauna, 163, 164
Melosira spp., 13
  M. varians, 306, 317
Mesismopedia sp., 120
microcysts, 382
microorganisms, removal of, 52, 55, 56–57
micropollutants
  mineral, 55
  organic, 51–52, 55, 56, 64, 257
  removal of, 50–51
microstraining, 314
microturbellarian flatworms, 163
midges, 323
modelling, 97, 99, 100, 148, 192–193, 349–366
  computer, 340
  mathematical, 97, 350
Monhystera vulgaris, 168
Mont Valérien (France), 265–280
multiple well equation, 97

Nais sp., 168, 187

Nauplii, 259
Navicula spp., 164
nematodes, 163, 169, 187, 259
Nephelometric Turbidity Units (NTUs), 29
New Haven (USA), 282, 283
Niphargus sp., 13
nitrate, 202
nitrifying bacteria, 202
nitrite, 202
Nitzschia acicularis, 164, 317
  N. linearis, 164
nonpurgeable dissolved organic carbon
  (NPDOC), 282, 284, 289–291
non-woven synthetic fabrics (NWF), 305, 306–308
normalised head-loss (NSH), 17, 20, 25
NPDOC, 282, 284, 289–291
NSH, 17, 20, 25
NWF, 305, 306–308

oligochaetaes, 163
Ormesby (UK), 182
Ostracoda, 259
Oswestry (UK), 154, 155, 162
outlet chamber, 4
outlet control, 4
Oxytrichia sp., 168, 169
ozonation, 153–162
ozone, 56, 82, 86, 332, 343–346

Panagrolaimus rigidus, 168
Paris (*see also* Ivry), 47, 60, 265
particle size analysis, 13, 22, 320
particulate organic carbon, *see* POC
pebble bed, 150
pebble matrix filtration, 141–152
Peru, 113, 116, 393–411
phosphate, 202, 203
photosynthesis, 179
plate counts, 275, 276, 279
POC, 13–15, 22, 27, 164, 170, 172, 314, 315, 320, 335, 338
  input loadings, 175–179
poliovirus, 208, 209
polyurethane, 367–378
Portsmouth (USA), 283, 292
pre-chlorination, 342, 343
prefilters, 104, 110
prefiltration, 104
  gravel, 128
  upflow, 131
pre-ozonation, 162, 342, 344, 345
primary filtration, 336
Pristina sp., 168
protozoa, 163, 164, 169–179, 207–209, 368
Puech–Chabal filter system, 105–106

rapid filtration, 48
rapid gravity (sand) filters, 183–184, 189, 208, 379
reovirus, 209–210
re-sanding, 10, 16, 20, 22, 40
Rhizophydium cyclotellae Zopf, 188
ripening, 9

Index

Rotatoria, 259
rotavirus, 209, 212–214, 222–223, 227
  simian, 211, 214
rotifers, 163, 170, 314, 368
roughing gravel filters, 103–121
Ruhr, River, 98, 197–199

SANNET, 360–366
SAP, 368, 374–376
Scenedesmus, 184
security, water supply, 61, 64–67
Selenastrum, 382
Serratia marcescens, 212, 218, 219, 226
shading, 342–343
shake flask adsorption, 214
silt content, 25, 323
simian rotavirus, 211, 214
skeleton-fill filter, 141
skimming (*see also* cleaning), 15
Spirogyra, 306
spirotrichs, 169
Springfield (USA), 282, 283
Staurastrum sp., 120
Stenosoma sp., 168
Stephanodiscus, 181, 306
  S. 'astraea', 183, 184, 317
  S. hantzschii, 317
  S. neoastraea, 184
stones, utilisation of, 148
Streptococcus faecalis, 367, 370, 375, 377, 378
Stylaria fossularis, 168
Stylonichia sp., 168, 169
submerged aerated prefilter (SAP), 368, 374–376
Sudan, 119
supernatant water layer, 2
Switzerland, 107
Synchaeta pectinata E., 188

Tachysoma sp., 168, 169
Testaceae, 259
Thalassiosira, 181
  T. fluviatilis, 185
Thames, River, 331
THM, 39, 281, 331
THM precursors, 39, 41, 281–304

THMFP, 33, 281–304
TOC (total organic carbon), 33, 76, 81, 82, 162, 231, 234, 242–250, 270, 274–275, 280
TPN (total particle numbers), 14, 15, 27
TPV (total particle volume), 14, 15, 27
Tracholomonas sp., 120
trenching, 16
Tribonema, 309
Trichoptera, 323
trihalomethane formation potential (THMFP), 33, 281–304
trihalomethane precursors, 39, 41, 281–304
trihalomethanes (THM), 39, 281, 331
Trilobus gracilis, 168

underdrain, 4, 152
uniformity coefficient, 393
United States, 29–44
University of Dar es Salaam (UDSM), 118–119
upflow coarse-grained prefilter, 123–139
upflow roughing filter, 114
upgrading SSFs, 331–347
uprating, 335, 339
upstream improvements, 54, 57
UV absorbance, 282, 284, 289

viruses, 207–229, 257–258
Volvox, 309
Vorticella sp., 164, 168, 169
Vyrnwy, Lake, 153

Walton (UK), 163, 346
washing, sand, 23, 25
Weesperkarspel (Netherlands), 254–255, 256, 260–261, 263–264
West Hartford (USA), 282, 289–290
wet cleaning, 340–342
Wraysbury (UK), 172

yellow-green algae, 309

Zimbabwe, 379–392
zoogloeal slime, 225, 351
zooplankton, 351